Advances in Experimental Medicine and Biology

AF806550

Proteomics, Metabolomics, Interactomics and Systems Biology

Series Editor

Daniel Martins-de-Souza
University of Campinas (UNICAMP)
Institute of Biology
Laboratory of Neuroproteomics
Campinas, Brazil

This series of volumes focuses on concepts, techniques and recent advances in the field of proteomics, interactomics, metabolomics and systems biology. Recent advances in various 'omics' technologies enable quantitative monitoring of myriad various biological molecules in a high-throughput manner, and allow determination of their variation between different biological states on a genomic scale. Now that the sequencing of various genomes, from prokaryotes to humans, has provided the list and linear sequence of proteins and RNA that build living organisms, defining the complete set of interactions that sustain life constitutes one of the key challenges of the postgenomic era. This series is intended to cover experimental approaches for defining protein-protein, protein-RNA, protein-DNA and protein-lipid interactions; as well as theoretical approaches dealing with data analysis, integration and modeling and ethical issues.

More information about this series at http://www.springer.com/series/15040

Marco Aurélio Zezzi Arruda
Editor

Metallomics

The Science of Biometals

Springer

Editor
Marco Aurélio Zezzi Arruda
Universidade Estadual de Campinas – Unicamp
Institute of Chemistry
São Paulo, Brazil

ISSN 0065-2598 ISSN 2214-8019 (electronic)
Advances in Experimental Medicine and Biology
ISBN 978-3-030-07943-7 ISBN 978-3-319-90143-5 (eBook)
https://doi.org/10.1007/978-3-319-90143-5

Printed on acid-free paper

This Springer imprint is published by the registered company Springer International Publishing AG part of Springer Nature.
The registered company address is: Gewerbestrasse 11, 6330 Cham, Switzerland

Preface

Metallomics is an emerging area of the omics-science that is growing since its recent conception. This area integrates the researches related to biometals, in symbiosis with genomics, proteomics, and metabolomics. In fact, some proteins need a metal in their activity centers to develop their precise functions, thus forming metalloproteins or metalloenzymes. They are important in a diversity of biological functions, and, however, some key studies involving how the element (metal or metalloid) is distributed in cellular compartments of a cell, its coordination environment, in which biomolecule is complexed, and, finally, the individual concentrations of the metallic species have important role in providing an integrated information connecting metallomics with other omics.

For attaining this task, a diversity of analytical and biological platforms are then required for the studies to make sense from a biological point of view. Then, this volume of *Proteomics, Metabolomics, Interactomics and Systems Biology* series, wrote by some of the most renowned experts in this area, provides a comprehensive view of metallomics from basic concepts to the most recent developments. In fact, it is organized into 10 chapters, being the concepts, and the opportunities arising from metallomics for a better understanding of human biology and health discussed in Chap. 1. Inside this context, Chap. 2 describes the recent advances in the metallomics area applied to neurodegenerative and mental disorders, and Chap. 3 points out environmental toxicological from a metallomics and metabolomics point of view. While Chap. 4 presents some studies related to biotransformation applied to plant metallomics, Chap. 5 demonstrates recent techniques and applications regarding to animal metallomics, which include fish, rat, and bovines. Chapter 6 focuses on isotopically enriched metalloproteins for exploring some biological pathways, which demonstrate the usefulness of isotopes in such studies. The state of the art in bioimaging metallomics including isotopes and elements bioimaging is demonstrated in Chap. 7, not only for cells but also for several tissues, including brain, live, hair, and others. As metallomics is an inherent part of chemical speciation, Chap. 8 demonstrates some examples involved in the human health risk, food and human diet, drugs, forensic, nanoscience and geological, including some aspects regarding sample preparation focusing on chemical speciation for metallomics purposes.

Chapter 9 comprises nuclear techniques applied to metallomics and nanometallomics, and, finally, Chap. 10 elegantly shows new frontiers in the metallomics applied to single cells.

For those professionals involved in omics-science, analytical, biological, toxicological, medical, among others areas, this book will be a useful reference. For graduate/undergraduate students or beginners in the metallomics area, this edition will provide valuable information from concepts to applications of metallomics from a transdisciplinary point of view.

Campinas, São Paulo, Brazil Marco Aurélio Zezzi Arruda

Contents

Contributors

Jiri Adamec Department of Biochemistry, University of Nebraska-Lincoln, Lincoln, NE, USA

Ana Arias-Borrego Department of Chemistry, Faculty of Experimental Sciences, University of Huelva, Campus de El Carmen, Huelva, Spain

Research Center on Natural Resources, Health and the Environment (RENSMA), University of Huelva, Huelva, Spain

International Campus of Excellence on Agrofood (ceiA3), University of Huelva, Huelva, Spain

Marco Aurélio Zezzi Arruda Universidade Estadual de Campinas – Unicamp, Campinas, São Paulo, Brazil

Spectrometry, Sample Preparation and Mechanization Group (GEPAM), Department of Analytical Chemistry, Institute of Chemistry, University of Campinas (UNICAMP), Campinas, São Paulo, Brazil

National Institute of Science and Technology for Bioanalytics – INCTBio, Institute of Chemistry, University of Campinas (UNICAMP), Campinas, São Paulo, Brazil

Jörg Bettmer University of Oviedo, Department of Physical and Analytical Chemistry, Oviedo, Spain

Katharina Bluemlein Fraunhofer Institute for Toxicology and Experimental Medicine, Hannover, Germany

Camila Pereira Braga Department of Biochemistry, University of Nebraska-Lincoln, Lincoln, NE, USA

Zhifang Chai CAS Key Laboratory for Biomedical Effects of Nanomaterials and Nanosafety, and Laboratory for Metallomic and Nanometallomics, Institute of High Energy Physics, Chinese Academy of Sciences, Beijing, China

Chunying Chen CAS Key Laboratory for Biomedical Effects of Nanomaterials and Nanosafety, National Center for Nanoscience and Technology, Beijing, China

Luana Ferreira da Costa Universidade Estadual de Campinas – Unicamp, Campinas, São Paulo, Brazil

Spectrometry, Sample Preparation and Mechanization Group (GEPAM), Department of Analytical Chemistry, Institute of Chemistry, University of Campinas (UNICAMP), Campinas, São Paulo, Brazil

National Institute of Science and Technology for Bioanalytics – INCTBio, Institute of Chemistry, University of Campinas (UNICAMP), Campinas, São Paulo, Brazil

Jemmyson Romário de Jesus Universidade Estadual de Campinas – Unicamp, Campinas, São Paulo, Brazil

Spectrometry, Sample Preparation and Mechanization Group (GEPAM), Department of Analytical Chemistry, Institute of Chemistry, University of Campinas (UNICAMP), Campinas, São Paulo, Brazil

National Institute of Science and Technology for Bioanalytics – INCTBio, Institute of Chemistry, University of Campinas (UNICAMP), Campinas, São Paulo, Brazil

Pedro de Magalhães Padilha Department of Chemistry and Biochemistry, Institute of Bioscience, São Paulo State University (UNESP), Botucatu, São Paulo, Brazil

Valderi Luiz Dressler Universidade Federal de Santa Maria, Santa Maria, Rio Grande do Sul, Brazil

Jörg Feldmann TESLA (Trace Element Speciation Laboratory), Department of Chemistry, University of Aberdeen, Scotland, UK

Rodrigo Moretto Galazzi Universidade Estadual de Campinas – Unicamp, Campinas, São Paulo, Brazil

Spectrometry, Sample Preparation and Mechanization Group (GEPAM), Department of Analytical Chemistry, Institute of Chemistry, University of Campinas (UNICAMP), Campinas, São Paulo, Brazil

National Institute of Science and Technology for Bioanalytics – INCTBio, Institute of Chemistry, University of Campinas (UNICAMP), Campinas, São Paulo, Brazil

Yuxi Gao CAS Key Laboratory for Biomedical Effects of Nanomaterials and Nanosafety, and Laboratory for Metallomic and Nanometallomics, Institute of High Energy Physics, Chinese Academy of Sciences, Beijing, China

Tamara García-Barrera Department of Chemistry, Faculty of Experimental Sciences, University of Huelva, Campus de El Carmen, Huelva, Spain

Research Center on Natural Resources, Health and the Environment (RENSMA), University of Huelva, Huelva, Spain

International Campus of Excellence on Agrofood (ceiA3), University of Huelva, Huelva, Spain

José Luis Gómez-Ariza Department of Chemistry, Faculty of Experimental Sciences, University of Huelva, Campus de El Carmen, Huelva, Spain

Research Center on Natural Resources, Health and the Environment (RENSMA), University of Huelva, Huelva, Spain

International Campus of Excellence on Agrofood (ceiA3), University of Huelva, Huelva, Spain

Rachel Ann Hauser-Davis Centro de Estudos da Saúde do Trabalho e Ecologia Humana (CESTEH), Escola Nacional de Saúde Pública Sérgio Arouca (ENSP), FIOCRUZ, Rua Leopoldo Bulhões, Rio de Janeiro, Brazil

Javier Jiménez-Lamana Institute of Analytical Sciences and Physico-Chemistry for Environment and Materials (IPREM), UMR 5254, CNRS-UPPA, Pau, France

Eva Maria Krupp TESLA (Trace Element Speciation Laboratory), Department of Chemistry, University of Aberdeen, Scotland, UK

Eraldo Luiz Lehmann Universidade Estadual de Campinas – Unicamp, Campinas, São Paulo, Brazil

Spectrometry, Sample Preparation and Mechanization Group (GEPAM), Department of Analytical Chemistry, Institute of Chemistry, University of Campinas (UNICAMP), Campinas, São Paulo, Brazil

National Institute of Science and Technology for Bioanalytics – INCTBio, Institute of Chemistry, University of Campinas (UNICAMP), Campinas, São Paulo, Brazil

Yu-Feng Li CAS Key Laboratory for Biomedical Effects of Nanomaterials and Nanosafety, and Laboratory for Metallomic and Nanometallomics, Institute of High Energy Physics, Chinese Academy of Sciences, Beijing, China

Ryszard Łobinski Institute of Analytical Sciences and Physico-Chemistry for Environment and Materials (IPREM), UMR 5254, CNRS-UPPA, Pau, France

Katherine Chacón Madrid Universidade Estadual de Campinas – Unicamp, Campinas, São Paulo, Brazil

Spectrometry, Sample Preparation and Mechanization Group (GEPAM), Department of Analytical Chemistry, Institute of Chemistry, University of Campinas (UNICAMP), Campinas, São Paulo, Brazil

National Institute of Science and Technology for Bioanalytics – INCTBio, Institute of Chemistry, University of Campinas (UNICAMP), Campinas, São Paulo, Brazil

Wolfgang Maret Metal Metabolism Group, Departments of Biochemistry and Nutritional Sciences, School of Life Course Sciences, Faculty of Life Sciences and Medicine, King's College London, London, UK

Maria Montes-Bayón University of Oviedo, Department of Physical and Analytical Chemistry, Oviedo, Spain

Martin Mueller TESLA (Trace Element Speciation Laboratory), Department of Chemistry, University of Aberdeen, Scotland, UK

Edson Irineu Müller Universidade Federal de Santa Maria, Santa Maria, Rio Grande do Sul, Brazil

Dirce Pozebon Universidade Federal do Rio Grande do Sul, Porto Alegre, Rio Grande do Sul, Brazil

Sara Ramírez-Acosta Department of Chemistry, Faculty of Experimental Sciences, University of Huelva, Campus de El Carmen, Huelva, Spain

Research Center on Natural Resources, Health and the Environment (RENSMA), University of Huelva, Huelva, Spain

International Campus of Excellence on Agrofood (ceiA3), University of Huelva, Huelva, Spain

Gema Rodríguez-Moro Department of Chemistry, Faculty of Experimental Sciences, University of Huelva, Campus de El Carmen, Huelva, Spain

Research Center on Natural Resources, Health and the Environment (RENSMA), University of Huelva, Huelva, Spain

International Campus of Excellence on Agrofood (ceiA3), University of Huelva, Huelva, Spain

Alessandra Sussulini Laboratory of Bioanalytics and Integrated Omics (LaBIOmics), Department of Analytical Chemistry, Institute of Chemistry, University of Campinas (UNICAMP), Campinas, São Paulo, Brazil

Spectrometry, Sample Preparation and Mechanization Group (GEPAM), Department of Analytical Chemistry, Institute of Chemistry, University of Campinas (UNICAMP), Campinas, São Paulo, Brazil

National Institute of Science and Technology for Bioanalytics – INCTBio, Institute of Chemistry, University of Campinas (UNICAMP), Campinas, São Paulo, Brazil

Joanna Szpunar Institute of Analytical Sciences and Physico-Chemistry for Environment and Materials (IPREM), UMR 5254, CNRS-UPPA, Pau, France

Barry Alan Wood Reading Scientific Services Ltd, The Reading Science Centre, Reading, Berkshire, UK

Jiating Zhao CAS Key Laboratory for Biomedical Effects of Nanomaterials and Nanosafety, and Laboratory for Metallomic and Nanometallomics, Institute of High Energy Physics, Chinese Academy of Sciences, Beijing, China

Chapter 1
Metallomics: The Science of Biometals and Biometalloids

Wolfgang Maret

Abstract Metallomics, a discipline integrating sciences that address the biometals and biometalloids, provides new opportunities for discoveries. As part of a systems biology approach, it draws attention to the importance of many chemical elements in biochemistry. Traditionally, biochemistry has treated life as organic chemistry, separating it from inorganic chemistry, considered a field reserved for investigating the inanimate world. However, inorganic chemistry is part of the chemistry of life, and metallomics contributes by showing the importance of a neglected fifth branch of building blocks in biochemistry. Metallomics adds chemical elements/metals to the four building blocks of biomolecules and the fields of their studies: carbohydrates (glycome), lipids (lipidome), proteins (proteome), and nucleotides (genome). The realization that non-essential elements are present in organisms in addition to essential elements represents a certain paradigm shift in our thinking, as it stipulates inquiries into the functional implications of virtually all the natural elements. This article discusses opportunities arising from metallomics for a better understanding of human biology and health. It looks at a biological periodic system of the elements as a sum of metallomes and focuses on the major roles of metals in about 30–40% of all proteins, the metalloproteomes. It emphasizes the importance of zinc and iron biology and discusses why it is important to investigate non-essential metal ions, what bioinformatics approaches can contribute to understanding metalloproteins, and why metallomics has a bright future in the many dimensions it covers.

Keywords Biometals · Biometalloids · Bioinformatics

W. Maret (✉)
Metal Metabolism Group, Departments of Biochemistry and Nutritional Sciences, School of Life Course Sciences, Faculty of Life Sciences and Medicine, King's College London, London, UK
e-mail: Wolfgang.Maret@kcl.ac.uk

M. A. Z. Arruda (ed.), *Metallomics*, Advances in Experimental Medicine and Biology 1055, https://doi.org/10.1007/978-3-319-90143-5_1

Abbreviations

AAS	Atomic absorption spectroscopy
BMP	Basic metabolic panel
CMP	Comprehensive metabolic panel
ICP-MS	Inductively coupled plasma mass spectrometry
MRI	Magnetic resonance imaging
SMA	Sequential multiple analysis

1.1 Introduction

Similar to genes/proteins, genomes/proteomes and genomics/proteomics, and any other trilogy applied to biomolecules, one can define the relationship among metals, metallomes, and metallomics. The late RJP "Bob" Williams used the word *metallome* to describe the variety of pathways for distributing individual chemical elements (metals) that are responsible for the specific character of an organism (Williams 2001). However, it was the inspiration and foresight of Hiroki Haraguchi, now professor emeritus, Nagoya University, to coin the word *metallomics* in order to establish an integration of biometal sciences and to highlight the significance of metal ions for life. Metallomics offers many opportunities to advance science by overcoming the limitations of individual and scattered disciplines that traditionally address only some aspects of metals in biology (Haraguchi 2004, 2017). It is a holistic approach that aims at describing functions of entire biological systems from an understanding of the complexities arising from the interactions of its parts. It emphasizes quantitative aspects and often includes multidisciplinary investigations.

The definitions of "omes and omics" have become a fad in biology. An editorial discussed the instances where the suffixes are useful, separating "goodomes" from "badomes" (Baker 2013). The former provide (i) a new focus and do not just rename existing fields, (ii) a comprehensive collection instead of describing an area with limited scope, and (iii) an easy-to-pronounce and easy-to-understand term rather than a garbled or obscure word. The editorial specifically mentions that the terms metallome and metallomics fulfil the criteria for goodomes/goodomics.

In an IUPAC (International Union of Pure and Applied Chemistry) publication, the following definitions and three notes have been given (Lobinski et al. 2010).

"Metallome Entirety of metal- and metalloid species present in a biological system, defined as to their identity and/or quantity.

1. The metallome can be determined in a bulk biological sample representative of the system [or its component(s)] or at specific location(s).
2. The metallome can be characterized with different degrees of approximation, such as a set of (i) total element concentrations; (ii) metal complexes with a

given class of ligands, e.g. proteins or metabolites, or (iii) all species of a given element, e.g. the copper metallome.

3. In contrast to the genome of which the analysis has a specific endpoint (…) the description of a metallome…can never be complete. In particular, the numerous … metal complexes with biological ligands can be described only in terms of kinetic constants and defined thermodynamic equilibria.

Metallomics Study of the metallome, interactions, and functional connections of metal ions and other metal species with genes, proteins, metabolites, and other biomolecules in biological systems. A metallomics study is expected to imply: (i) a focus on metals (…) or metalloids (…) in a biological system, (ii) a link between the set of element concentrations or element speciation with the genome. This link may be statistical (…), structural (…) or functional (…); or (iii) a systematic or comprehensive approach. The identification of a single metal species, however important, without specifying its significance and contribution to a system should not be referred to as metallomics."

Two aspects are worth adding:

(a) Per above definitions *metallomes* refer to biological systems. However, metallomes also can be investigated in relation to the environment of biological systems, in water (hydrosphere), soil (lithosphere), and air (atmosphere) (Haraguchi 2004), generating knowledge of how the connection between the geosphere and the biosphere comes to bear on biological systems.

(b) There is a semantic issue, namely, that the word *metal* refers to a chemical element. After all, the methods employed to analyse metals in biology, such as atomic absorption spectroscopy (AAS) and inductively coupled plasma mass spectrometry (ICP-MS), are methods of elemental analysis – in contrast to the more common use of mass spectrometry as a method of molecular analysis. In biology, however, the ionic state and not the elemental state of the metal is the one that is almost always functionally important. To overcome this limitation of the words metallome and metallomics, the terms *ionomes* and *ionomics* were introduced in work on systematic elemental profiling of minerals and trace elements in plants. The terms include metals, metalloids, and non-metals (Lahner et al. 2003). The ionome as one of the "four basic biochemical pillars of functional genomics", namely, the transcriptome, proteome, metabolome, and ionome, draws attention to the fact that the information of the elemental composition of an organism is encoded in the genome in terms of the proteins that coordinate the acquisition and distribution of the chemical elements. The functional interactions of the ionome with the other three basic pillars are critical for understanding the biochemical basis of life (Salt 2004). These interactions can be direct, for example, when a metal ion affects the structure and function of a protein, or indirect, for example, when a metal ion affects the expression of proteins.

Included in the above definition of metallomics are metals and metalloids; none of the latter has been shown to be essential for humans, though. Excluded are a number of non-metals that are also essential for life. Another word in lieu of ionomics is *elementomics*. Clearly, not all chemical elements occur in an ionic form, as they can also form covalent compounds as in the case of iodine or selenium, for example. It includes all the chemical elements (Li et al. 2008). Since metal ions and non-metals are metabolites, the discipline has also been referred to as elemental metabolomics (Zhang et al. 2017).

The information gathered in metallomics is not only that of the type of ions but also includes the identification of their valence states and coordination environments, namely, speciation analysis in qualitative and quantitative terms, which can be addressed by combining elemental and molecular mass spectrometry with hyphenated techniques (Szpunar 2004). Investigations, therefore, address the presence of metal ions in different oxidation states, the binding of metal ions to macromolecules, where interactions with proteins dominate (metalloproteomes), and the interaction of metal ions with low molecular weight compounds. For example, when we talk about "iron biology", we neither specify the valence state of iron nor its coordination environment. Iron in biology refers to iron ions, Fe^{2+} (ferrous), Fe^{3+} (ferric), or even higher oxidation states in intermediates of enzymatic reactions. This distinction is important because the chemistries of Fe(II) and Fe(III) are very different in terms of reactivity and the stability of complexes. In the case of chromium, Cr(III) is a cation and supposedly essential, while Cr(VI) occurs as chromate, an oxyanion, which is toxic and a carcinogen. Therefore Cr(III) is expected to compete with cations such as Fe(III), whereas Cr(VI) competes with anions such as sulphate or phosphate. In zinc biology, however, only one valence state, Zn(II), is important. Yet, the coordination chemistry of zinc in biology is quite exquisite and dynamic, and protein-bound zinc as well as non-protein bound zinc ions are functionally important (Maret and Li 2009).

Last but not least, a reflection on the meaning of the term *biometals* is useful. The term refers to the metals that are present in or important for organisms. It is appropriate because with the high sensitivity of modern ICP-MS instrumentation, it became possible to measure almost all chemical elements in organisms. This finding constitutes a certain shift of paradigm because it indicates that we should not focus only on the metals that are essential for a species but also on the ones that are non-essential because metal ions are reactive and their presence has functional consequences. When extending the concept of biometals to essential and non-essential non-metals, it is appropriate to use the term *bioelements*. A speciation and metalloproteomics approach amplified the notion that non-essential elements are functionally significant and indicated the extent of uncharted territory (Cvetkovic et al. 2010), namely, that the metalloproteome of the microorganism *Pyrococcus furiosus* is largely uncharacterized. Specifically, out of the 343 metalloproteins characterized, 158 were recently unknown, including new nickel and molybdenum proteins but also proteins that contain lead and uranium. This experimental approach also shows the limitations of bioinformatics approaches to predict whether or not proteins are metalloproteins as discussed later in this article. Such predictions are possible on the basis of knowledge for essential metal ions and their coordination environments but can only gauge the possible interaction of non-essential metal ions with proteins in sites that normally bind essential metal ions.

In this article, I will discuss why an integration of the different fields dealing with biometals is useful, and in fact desirable, and point out the potential of metallomics for advancing science, using human biology as an example. A more comprehensive text on how metallomics bridges the biology and chemistry of biometals has been published (Maret 2016a).

1.2 History of Biometal Sciences

Metal ions in biology are investigated in many different fields. In nutrition, the focus was – and remarkably still is – to define the elements that are essential for life. In toxicology, the adverse effects of metal ions on biological structure and function are investigated, and in pharmacology, their therapeutic effects. The three disciplines cover different ranges of the biological dose-response curve: nutrition, mainly the lower concentration range up to the range where functional responses are optimal; pharmacology, the range where an enhancement of functions is observed, generally supraphysiological concentrations; and toxicology, the range of even higher concentrations. Whereas nutrition is primarily concerned with essential metal ions, pharmacology and toxicology address non-essential metal ions as well. However, this separation is arbitrary. In nutrition, the presence of non-essential elements affects the function of essential metal ions, and in toxicology the presence of essential elements affects the adverse health effects of non-essential metal ions, notwithstanding the fact that essential metal ions also become toxic at higher concentrations.

A first attempt at integration occurred when a critical amount of information about the molecular structures of metalloproteins became available. The field was then called bioinorganic chemistry with permutations of the name expressing slightly different emphasis: inorganic biochemistry, inorganic biological chemistry, and inorganic chemical biology. The main focus of this field is the remarkable chemistry of coordination environments of metals in biomolecules, the discovery of which turned out to be very exciting because in many cases it set precedence for structures and functions hitherto unknown in chemistry. Bioinorganic chemistry includes investigations that may not have biological significance. By and large, it espouses an emphasis on molecules and, thus, is a reductionist approach. Clearly, after decades of research, in parallel with contemporaneous developments in other fields such as genomics and proteomics, a holistic approach became an attractive way of thinking. Metallomics is such an approach as it addresses the entirety of metal ions in a system rather than individual biomolecules. For example, the knowledge about genomes and proteomes allowed estimating the entirety of metalloproteins in a system, i.e. the metalloproteome.

As an approach to biological systems, metallomics draws attention to the overall significance of metal ions in biology. This development is welcome as the teaching of biochemistry in most textbooks fails to address many of the elements on which life is based and the specific metabolic pathways that handle and control the distribution of the elements in organisms. While this omission was perhaps appropriate when the knowledge about additional bioelements was in its infancy, the current knowledge no longer justifies omission. Historically, the field of biochemistry developed from organic to physiological to biological chemistry. The perceived foundation of biochemistry on organic chemistry was meant to express the fact that life is based on carbon chemistry together with hydrogen, oxygen, and nitrogen chemistry and to some extent sulphur and phosphorus chemistry. However, associating biochemistry with organic chemistry and non-metals only and thus identifying "organic" with the living world and "inorganic" with the inanimate world is not a useful concept in biochemistry as many more chemical elements than the ones traditionally considered in organic chemistry interact with life or are necessary for life and its evolution. A look at the history of the biometal sciences explains the rather slow evolution of the field and further supports this conclusion.

Investigations of the elemental composition of biological matter date back to at least the beginning of the nineteenth century. In the case of iron, they can be traced back even further to the identification of iron in blood in the eighteenth century (Menghini 1747). The essentiality of zinc for fungal growth was shown by Jules Raulin, a student of Louis Pasteur, in 1869 (Raulin 1869) in the process of developing industrial fermentation and understanding the nutrient requirements of yeast. However, even as late as in the 1920s, it was a matter of conjecture whether the presence of zinc in human tissues is the result of environmental exposure or reflects a functional requirement. Convincing arguments were made for a biological role of zinc in humans (Drinker and Collier 1926). However, its presence in an enzyme, carbonic anhydrase, was not demonstrated until the late 1930s (Keilin and Mann 1939), and it took until the 1960s, 100 years after discovering its essential role for fungal growth, to demonstrate that it is an essential element for humans (Prasad et al. 1961). Since then the field of human zinc biology has advanced to the point where zinc must be considered as important or even more important as iron. At least two reasons can be given why metallobiochemistry was considered a specialty and lagged behind other developments that shaped mainstream biochemistry. The analytical instrumental methods to measure metal ions with sufficiently low detection limits and the ability to address speciation have been developed only rather recently (Maret et al. 2015). Also, knowledge about the complexity of pathways that control cellular metal ion concentrations and redistribution emerged quite late. Only in the last two decades a critical mass of knowledge allowed appreciating the overall significance of biometals for biochemistry, thus starting to overcome the initially rather skewed assessment of the chemical basis of life.

1.3 Biological Periodic Systems of the Elements

Considering our detailed knowledge today about some aspects of biochemistry, e.g. having sequenced the entire human genome and many other genomes, it is remarkable that the final count on how many chemical elements are essential for life is open, even for humans. For example, bromine was added as an essential element only in 2014 and there are lingering questions as to whether or not chromium is an essential nutrient (McCall et al. 2014; Mertz 1993; Vincent 2014). It is not widely known that elements such as molybdenum are essential for humans. Only four enzymes use it in humans and they all require the biosynthesis of the pterin cofactor that binds molybdenum to form molybdopterin. Rare diseases leading to molybdenum cofactor deficiency are fatal. Which chemical elements are essential for life is not identical to the question which ones are essential for humans. Some specialized organisms use "unexpected" biometals such as tungsten, cadmium, or even lanthanides (Ljungdahl and Andreesen 1975; Lane and Morel 2000; Pol et al. 2014). The use of nickel and vanadium seems to be restricted to only a few enzymes in certain organisms, not including humans. Biological periodic systems of the chemical elements often do not acknowledge this variation in the usage of the elements in different organisms. For humans, at least 20 chemical elements are essential (Maret 2016b) (Fig. 1.1).

The definition what "essential" means in this context has undergone some changes. It is important to distinguish whether an element is essential for survival or only important for optimizing a particular function. In some articles, fluorine, which supports tooth health as fluoride, is given as an essential element. However, protecting teeth is not an essential function for life. Fluoride is toxic at rather low concentrations. Some elements, such as the metals nickel and vanadium and the non-metals silicon and boron, are bioactive in certain concentration ranges with some beneficial, but not necessarily essential, effects in humans. While molecular mechanisms for their actions are poorly characterized, it is discussed whether intake recommendations for additional not necessarily essential elements should be made (Nielsen 2014a, b, c). It may be relevant that we live with a few hundred different types of commensal bacteria on our mucosal surfaces and skin. In fact,

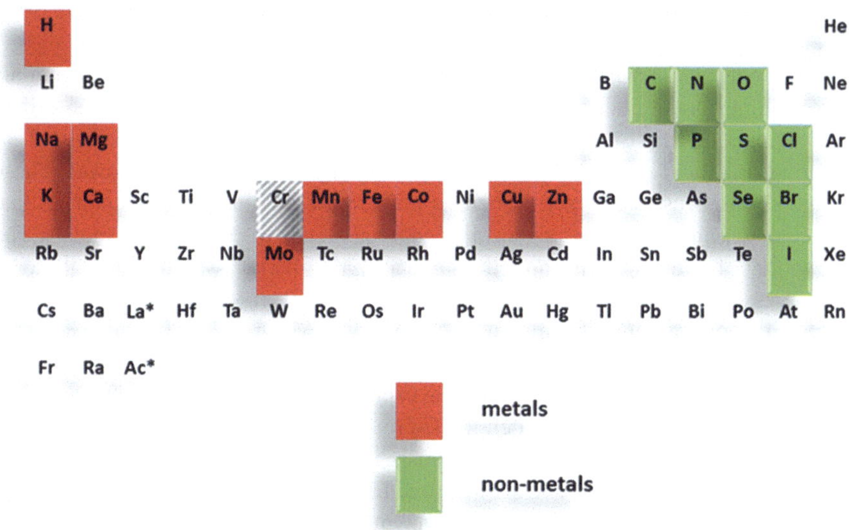

Fig. 1.1 Biological periodic system of the chemical elements that are essential for humans. Chromium is indicated on a background with grey stripes as its status as an essential micronutrient in the form of Cr(III) has been challenged recently

the number of our bacterial cells is higher than the number of our own cells. These bacteria have a different spectrum of essential chemical elements. Thus one reason why some elements are beneficial to us likely relates to the fact that they support the health of our microbiota.

The quantity and relative significance of chemical elements is another area that must be put into proper perspective. Some apparently non-essential elements are present at much higher concentrations than others that are essential (Table 1.1). It is not a foregone conclusion that the presence of these elements is without consequence. The same can be said about other non-essential elements that are present at lower concentrations. Employing highly sensitive ICP-MS, the presence of almost any element in a biological sample can be demonstrated: 74 out of the 78 natural elements investigated were measured in salmon egg cells (Haraguchi et al. 2008). Even radioactive elements such as thorium and uranium can be measured in these egg cells and in the human body at relatively high concentrations when compared to lanthanides, for example. In Table 1.1, both essential and non-essential metals are listed according to their abundance in humans.

A number of conclusions can be drawn from this compilation and knowledge about metal metabolism.

1. The quantities of essential metal ions cover approximately six orders of magnitude, from kg (Ca) to g (Na, K, Mg, Fe, Zn) to mg (Cu, Mn, Mo). Thus:

$$\mathbf{Ca > K, Na > Mg > Fe, Zn > Cu > Mn, Mo > Co.}$$

   The amounts of iron (Fe) and zinc (Zn) are in the gram range typical for *minerals* and thus are not really traces. Referring to them as trace metals is not entirely appropriate. In extension of the above terminology, for the studies of *biominerals*, the term *mineralomics* has been proposed (Yasuda et al. 2006). The amounts of the next two essential metals are two orders (copper, Cu) and three orders (manganese, Mn) of magnitude lower and followed by molybdenum (Mo)/chromium (Cr) and cobalt (Co), which are four orders of magnitude lower. These metals are present as traces. There is no clear definition of the boundaries that separate minerals from trace metals and trace metals from ultratrace metals. The amount of manganese is about the same as that of nickel, which is thought to be non-essential for humans. Remarkably, the amount of vanadium is even lower than that of uranium. The human biochemistry of Mo and Co is linked to cofactor chemistries – pterins and corrins, respectively. But higher amounts of Mo and Co, presumably exceeding the binding capacities of the cofactors, have a variety of adverse health effects as documented for cobalt (Leyssens et al. 2017) and for molybdenum in the form of molybdosis in ruminants. Mo and Co have very limited usage in humans and are acquired for specific functions as cofactors in as far as we know only in a few human enzymes, two for cobalt in the form of vitamin B_{12} and four for molybdenum in molybdopterin. Manganese is a cofactor in a number of enzymes, but it is unknown how many human proteins require it. Even as late as in 1996, Mn was listed only as "probably essential" for humans

Table 1.1 Metal composition of the human body (70 kg human) and metal concentrations in the liver and in whole blood. Essential elements are in bold

	Total amount	Concentration[b,c]	Concentration[b,c]
		Liver (μg kg^{-1})/(ppm)	Whole blood (μg L^{-1})
Kilogram range[a]			
Calcium	1 kg	–/5100	–/60.5 mg L^{-1}
Gram range			
Potassium	140	–/207,000	–/1.62 g L^{-1}
Sodium	110/100	–/–	–/1.97 g L^{-1}
Magnesium	19	–/13,000	–/37.8 mg L^{-1}
Iron	4.2	150–250 mg/kg/16,769	425–500/447 mg L^{-1}
Zinc	2.3	40–60 mg/kg/5543	6–7/7 mg L^{-1}
Milligram range[d]			
Rubidium	320/680	400–6000/–	2–4/2.5 mg L^{-1}
Strontium	320	–/4	–/31
Lead	120	300–600/122	50–150/214
Copper	72	5000–7000/882	0.8–1.1(m); 1–0.4(f)/1.01 mg L^{-1}
Aluminium	61/60	0.3–2/65–500	2–8/390
Cadmium	50	500–2000/203	0.3–1.2(ns); 1–4(s)/5.2
Cerium	–/40	–/–	–/–
Barium	22	–/0.2–10	0.5–2.5/<100
Tin	<17/20	100–1000/5–23	<1/<300
Titanium	–/20	–/–	–/<100
Manganese	12	1000–2000/138	8–12/1.6–7.5
Nickel	10/15	10–50/–	1–5?/4.8–106
Gold	10/0.2	–/–	–/0.04–0.42
Molybdenum	9.3/5	400–800/86	1–3/0.95–75
Chromium	1.8/14	5–50/0.7–12.7	<5?/6.5–107
Lithium	–/7	–/–	0.4–1/<40
Cesium	1.5/6	5–20/–	1.5–4.5/<5
Mercury	–/6	30–150/–	2–20/5–20
Germanium	–/5	–/–	–/440–5000
Cobalt	1.5/3	30–150/2–13	5–10/0.3–9.9
Antimony	2	–/–	–/<5
Silver	2	–/0.1–1.7	–/3.4–120
Microgram range			
Uranium	900	–/–	–/<1
Beryllium	360/36	–/–	–/<3.8
Vanadium	–/110	5–20/<1	0.1–0.5/13.6

ns non-smokers, *s* smokers, *m* male, *f* female, a human liver weighs 1.2–1.5 kg

[a]Sources: web2.airmail.net/uthman/elements_of_body, accessed 10/5/2017, which is based on data in Emsley (1998), and for the second values after the slash www.random-science-tools.com/chemical_comp_of_body, accessed 10/5/2017, which is based on data in the *CRC Handbook of Chemistry and Physics* (1998)

(continued)

Table 1.1 (continued)

[b]Source: WHO (1996); given as μg kg^{-1} = ppb for liver; aside from the different units, comparison with the values after the slash cannot readily be made as the former values are not specified with regard to the procedure, i.e. ashed, dried, or fresh liver

[c]Source: Iyengar et al. (1978) for mean values after the slash and given as mg kg^{-1} = μg g^{-1} = ppm for ashed liver

[d]Additional metals have been measured in the milligram and microgram range; depending on values given from the second source after the slash, the order of abundance may change after titanium; since measurements are from ashed corpses, elements such as mercury and gold used in tooth fillings may contribute to the variation. In addition, there seem to be occasional copying errors and/or issues with the correct units/detection limits, particularly in the case of some of the lower abundance metals

(WHO 1996), the reason being that a critical experiment, namely, demonstrating essentiality by depleting Mn sufficiently to elicit adverse health effects, has been performed in animals but not in humans. Nevertheless, low Mn status is associated with metabolic syndrome, diabetes, and poor birth outcome in humans and is becoming increasingly an issue due to changes in diets as the primary dietary source of Mn is from plants (Freeland-Graves et al. 2016). On the other hand, Mn toxicity is a concern from natural or contaminated water with high Mn content and from other environmental sources. There seems to be a correlation between the abundance of an essential element and the number of functions it is involved in, giving the field of iron and zinc biology some dominance while making the biology of other, less abundant metal ions more of a specialty, at least for humans.

2. The quantities of non-essential metal ions in biological tissues span a similarly wide range. None are in the range of grams but several are in the milligram range, and importantly, several are at concentrations that are higher than those of some essential metal ions. Noteworthy are the relatively high amounts of rubidium (Rb) and strontium (Sr), where we have very little information about functions, but also those of several others such as alumin(i)um (Al) and titanium (Ti). Thus, with reference to essential elements given in italics:

$$\mathbf{Rb, Sr} > \mathbf{Pb} > \boldsymbol{Cu}, \mathbf{Al, Cd} > \mathbf{Ba, Sn, Ti} > \boldsymbol{Mn}, \mathbf{Ni} > \boldsymbol{Co}, \mathbf{U}.$$

"Non-essential" should not be interpreted as meaning "non-functional". What is needed is a better understanding of the functional responses to the presence of non-essential elements in organisms, when, where, and how they interfere with the functions of essential elements, and which range of their concentrations is acceptable without compromising either our short-term or long-term health. Non-essential metals have their own metalloproteomes, which may partially overlap with the metalloproteomes of essential metals because some non-essential metal ions can bind at sites otherwise occupied by essential metal ions.

3. The uneven distribution of metal ions in biological tissues is a major incentive for research to determine functions, making imaging an attractive approach in metallomics. The uneven distribution is both at the subcellular and organismal levels as, for example, shown in imaging metals in the water flea or the zebraf-

ish (De Samber et al. 2008; Bourassa et al. 2014). Specific tissues need particular metal ions. Examples include calcium in the bone and iron in the blood, where 90% of all iron is in the hemoglobin of erythrocytes. In contrast to iron, only 1% of zinc is in the blood, making it mostly a cellular ion. Subcellularly, a large part of iron metabolism is linked to mitochondria. The highest concentration of calcium is in the cytosol, whereas the highest concentration of zinc is in the nucleus.

4. The distribution of elements between blood and organs shows evidence of active processes in taking up certain metal ions and excluding others. Bioaccumulation factors measured for salmon eggs versus seawater vary over six orders of magnitude (Haraguchi et al. 2008). The essential metal ions (Fe, Zn, Cu, Mn, Co) all have accumulation factors in excess of 10,000 up to half a million. Some elements that are present at high concentrations in sea water, such as sodium and chloride, have negative accumulation factor, i.e. they are excluded in marine (salt water) organisms. Remarkably, the toxic mercury ion also has a bioaccumulation factor of 10,000. Thus, *if present*, it accumulates in biological matter due to the presence of ligands with sulphur donors that have high affinity for mercury ions. There are gradients of metal ions across the plasma membrane: for potassium the gradient is inward while for sodium the gradient is outward. It is not known how specific these processes of distribution are and whether the presence of non-essential metal ions in cells simply reflects a lack of absolute discrimination, i.e. non-essential metal ions piggybacking on transport proteins for essential metal ions. Matters of great importance in evaluating the significance of the presence of non-essential metal ions are whether promiscuity was encouraged evolutionarily so that when one metal ion becomes limiting, another one can substitute for it and maintain function, whether such metal swapping in proteins is even accompanied with intended functional changes that signal the organism limited availability of the correct metal, or metal substitution merely compromises function.
5. Control of systemic and cellular homeostasis maintains concentrations of essential metal ions in a rather narrow range. Such control minimizes overlap between the functions of different metal ions, is extremely tight for iron and zinc, and is necessary as the specificity of biological coordination environments is not high enough to select only the required metal ion. Thus, in vitro, zinc can bind in iron-binding sites of proteins and iron can bind in zinc-binding sites of proteins. The control of availability of metal ions in biological systems provides additional selectivity and avoids that more competitive metal ions bind in sites that need to be populated by less competitive metal ions. For some metal ions, e.g. copper, specific metallochaperones transfer the metal ions to the sites where they are required. The mechanisms of control are quite elaborate and metal-specific and are part of the many pathways that incorporate chemical elements into biomolecules (Foster et al. 2014). In addition to the metalloproteins that use metal ions for catalytic or structural functions, many metalloproteins are involved in handling and controlling metal ions. The large number of these proteins is further evidence for the impor-

tance of pathways for acquisition and distribution of chemical elements and thus the significant role of metallobiochemistry in general biochemistry (Maret and Wedd 2014).

Other elements that do not underlie such strict homeostatic control as iron and zinc cover a wider range of concentrations in tissues. It appears that essential metal ions are controlled in a narrow range, whereas non-essential metal ions, due to a lack of homeostatic control, can vary over a much wider range.

1.4 Metallomics and Human Health

One aim of metallomics is to establish the normal distribution and concentrations of metal ions in tissues and cells. Such reference values then can serve to understand changes that cause disease or are a result of disease. Control of optimal concentrations of metal ions in the human body has huge implications for human health and requires further scrutiny to establish additional guidelines to protect human populations and to prevent and treat disease. Disease can result from a deficiency or an overload of an essential element or the presence of a non-essential and toxic element. Not only the nutritional availability of metal ions determines risk for disease but also the interactions among metal ions, the presence of metal-binding (chelating) agents in the diet, and, when patients are medicated, the interactions of metal ions with therapeutic drugs. In addition, genetic variations in the proteins that handle the metal ions affect metal metabolism and utilization. For example, genetic and non-genetic factors can cause iron overload. For zinc, there is no known condition of cellular overload aside from cellular injury. However, there are many genetically determined changes in zinc metabolism, including a condition of high zinc in the blood (hyperzincaemia) due to increased levels of the protein calprotectin (Hogstrand and Maret 2016).

The consequences of deficiencies of essential biominerals and trace and ultratrace biometals on health have been the major focus of investigations, but we know comparatively little what the consequences for human health are, either acutely or chronically, as a result of the presence or accumulation of non-essential metals. It is a very important area of research, as the presence of non-essential metals and the majority of essential metals are never examined clinically on a routine basis. A blood test such as the basic metabolic panel (BMP) contains seven determinations, hence its name CHEM-7 or SMA-7 (sequential multiple analysis). Only sodium and potassium are measured, and sometimes calcium (CHEM-8). While there are seven rather routine tests for iron (ferritin, serum iron, transferrin saturation, total iron-binding capacity, transferrin, transferrin receptor), there is not a single test for zinc or any of the other essential elements, even in extended tests that include liver function, in the so-called comprehensive metabolic panel (CMP), in which 14 parameters are determined with SMAC autoanalyzers. Magnesium deficiency is associated with some chronic diseases; there is now evidence for subclinical deficiencies even if the serum magnesium status is normal within the currently accepted reference interval (Costello et al. 2016). We have recommendations for optimal concentrations of the essential elements. However, we often do not have biomarkers that inform us whether the concentrations in tissues and the associated functions are optimal. There should be investigations on the health effects of many more elements aside from the essential and highly toxic ones.

Homeostatic mechanisms control essential but not non-essential elements. The concentrations of non-essential elements are more variable and depend on exposure, interactions, and genetic factors. While UL (upper limit) values have been estimated on the basis of overt signs of toxicity, we do not know all the consequences of the presence of these elements at lower concentrations. Many exposure models are based on a linear relationship between dose and response. However, there are many examples of non-linear responses with hormetic effects. Exposure to some non-essential elements through water, soil, and air has changed drastically in recent years due to new manufacturing practices and using metal compounds and novel nanoparticles in various high-tech applications. Aside from lanthanides and chemically very reactive tellurium compounds, it includes elements such as hafnium, tantalum, and indium, for which biological investigations of exposure are extremely limited or even non-existent (Ridgway and Webb 2015). Also, metal ions or metal compounds are used as metallodrugs for therapeutic or diagnostic purposes, introducing metals that are usually virtually absent, e.g. Pt-based anticancer drugs, or present at low concentrations, e.g. Li salts for treating manic depression. Gadolinium deposits from gadolinium-based contrast agents used in magnetic resonance imaging (MRI) have been detected in the human brain and bone, resulting in a "gadolinium deposition disease"/"gadolinium toxicity disorder", which is presently being further evaluated (Ramalho et al. 2016). On the other hand, we are exposed to chelating agents such as some nutritional supplements or food additives and pharmaceuticals that bind metal ions and interfere with metal metabolism. Also, changing diets in many countries due to new ways of food production and processing and changes in food preferences and availability have additional effects on the nutritional status of essential and non-essential elements.

Some mechanisms of detoxification for non-essential elements have been discussed. There is a wide range of toxicities with different modes of action. Lead (Pb), for example, is a very potent inhibitor for δ-aminolevulinic acid dehydratase interfering with heme biosynthesis. In contrast, the actions of cadmium (Cd) seem to be pleiotropic as a large number of proteins interact with it (Maret and Moulis 2013). The amounts of cadmium and lead in the human body are remarkably high, in the order of those of essential copper and presumably non-essential nickel. For lead, it is now becoming accepted that there is no threshold value for its toxicity and that no level of exposure can be considered safe (Maret 2017). For cadmium, such a relationship between exposure and health effects has not been established.

The description of functional effects would be incomplete without pointing out the significance of interactions. Thus, in many instances the presence or absence of one biometal influences the presence or absence of another biometal. For example, under iron deficiency, more cadmium is taken up, making deficiencies of essential metal ions a risk factor for toxic metal ions. The availability of a metal ion may be influenced by the nutritional status of a non-metal, e.g. selenium. Requirements for essential metal ions and vulnerability to the toxic effects of metal ions depend on life cycle stages. Toxic metals such as lead and cadmium accumulate with age. There is evidence of dyshomeostasis of essential metal ions in older people. Apparently, homeostatic control works less efficiently and uncomplexed (free) metal ions accumulate and interfere with protein functions. One effect is pathological protein aggregation triggered by metal binding to proteins. Neurodegenerative disease such as Alzheimer's disease is one example where such a role of metal ions has been demonstrated.

Each element requires biochemical investigations to determine the range of concentrations in healthy and diseased human tissue. There should be increased monitoring, which is relatively inexpensive for metals compared to organic molecules, as metal ions are relatively easy to identify and quantify without further separation techniques.

1.5 Bioinformatics: Predicting Metalloproteomes

Most of our present knowledge of metallobiochemistry relates to the interaction of metal ions with proteins, with the notable exception of magnesium biochemistry pertaining to the interaction with RNA. Accordingly, a major emphasis in metallomics is on metalloproteomes and metalloproteomics (Maret 2010). The combination of experimental and computational approaches has been very successful and together with advances in analytical techniques has laid the foundations for the field of metallomics as a systems biology approach. Once a sufficient number of sequences and 3D structures of metalloproteins was determined, analyses of the metal-binding sites revealed some general features. It was recognized, for example, that the binding of calcium in EF hands and zinc in zinc fingers is similar in different proteins, as reflected in the spacing and types of ligand donors in the primary structure (sequence) (Kretsinger and Barry 1975; Miller et al. 1985). Recurring patterns or "signatures" of amino acids in sequences were recognized, which then could be employed as templates for identifying unknown metal-binding sites in other proteins. One factor is the type of amino acid that provides the ligand donor to the metal. Only a few amino acids have such donors in their site chain: oxygen from Glu, Asp, nitrogen from His, and sulphur from Cys and in some instances oxygen from Tyr (Fe) and sulphur from Met (Cu). Rarely others, such as oxygen from serine, serve as ligand donors. Another critical observation was that short spacers of 1–4 amino acids – or no spacer at all – occur between amino acids that provide the donors and form the metal-binding sites (Fig. 1.2). In metalloenzymes, usually a long spacer follows the short spacer to reach the amino acid that provides the third ligand donor (Vallee and Auld 1989).

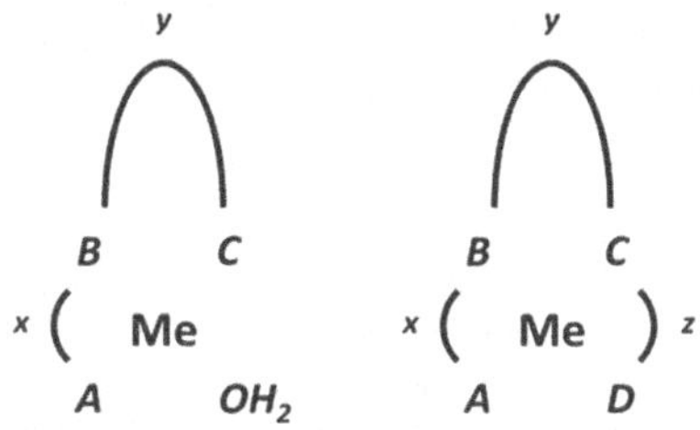

Fig. 1.2 Metal signatures in proteins. The amino acids that provide the ligand donors (A, B, C, D) to the metal (Me) are separated by characteristic spacers (x, y, z). Left: catalytic metal site in enzymes, where the fourth ligand donor is a water molecule in the substrate binding site. Right: structural metal site in proteins. The presentation should be understood as an abstract cartoon only to illustrate the concept of how spacers and ligand donors are employed to predict metal sites in proteins. In reality, there is considerable variation in the number of ligand donors (coordination number) and the lengths of spacers

In structural metal sites, the spacers are usually shorter. The number of amino acids in the spacers is often conserved, as are, to some extent, the types of amino acids. There are exceptions to these rules, though. When inspecting the tertiary structures of proteins, it was noticed that domains for metal binding are conserved in a large number of metalloproteins. This modular nature of metal binding sites in proteins is also employed for prediction, either in queries for unknown metalloproteins when using protein or nucleic acid sequences or in queries of entire genomes, which became available in increasing numbers (Andreini et al. 2011). Several web servers are available for performing predictions.

While this novel way of searching for metalloproteins employing bioinformatics and mining databases turned out to be remarkably successful, the classical and labour-intensive approach of isolating and characterizing metalloproteins is still needed for experimental verification of the predictions. The bioinformatics approach made it possible to predict the total number of metalloproteins for some essential metals, the metalloproteomes, in different organisms. It allowed estimating the number of iron, copper, and zinc proteins, but not manganese proteins due to the lack of suitable consensus signatures. The bioinformatics approach made an enormous contribution to the field because it increased the number of known metalloproteins significantly, at least by an order of magnitude, and thus, for the first time, showed the full impact of metals in biochemistry: 1% of all human proteins contain non-heme iron plus about 70 proteins Fe-S clusters, 9% zinc, and 1% copper (Andreini et al. 2006; Andreini et al. 2016). If one adds calcium, magnesium, and heme proteins, an estimated 30–40% of all proteins require a metal cofactor. Thus, in humans, there are more than 3000 zinc proteins and 54 copper proteins (Blockhuys et al. 2017). The mining of genomes using metal signatures introduced an omics approach to metallobiochemistry for investigating the entirety of metalloproteins. A variant of this approach is the search for selenoproteins in genomes. Selenium, a chemical homologue of sulphur, can be present in the inorganic ionic form or covalently bound in many chemical species and with various valence states. It is incorporated into selenoproteins as selenocysteine by recoding the stop codon and using a stem-loop structure in the mRNA and a couple of ancillary proteins. Twenty human selenoproteins are known (Lu and Holmgren 2009). Selenium from selenocysteine can also be a ligand donor to metal ions. Selenomethionine, on the other hand, is incorporated into proteins randomly.

The specificity of signatures is linked to the factors that determine coordination environments of biometals in proteins, which is a mainstay area of bioinorganic chemistry/structural biology. For calcium, which is an important cellular signalling ion (second messenger), remarkably specific sites have evolved. To perform the functions of calcium, biology had to make coordination environments that bind calcium with higher affinity than magnesium, opposite to general chemical principles according to which the affinity of magnesium for the same ligands is higher. Selectivity was achieved by using up to seven donor atoms from ligands in sites and also a unique amino acid, γ-carboxyglutamic acid (Gla), which is synthesized in a vitamin K-dependent reaction.

The bioinformatics approach to metalloproteomes of essential metal ions has some limitations as discussed for the zinc proteome (Maret 2004, 2008): (i) it is a prediction and not a chemical analysis; (ii) it can employ only known signatures for prediction, clearly leaving out metalloproteins for which signatures are yet unknown or are not recognizable in sequences because the ligands differ or have different spacer lengths; (iii) it *assumes* that the signature binds a specific metal ion. The latter point is critical because the metal ion may not be present at all, as the availability of metal ions is determined by the entire regulatory systems of the organism responsible for distributing and allocating metal ions and not just the binding capacity of a site. The availability of metal ions is variable and hence the metalloproteome is variable. At low availability, some sites will not be occupied while at high availability additional sites will bind metal ions. In some sites, the metal binds only transiently and hence the metal may not be present when the protein is isolated. There is also ambiguity concerning which metal ion is bound. A particular signature can bind one metal in one instance and another one in another instance. This promiscuity is known for manganese and iron superoxide dismutase and other cambialistic proteins. Another example is metallothionein, a family of proteins in humans, binding zinc, copper, cadmium, and less abundant metal ions (Krężel and Maret 2017). Other limitations include: (iv) prediction is possible for certain structural and catalytic sites but not for inhibitory sites, intersubunit sites, where the ligand donors are on different proteins, or sites where the ligand donors are far apart from each other in the sequence. Signatures are linear, sequential motifs of amino acids. However, there are instances where the ligand donors are arranged non-sequentially or where the donors form ligand bridges. Such binding is difficult or impossible to predict. Lastly, (v) predictions are looking only at metal binding to proteins. The metalloproteome is not identical with the metallome. The difference between metals bound to proteins and metals bound to other biomolecules increases when the metal ions become less competitive. Less competitive metal ions are distributed in a wider pool of biomolecules because the equilibrium between protein-bound and free metal ion leaves much more free (non-protein bound) metal ions to bind to other species. Even for a highly competitive metal ion such as Zn^{2+}, the pool of non-protein bound cellular zinc is not negligible and has a significant role in cellular regulation as a signalling ion. Accordingly, it requires regulatory mechanisms in addition to the control of cellular homeostasis (Maret 2013, 2014). Metal buffering involving metal-binding biomolecules and transport of metal ions through the cytoplasmic membrane or subcellular membranes controls free metal ion concentrations. The dynamic biological transport processes contributing to biological metal buffering are referred to as muffling (Colvin et al. 2010).

1.6 The Multidimensionality of Metallomics

Metallomics can be subdivided into structural, functional, and quantitative metallomics. Structural metallomics addresses the chemical structures of metal ions and metalloids. Functional metallomics aims at annotating these structures in terms of their biological functions. Quantitative metallomics determines the concentrations of metal ions and metalloids in tissues and cells, including imaging their distribution in two and three dimensions and overlaying such images with physical maps obtained with microscopy and computer tomography techniques. Comparative genomics of metallomes focuses on the utilization of metals and evolutionary dynamics in different forms of life (Gladyshev and Zhang 2013).

Metallomes are inherently incomplete and that there are an infinite number of metallomes depending on the biological context. Biological systems are dynamic, the essence of being alive, with active metabolism and changes occurring in biological space and time. Every individual is slightly different. We observe this in genetics with the *variome*, a set of genetic variations in populations of one species, some of which affect metal metabolism and distribution, and we see this in non-genetic factors, the *phenome*, a sum of all phenotypic traits. Individual variation is of course present in metallomes due to environmental factors such as nutrition, exposure, and disease. This is the area of *metametallomics*, analogous to the field of metagenomics, the study of genetic material dependent on all environmental influences.

1.7 Conclusion

Metallomics emphasizes the importance of metals in biology through the integration of research fields and methodologies. It is an emerging multidisciplinary field of research and when applied properly has far-reaching implications and future potential. The interpretation of metallomics data requires training in biology and medicine, which is not part of the traditional education of a chemist. Thus, additional curricular activities and training opportunities are necessary for the field to prosper. It took considerable time and effort to arrive at this point and will require such for further development.

References

Andreini C, Bertini I, Rosato A (2006) Metalloproteomes: a bioinformatic approach. Acc Chem Res 42:1471–1479

Andreini C, Banci L, Rosato A (2016) Exploiting bacterial operons to illuminate human iron-sulfur proteins. J Proteome Res 15:1308–1322

Andreini C, Bertini I, Cavallaro G et al (2011) A simple protocol for the comparative analysis of the structure and occurrence of biochemical pathways across superkingdoms. J Chem Inf Model 51:730–738

Baker M (2013) The 'omes puzzle. Nature 494:416–419

Blockhuys S, Celauro E, Hildesjö C et al (2017) Defining the human copper proteome and analysis of its expression variation in cancers. Metallomics 9:112–123

Bourassa D, Gleber S-C, Vogt S et al (2014) 3D imaging of transition metals in the zebrafish embryo by X-ray fluorescence microtomography. Metallomics 6:1648–1655

Colvin RA, Holmes WR, Fontaine CP et al (2010) Cytosolic zinc buffering and muffling: their role in intracellular zinc homeostasis. Metallomics 2:306–317

Costello RB, Elin RJ, Rosanoff A et al (2016) Perspective: the case for an evidence-based reference interval for serum magnesium: the time has come. Adv Nutr 7:977–993

Lide DR (ed) (1998) CRC Handbook of Chemistry and Physics, 79th edn. CRC Press, Boca Raton

Cvetkovic A, Menon AL, Thorgersen MP et al (2010) Microbial metalloproteomes are largely uncharacterized. Nature 466:779–784

De Samber B, Evens R, De Schamphelaere G et al (2008) A combination of synchrotron and laboratory X-ray techniques for studying tissue-specific trace level metal distributions in *Daphnia magna*. JAAS 23:829–839

Drinker KR, Collier ES (1926) The significance of zinc in the living organism. J Industr Hygiene 8:257–269

Emsley J (1998) The elements, 3rd edn. Clarendon Press, Oxford

Foster AW, Osman D, Robinson NJ (2014) Metal preferences and metallation. J Biol Chem 289:28095–28103

Freeland-Graves JH, Mousa TY, Kim S (2016) International variability in diet and requirements of manganese: causes and consequences. J Trace Elem Med Biol 38:24–32

Gladyshev VN, Zhang Y (2013) Comparative genomics analysis of the metallomes. In: Metallomics and the cell, Banci L, (Guest ed.) vol 12 of Metal Ions in Life Sciences, A. Sigel, H. Sigel, R. K. O. Sigel (eds). Springer Science + Business Media B.V., Dordrecht, pp 529–580

Haraguchi H (2004) Metallomics as integrated biometal science. JAAS 19:5–14

Haraguchi H, Ishii A, Hasegawa T et al (2008) Metallomics study on all-elements analysis of salmon egg cells and fractionation analysis of metal in cell cytoplasm. Pure Appl Chem 80:2595–2608

Haraguchi H (2017) Metallomics: the history in the last decade and the future outlook. Metallomics 9:1001–1013

Hogstrand C, Maret W (2016) Genetics of human zinc deficiencies. In: eLS. John Wiley & Sons, Ltd, Chichester. https://doi.org/10.1002/9780470015901.a0026346

Iyengar GV, Kollmer WE, Bowen HJM (1978) The elemental composition of human tissues and body fluids. Weinheim, Verlag Chemie

Keilin D, Mann T (1939) Carbonic Anydrase. Nature 144:442–443

Kretsinger RH, Barry CD (1975) The predicted structure of the calcium-binding component of troponin. Biochim Biophys Acta 405:40–52

Krężel A, Maret W (2017) The functions of metamorphic metallothioneins in zinc and copper metabolism. Int J Mol Sci 18:1237

Lahner B, Gong J, Mahmoudian M et al (2003) Genomic scale profiling of nutrient and trace elements in *Arabidopsis thaliana*. Nat Biotechnol 21:1215–1221

Lane TW, Morel FMM (2000) A biological function for cadmium in marine diatoms. Proc Natl Acad Sci U S A 97:4627–4631

Leyssens L, Vinck B, Van Der Straeten C et al (2017) Cobalt toxicity in humans-a review of the potential sources and systemic health effects. Toxicology 387:43–56
Li Y-F, Chen C, Qu Y et al (2008) Metallomics, elementomics, and analytical techniques. Pure Appl Chem 80:2577–2594
Ljungdahl LG, Andreesen JR (1975) Tungsten, a component of active formate dehydrogenase from *Clostridium thermoaceticum*. FEBS Lett 54:279–282
Lobinski R, Becker JS, Haraguchi H et al (2010) Metallomics: guidelines for terminology and critical evaluation of analytical chemistry approaches (IUPAC technical report). Pure Appl Chem 82:493–504
Lu J, Holmgren A (2009) Selenoproteins. J Biol Chem 284:723–727
Maret W (2004) Exploring the zinc proteome. JAAS 19:15–19
Maret W (2008) Zinc proteomics and the annotation of the human zinc proteome. Pure Appl Chem 80:2679–2687
Maret W, Li Y (2009) Coordination dynamics of zinc in proteins. Chem Rev 109:4682–4707
Maret W (2010) Metalloproteomics, metalloproteomes, and the annotation of metalloproteins. Metallomics 2:117–125
Maret W (2013) Zinc and the Zinc Proteome. In Metallomics and the cell, Banci L, (Guest ed.) vol. 12 of Metal ions in life sciences, A. Sigel, H. Sigel, R. K. O. Sigel (eds). Springer Science + Business Media B.V., Dordrecht, pp 479–501
Maret W, Moulis J-M (2013) The bioinorganic chemistry of cadmium in the context of its toxicity. In Cadmium: from toxicity to essentiality, vol. 11 of Metal ions in life sciences, A. Sigel, H. Sigel, R. K. O. Sigel (eds). Springer Science + Business Media B.V., Dordrecht, pp 1–29
Maret W (2014) Molecular aspects of zinc signals. In: Fukada T, Kambe T (eds) Zinc signals in cellular functions and disorders. Springer, Tokyo, pp 7–26
Maret W, Wedd AG (eds) (2014) Binding, Transport and Storage of Metal Ions in Biological Cells. Royal Society of Chemistry, Cambridge
Maret W, Caruso JA, Contag CH et al (2015) In: Nriagu JO, Skaar EP (eds) Trace Metals and Infectious Diseases. MIT Press, Cambridge, pp 341–401
Maret W (2016a) Metallomics, a primer to integrated biometal sciences. Imperial College Press, London
Maret W (2016b) The metals in the biological periodic system of the elements: concepts and conjectures. Int J Mol Sci 17:66
Maret W (2017) The bioinorganic chemistry of lead in the context of its toxicity. In Lead – its effects on environment and health, vol 17 of Metal Ions in Life Sciences, A. Sigel, H. Sigel, R. K. O. Sigel (eds), W. de Gruyter, Berlin, pp 1–20
McCall AS, Cummings CF, Bhave G et al (2014) Bromine is an essential trace element for assembly of collagen IV scaffolds in tissue development and architecture. Cell 157:1380–1392
Menghini V (1747) De Ferreorum particulerum sede in sanguis. Commentar. Bononiens (Bologna) ii, p 475
Mertz W (1993) Chromium in human nutrition. J Nutr 123:626–633
Miller J, McLachlan AD, Klug A (1985) Repetitive zinc-binding domains in the protein transcription factor IIIA from Xenopus oocytes. EMBO J 4:1609–1624
Nielsen FH (2014a) Should bioactive trace elements not recognized as essential, but with beneficial health effects, have intake recommendations. J Trace Elem Med Biol 28:406–408
Nielsen FH (2014b) Update on the possible nutritional importance of silicon. J Trace Elem Med Biol 28:379–382
Nielsen FH (2014c) Update on human health effects of boron. J Trace Elem Med Biol 28:383–387
Pol A, Barends TRM, Dietl A et al (2014) Rare earth metals are essential for methanotrophic life in volcanic mudpots. Env Microbiol 16:255–264
Prasad AS, Halsted JA, Nadimi M (1961) Syndrome of iron deficiency anemia, hepatosplenomegaly, hypogonadism, dwarfism and geophagia. Am J Med 31:532–546
Ramalho J, Ramalho M, Jay M et al (2016) Gadolinium toxicity and treatment. Magn Reson Imaging 34:1394–1398

Raulin J (1869) Etudes chimiques sur la végétation. Ann Sci Nat Bot Biol Veg 11:92–299
Ridgway A, Webb R (2015) The materials bonanza. New Sci 225:35–41
Salt DE (2004) Update on ionomics. Plant Physiol 136:2451–2456
Szpunar J (2004) Metallomics: a new frontier in analytical chemistry. Anal Bioanal Chem 378:54–56
Vallee BL, Auld DS (1989) Short and long spacer sequences and other structural features of zinc binding sites in zinc enzymes. FEBS Lett 257:138–140
Vincent JB (2014) Is chromium pharmacologically relevant? J Trace Elem Med Biol 28:397–405
WHO (1996) Trace elements in human nutrition and health. World Health Organization, Geneva, pp 258–259
Williams RJP (2001) Chemical selection of elements by cells. Coord Chem Rev 216:583–595
Yasuda H, Yonashiro T, Yoshida K et al (2006) Relationship between body mass index and minerals in male Japanese adults. Biomed Res Trace Elem 17:316–321
Zhang P, Georgiou CA, Brusic V (2017) Elemental metabolomics. Brief Bioinform. https://doi.org/10.1093/bib/bbw131

Chapter 2
Metallomics Applied to the Study of Neurodegenerative and Mental Diseases

Alessandra Sussulini and Rachel Ann Hauser-Davis

Abstract Biochemical imbalances, provoked by aging or a secondary illness, might directly affect the brain, causing severe problems, such as loss of memory or alteration of behavior patterns. Brain disorders are usually classified as injuries (such as stroke, hematomas, and concussions), tumors, and neurodegenerative (such as Parkinson's and Alzheimer's diseases) and mental (such as depression, bipolar disorder, schizophrenia) diseases. As the pathophysiology of these illnesses is not completely established and multiple factors are involved, metallomics, a bioanalytical strategy that allows the detection of metal ions and metalloproteins in diverse biological matrices, is of extreme relevance in identifying which elements are affected by a disease and/or treatment. Thus, determining which element ions suffer disturbances in their homeostasis during the disease progress is relevant to understand the biochemical changes and propose new drug targets. In addition, it is well known that oxidative stress plays an important role in the development of pathological neurodegenerative and mental diseases, which may be caused by metal ion dyshomeostasis, so it is also important to understand endogenous antioxidant metalloprotein and metalloenzyme mechanisms in this regard. In this context, recent applications of metallomics in the study of neurodegenerative and mental disorders are discussed in this chapter, as well as future trends in this research area.

A. Sussulini (✉)
Laboratory of Bioanalytics and Integrated Omics (LaBIOmics), Department of Analytical Chemistry, Institute of Chemistry, University of Campinas (UNICAMP), Campinas, São Paulo, Brazil

Spectrometry, Sample Preparation and Mechanization Group (GEPAM), Department of Analytical Chemistry, Institute of Chemistry, University of Campinas (UNICAMP), Campinas, São Paulo, Brazil

National Institute of Science and Technology for Bioanalytics – INCTBio, Institute of Chemistry, University of Campinas (UNICAMP), Campinas, São Paulo, Brazil
e-mail: sussulini@iqm.unicamp.br

R. A. Hauser-Davis
Centro de Estudos da Saúde do Trabalho e Ecologia Humana (CESTEH), Escola Nacional de Saúde Pública Sérgio Arouca (ENSP), FIOCRUZ, Rua Leopoldo Bulhões, Rio de Janeiro, Brazil

M. A. Z. Arruda (ed.), *Metallomics*, Advances in Experimental Medicine and Biology 1055, https://doi.org/10.1007/978-3-319-90143-5_2

Keywords Metallomics · Brain · Mass spectrometry imaging · ICP-MS · Neurodegenerative diseases · Mental disorders · Metalloproteins · Metalloenzymes · Oxidative stress

Abbreviations

6-OHDA	6-hydroxydopamine
AAS	Atomic absorption spectrometry
AD	Alzheimer's disease
ALS	Amyotrophic lateral sclerosis
Aβ	β-amyloid
BD	Bipolar disorder
CNS	Central nervous system
CSF	Cerebrospinal fluid
HD	Huntington's disease
ICP-MS	Inductively coupled plasma mass spectrometry
LA-ICP-MS	Laser ablation inductively coupled plasma mass spectrometry
MPTP	1-methyl-4-phenyl-1,2,3,6-tetrahydropyridine
MRI	Magnetic resonance imaging
MS	Multiple sclerosis
MT	Metallothionein
PD	Parkinson's disease
SOD	Superoxide dismutase
WD	Wilson's disease

2.1 Introduction

The role of metal ions in the brain is directly connected to the development and maintenance of enzymatic activities, learning and memory, mitochondrial function, myelination, and neurotransmission. The disruption of any of these mechanisms or the absorption of toxic metals generates a metal homeostasis imbalance in the brain, which initiates a cascade of events that leads to neurodegeneration and cell death, resulting in a diseased state (Prakash et al. 2017). Aging is an unavoidable risk factor for neurodegenerative disorders, and an age-dependent deterioration of the metal homeostatic system hypothesis has been proposed (Barnham and Bush 2014).

The etiology of most neurodegenerative and mental disorders is linked to different biological processes, including oxidative stress and proteinaceous aggregations, in which metal ions and metalloproteins are involved. Based on this fact, metallomics is a key bioanalytical strategy, which is able to provide relevant information on the distribution and concentrations of metal ions and/or metalloproteins in the brain, as well as in biological fluids, such as serum, plasma, or cerebrospinal fluid (CSF).

In this chapter, selected applications of the metallomics strategy for studying neurodegenerative and mental disorders will be discussed. First, we shall discuss the analysis of brain samples through elemental mass spectrometry imaging, followed by metallomic studies of biofluid samples, including metalloproteins and metalloenzyme investigations.

2.2 Metal Dyshomeostasis and Oxidative Stress

Metal dyshomeostasis is present in many neurodegenerative and psychiatric disorders of the brain, although reports are extremely conflicting, since many different elements have been shown to be both increased or decreased, or even not altered, in these conditions. Differential levels have also been reported during several different stages of progression of many disorders, and confounding factors are usually present in human studies. Thus, metallomic investigations are paramount in order to identify possible biomarkers for these diseases and further advance knowledge in both essential and nonessential element homeostasis. Metal determinations in this regard, in different biological matrices such as the brain and biofluids, are usually conducted either by inductively coupled plasma mass spectrometry (ICP-MS) or by atomic absorption spectrometry (AAS) (Subramanian 1996; Bolann et al. 2007).

Oxidative stress is also an important symptom of these conditions, since many, if not most, present neuroinflammation aspects in conjunction with other pathophysiological findings. Also, the brain, with its extensive capacity to consume large amounts of oxygen and produce free radicals, is considered particularly sensitive to oxidative damage (Halliwell 2006). This condition results from the incomplete reduction of oxygen, leading to the formation of reactive oxygen species, decreases in the degradation of these species by endogenous antioxidant agents, or both (Halliwell 2001; Sas et al. 2007). Thus, many antioxidant proteins and enzymes are of interest in the study of these diseases, in both the brain itself and in biofluids.

In the following sections, we will deal with both aspects, metal dyshomeostasis and oxidative stress, in some important brain diseases. Enzyme and protein determinations, both in the brain and in biofluids, are mainly conducted through mass spectrometry techniques.

2.3 Elemental Mass Spectrometry Imaging in Brain Samples

Elemental imaging by laser ablation inductively coupled plasma mass spectrometry (LA-ICP-MS) provides spatial information of the distribution of elements in brain samples that allows for the correlation of neuroanatomical regions (and their known function) with metal-dependent biological processes (Hare et al. 2017). Complementary information regarding LA-ICP-MS can be found in the Chap. 7 of this book. However, a review focusing on the current capabilities and applications of LA-ICP-MS imaging in biomedical research, including brain diseases, contrast agents and metallodrug distribution, and metalloproteomics, has been recently published (Sussulini et al. 2017). Herein, advances in the study of neurodegenerative diseases using LA-ICP-MS will be discussed.

The LA-ICP-MS technique was applied in the evaluation of Parkinson's disease induced by unilateral 6-hydroxydopamine (6-OHDA) in mouse models (Sussulini et al. 2012). An increase of periventricular and decreased striatal parenchymal copper concentrations were found, which are consistent with increased copper levels in human CSF and a degenerative pathophysiology. In another study using 6-OHDA-induced Parkinson's disease mouse models (Matusch et al. 2012), increased concentrations of iron, manganese, and copper in substantia nigra were reported.

A different model of Parkinson's disease, induced by 1-methyl-4-phenyl-1,2,3,6-tetrahydropyridine (MPTP), was also evaluated by LA-ICP-MS imaging (Matusch et al. 2010). In this case, copper, iron, zinc, and manganese in native form were quantified in brain sections. Significant decreases of copper concentrations were observed in the periventricular zone and the fascia dentata. In the cortex, copper concentration decreased, and iron was increased in the interpeduncular nucleus, but not in the substantia nigra. These results emphasize the value of microlocal analytical techniques, such as LA-ICP-MS, in studying brain disorders and affirm a role of copper availability in Parkinson's disease.

Regarding Alzheimer's disease (AD), altered concentrations of metal ions present in the β-amyloid (Aβ) containing plaques in brains suffering from this condition have been reported. In a study using transgenic mouse model of AD (TASTPM, which overexpresses human mutant APP and presenilin-1 transgenes and develops senile plaque-like deposits of Aβ protein), LA-ICP-MS was employed for the imaging of Aβ deposits based on measurement of labeled (Eu/Ni) monoclonal antibodies (Hutchinson et al. 2005). Magnesium, calcium, copper, zinc, iron, and aluminum were mapped and a correlation between trace elements and Aβ deposits was suggested, but no precise conclusions were presented.

Wilson's disease (WD) is known to be caused by a genetic defect of the copper efflux pump ATPase7B. A model of WD based on mouse strains with natural or engineered transgenic defects of the Atp7b gene was studied using LA-ICP-MS imaging (Boaru et al. 2014). Copper accumulation was detected in brain parenchyma in WD mice compared to controls. In the periventricular regions, copper concentration was reduced and zinc concentration was increased. The concentrations of iron and manganese were constant throughout all evaluated regions.

Multiple sclerosis (MS) is a neurodegenerative disorder characterized by active demyelination and iron deposition in lesions. Hence, iron detection might provide a biomarker for chronic inflammation within lesions, and a study using LA-ICP-MS imaging combined with magnetic resonance imaging (MRI) was performed (Wisnieff et al. 2015). Two lesions of the same individual were analyzed, and the lesion borders showed differing amounts of iron and myelin contributing to the total susceptibility. One of the lesions had high iron and low myelin content, whereas the other lesion showed low iron contribution and high myelin concentration. This was a preliminary study involving the characterization of MS lesions, which needs to be further explored.

A model of amyotrophic lateral sclerosis (ALS) using transgenic mice expressing human mutated superoxide dismutase 1 (SOD1) linked to familial forms was studied by elemental mass spectrometry imaging of spinal cord samples (Steinacker et al. 2010). The quantification of copper, phosphorus, and zinc was performed to compare control (wild type), prion knockout (Prp−/−), and terminal ALS ($SOD1^{G93A}$) models. Cellular prion protein (Prp) was selected due to its proposed neuroprotective function. This study reported that elemental distribution was similar in Prp−/− and $SOD1^{G93A}$ mice, with a decrease in copper and an increase in zinc concentrations in comparison with controls.

To the best of our knowledge, mental disorders, such as schizophrenia and bipolar disorder, have not yet been studied using elemental mass spectrometry imaging strategy. In this context, this is an interesting future trend in the area of metallomics, with high discovery potential, since this is a completely unexplored research topic.

2.4 Metallomics in Biofluids

One of the most desirable goals for healthcare research and delivery is the capability to assess physiological state and detect morbidity initiation and progression through non-invasive approaches (Lee and Wong 2009). In this regard, the analysis of biofluids to detect early stages of brain diseases has been increasingly applied with metallomics playing an important role, since metallomic evaluations of biofluids in patients suffering several different brain diseases in the search for disease biomarkers can been conducted *antemortem*.

2.4.1 Parkinson's Disease

Metal dyshomeostasis of different elements in PD patients has been observed in different biofluids, although there is no consensus regarding which element can be used as an indicative of the disease, since differential levels have been reported for several elements, with many discrepancies. For example, significantly higher calcium, chromium, and magnesium and significantly lower aluminum, cobalt, iron, manganese, silicon, and zinc have been observed in CSF from PD patients, whereas significantly higher aluminum, calcium, and magnesium and significantly lower copper and iron have been detected in serum (Sanyal et al. 2016). Other studies indicate significantly higher whole-blood manganese concentration, higher iron concentration in serum and urine, and significantly lower serum copper concentration in PD patients (Fukushima et al. 2013; Forte et al. 2004). A decreasing trend for aluminum levels in PD patients' urine, serum, blood, and CSF; significantly higher calcium levels in urine, serum, and blood; significantly lower magnesium level in serum; and increased zinc levels in whole blood have also been reported (Forte et al. 2004). However, other studies indicate significant increases of vanadium concentrations in blood and urine and decreases in aluminum, cadmium, mercury, and lead in serum and of cadmium, cobalt, chromium, mercury, and lead in CSF (Bocca et al. 2004).

Alterations in several inter-elemental interactions in both CSF and serum have also been reported (Sanyal et al. 2016; Ahmed and Santosh 2010). Interestingly, in some cases serum has shown relatively high accuracy compared to CSF, indicating the possibility of using only serum analyses in detriment of CSF samples, which are more invasive to obtain (Sanyal et al. 2016).

In addition to metal dyshomeostasis, several metal-binding proteins, or metalloproteins, found in patient biofluids have also been implicated in PD progression and etiology, although it is not known if alterations in protein synthesis lead to metal dyshomeostasis or if metal dyshomeostasis then leads to modifications in protein synthesis. For example, decreased levels of serum ceruloplasmin, the major copper-carrying protein in the blood, which also plays a role in iron metabolism, have been hypothesized to exacerbate nigral iron deposition in the brain of patients with PD, a prominent pathophysiological feature of this condition (Jin et al. 2011). In addition, its role as an antioxidant protecting the CNS from iron-mediated free radical injuries has also been postulated, since serum ferritin levels, an iron-binding protein, have been verified as significantly lower in PD patients when compared to controls (Rohani et al. 2013). Moreover, a negative correlation between iron accumulation in the brain and copper-dependent ferroxidase activity (a protein maintained via ceruloplasmin) in plasma has been observed (Martínez-Hernández et al. 2011; Ayton et al. 2013). Interestingly, a positive correlation between ceruloplasmin and the age of onset of PD has been observed, with younger-onset PD patients presenting significantly lower levels of serum ceruloplasmin compared to those with older-onset PD (Bharucha et al. 2009). Because of this, the administration of ceruloplasmin has been proposed as potential therapeutic treatment of PD (Barbariga et al. 2015).

High aggregation and deposits of alpha-synuclein in the brain, a copper-binding protein with two copper-binding sites per monomer at nanomolar and micromolar concentrations, are a characteristic of PD (Rasia et al. 2005). It is thought that copper binding to this protein facilitates alpha-synuclein fibrillation and aggregation in the brain, leading to changes in the redox properties of this metal, thus leading to increased radical oxygen species in the brain environment (Chahine et al. 2014). In addition, copper-bound alpha-synuclein has been shown to promote the contribution of iron in the biosynthesis of free radicals, and the triplication of the gene encoding this metal-binding protein is a known cause for the familial form of PD (Singleton et al. 2003). As it is an extracellular protein, it is no surprise that it has also been identified in the CSF and plasma in PD patients (El-Agnaf et al. 2003), although results regarding its evaluation as a PD biomarker are still conflicting, since both increases and decreases in plasma alpha-synuclein in PD patients compared to controls have been reported (Chahine et al. 2014; Lee et al. 2006; Li et al. 2007; Kim 2013). Further studies attempting to make use of this protein as a biomarker for PD are currently underway.

Superoxide dismutase (SOD), a metalloprotein that plays an important antioxidant role in the organism, has also shown significance in PD. The different isoforms of this protein require copper (cytoplasmic type I), zinc (extracellular type III), and manganese (mitochondrial type II) as cofactors (Marklund 1982). Significantly lower Cu/Zn-SOD levels in whole blood have been reported in PD patients (Ihara et al. 1999), as well as significantly decreased SOD activity with increased duration of PD, suggesting faster deterioration of the antioxidant ability of Cu/Zn-SOD in this condition (Tórsdóttir et al. 1999).

Metallothionein, an important metalloprotein that shows both metal-binding capability and antioxidant activity, has been shown to be increased in glial cells in the substantia nigra and cortex of patients with PD (Michael et al. 2011), while, in rodent models for PD, transgenic mice overexpressing metallothionein I were shown to be more resistant to this disease than non-transgenic controls (Ebadi and Sharma 2006). However, not many studies are available concerning the expression of this protein in biofluids in PD patients, mainly because interest is high in a specific isoform, MT-III, which is mainly found in the central nervous system (Aschner 1996). As metal dyshomeostasis in this condition, described above, is well known, this protein, which shows high affinity to essential ions such as zinc and copper, may be an interesting biomarker in this regard, although it may also be induced by oxidative stress and other biotic parameters, and not only by the simple presence of metals.

2.4.2 *Alzheimer's Disease*

Metal dyshomeostasis has also been demonstrated for several elements in AD and has been postulated as impacting not only AD onset but also progression (Squitti and Polimanti 2013). In addition to metal imbalances in the AD brain itself, biofluid metal levels are also altered in AD patients. For example, increased copper levels have been observed in AD patient serum and CSF (Basun et al. 1991; Squitti et al. 2002), while plasma/serum copper ratio levels are also significantly higher in AD patients (Ventriglia et al. 2012). Since high copper levels in CSF are consistent in AD patients, they have been proposed as a biomarker for the disease (Boll et al. 2008). Decreased serum iron and zinc (Loeffier et al. 1995; Molina et al. 1998) have also been observed. Interestingly, these are opposite trends to what is found in the AD brain, suggesting a redistribution of metals between the brain and the biofluids responsible for supplying and clearing excreted metals from the brain (González-Domínguez et al. 2014). Due to this apparent metal dyshomeostasis condition in AD, some clinical trials with some metal modulators are in progress, in order to assess the effects of therapies redistributing metals among different organism compartments (Crouch and Barnham 2012).

Due to the fact that a main neurological finding in AD brains is the accumulation and deposition of amyloid plaques in the brain, composed of the amyloid beta protein (Aβ) (Murphy and LeVine 2010), which presents high levels of precipitated copper and zinc, studies have mainly focused on the role of these metals in AD pathology. However, although the focus of AD investigations is usually copper and zinc homeostasis, many other elements are differentially observed in AD patient biofluids. For example, in addition to copper and zinc, iron and aluminum have also been demonstrated as suffering alterations in AD patients, and cadmium concentrations in whole blood have been found in significantly higher levels in AD patients, while lead and mercury levels in serum were reported as higher (Lee et al. 2012). Lower levels of cesium in both plasma and CSF have also been reported in AD patients (Gerhardsson et al. 2008).

As in PD, up- and downregulation of several metalloproteins have also been observed in AD. As stated previously, one of the main neurological findings in AD brains is the accumulation and deposition of amyloid plaques in the brain, composed of the Aβ peptide that has the ability to bind to several metals, including zinc, copper, and iron. Serum levels of this peptide are now being considered as an early biomarker in the diagnosis of AD, since serum Aβ peptide levels may be peaked during early stages of AD, as demonstrated in transgenic mouse models for AD (He et al. 2013). $A\beta_{42}$ (one of the Aβ forms) levels in CSF have been used as an adequate biomarker to distinguish subjects with MCI who are likely to progress to AD from non-converters (Hampel et al. 2008), while plasma total Aβ and Aβ42 were found increased in subjects with familial AD (Scheuner et al. 1996; Kosaka et al. 1997). However, as in many other brain diseases, results are still contradictory in this regard, with some studies observing no significant differences in Aβ plasma levels when comparing AD patients and control cases (Scheuner et al. 1996; Kosaka et al. 1997). Others suggesting that high plasma $A\beta_{42}$ levels in non-demented elderly people are a risk factor for developing cognitive decline related to AD (Pomara et al. 2005; Schupf et al. 2008), while yet others have reported both associations and non-associations between plasma $A\beta_{40}$ and $A\beta_{42}$, and cognitive impairment, risk of dementia, or level of dementia severity (van Oijen et al. 2006; Mehta et al. 2000). However, its importance in the AD condition is undisputed, and therapy options have included attempts at decreasing plasma Aβ plasma levels by treatment with compounds that inhibit the β-secretase or γ-secretase enzymes, which are known to produce Aβ (Schneider et al. 2009).

Tau is a cytosolic protein that plays a role in microtubule assembly and stability and, when hyperphosphorylated, aggregates and forms toxic deposits in the brain, leading to the characteristic neurofibrillary tangles present in AD patients (Maeda et al. 2007; Braak and Braak 1991). This process is modulated by metal ions, such as zinc, iron, and copper (Cristóvão et al. 2016). Elevated tau and phosphorylated tau concentrations in CSF are well-established hallmarks of AD (Shekhar et al. 2016) and have been shown as significantly upregulated by up to an impressive 300% in CSF samples from AD patients, with about 80 to 90% sensitivity and specificity (Blennow and Zetterberg 2009). High T-tau levels in this biofluid have been linked to fast progression from mild cognitive impairment to the AD condition, as well as with rapid cognitive decline and high mortality rates in patients presenting this condition (Blom et al. 2009; Samgard et al. 2010). Tau is also found in plasma, although plasma tau levels are normal in the mild cognitive impairment stage of the AD condition, suggesting that plasma tau is a late biomarker for this disease, requiring substantial axonal injury before increasing to abnormal levels (Zetterberg et al. 2013). In addition, there is a lack of correlation between tau levels in plasma and in CSF, indicating that steady-state concentrations of tau in these biofluids are differentially regulated (Zetterberg et al. 2013).

As in other brain diseases, both elevated and decreased serum ceruloplasmin levels have been reported in AD (Park et al. 2014; Arnal et al. 2010), although some authors have suggested that AD patients may have significantly more "defective" ceruloplasmin, apo-ceruloplasmin lacking its copper, than in healthy controls (Park et al. 2014).

2.4.3 *Huntington's Disease*

Significant and consistent alterations in iron homeostasis have been frequently reported in Huntington's disease (HD) (Muller and Leavitt 2014), and serum ferritin deficiency has been postulated as being the cause for elevated brain iron levels observed in HD patients (Bonilla et al. 1991; Morrison and Nevin 1994). However, metallomic studies in biofluids from HD patients are scarce.

The main implicated protein in HD is huntingtin that, although not a metalloprotein per se, is an iron-responsive protein involved in brain development that also regulates receptors for transferrin, one of the key iron metabolism proteins. The presence of huntingtin in biofluids has been proposed as a biomarker for brain huntingtin, since there is a need for biomarkers that could more accurately predict HD onset or progression (Southwell et al. 2015), and it has been demonstrated that huntingtin levels in cerebrospinal fluid reflect brain levels, increasing with disease stage (Southwell et al. 2015).

In HD, polyQ is a polyglutamine expansion in the huntingtin protein that results in huntingtin misfolding and aggregation. Metallothionein, in the form of MT-III, has been found to significantly reduce polyQ aggregation and toxicity when overexpressed in mammalian cells. Thus, this metalloprotein has been proposed as a potential therapeutic target for HD (Hands et al. 2010). However, once again, no interest is seemingly present in evaluating these metalloproteins in biofluids, only in the MT-III isoform in the brain. It is expected that MT should be overexpressed in CSF, at least, due to the elevated brain iron levels observed in HD patients.

2.4.4 *Wilson's Disease*

Wilson's disease (WD) is an autosomal recessive inherited disorder of the hepatic copper metabolism, resulting in the accumulation of copper in many organs and tissues (Ferenci 2004). This condition is characterized by very low copper levels in the blood, while urinary copper levels are typically increased and are used as a biomarker of the disease (Manto 2014). Interestingly, total serum copper levels are usually low, reflecting low ceruloplasmin levels, since 95% of plasma copper in WD is bound to ceruloplasmin, while serum copper unbound to ceruloplasmin is increased in this condition (Ogihara et al. 1995).

Metallothionein concentrations in livers from patients suffering this condition have been shown to be increased in several studies (Mulder et al. 1992; Nartey et al. 1987), most certainly in response to copper accumulation in this organ, since about 73% of subcellular copper was present in the cytoplasmic fraction and most of it was in association with MT (Nartey et al. 1987). In the early stages of the disease, copper is mainly in the cytoplasm bound to metallothionein (Goldfischer and Sternlieb 1968). Interestingly, only in degenerating hepatocytes is MT found in hepatocyte nuclei, as well as the cytoplasm, indicating differential locations in differentially affected tissues. Unfortunately, scarce studies regarding this protein in the biofluids of WD patients are available. One indirect finding in this regard is that zinc supplementation in WD patients is an efficacious therapy, achieving normalization of free copper levels in the blood, since this causes the induction of metallothionein and its binding of free copper by zinc (Rodriguez-Castro et al. 2015).

2.4.5 *Bipolar Disorder*

Metallomic analyses have also been conducted in biofluids from bipolar disorder (BD) patients, although studies are still scarce. One study found increased lead and cadmium levels in both blood and urine of BD patients, as well as increased serum zinc and thallium in urine (González-Estecha et al. 2011), and the authors suggested that BD may reflect a past high lead exposure in childhood and adolescence and that oxidative stress seems to play a role in BD pathophysiology.

Lead and cadmium are both toxic metals that are known to disrupt mitochondrial functions through many processes leading to energy metabolism and malfunctions in mitochondrial biochemical cascade, mechanisms directly linked to BD (Fattal et al. 2006). Many studies have linked BD with increased lead body burden, and some authors have postulated that specifically reducing blood lead levels in patients presenting BD may contribute to an improvement in symptoms (Orisakwe 2014).

Regarding essential metals, some studies have indicated no difference, for example, in serum copper levels in patients in different BD phases or stages and healthy volunteers, although levels of this metal in the first stage of the disease were significantly higher than in advanced stage. In addition, serum copper concentrations were higher in later-onset BD, indicating that different physiological processes, such as inflammatory responses and oxidative stress, occur during different BD stages (Siwek et al. 2017).

As for metalloproteins in BD, studies are even scarcer. In one study, 32 serum metalloproteins were observed, showing differential expression when comparing control and BD patients treated with lithium. The main metals found in all groups were sodium, magnesium, zinc, calcium, and iron with cobalt only observed in the control group and BD patients treated with lithium. Potassium and titanium were only found in the BD patient groups, and phosphorus was only observed in control and BD patients not treated with lithium. However, the metalloproteins were not identified (Sussulini et al. 2010).

SOD has also been evaluated in this condition in different biofluids, such as plasma, red blood cells, and serum, and results are still contradictory, which has been postulated as being the reflection of fluctuations in response to state-dependent changes in oxidant levels in BD, perhaps linked to different manic states of this condition (Savas et al. 2006; Selek et al. 2008; Abdalla et al. 1986).

2.4.6 *Schizophrenia*

Studies regarding metals in schizophrenia have reported conflicting results. Some studies report the decrease of essential trace metals such as iron and selenium, while certain toxic metals (lead, chromium, and cadmium) were increased in schizophrenic patient plasma samples (Arinola et al. 2010). Copper has been reported as increased in both serum and plasma (Sharma et al. 2013; Olatunbosun et al. 1975), although decreased CSF copper has also been observed (Tyrer et al. 1979), which has been suggested as corroborating the hypothesis of decreased central copper-dependent enzymes activity in schizophrenia, which are involved in the synthesis and catabolism of dopamine (Tyrer et al. 1979). Zinc alterations are inconsistently reported, both in low and high levels in various biofluids (Sharma et al. 2013; Yanik et al. 2004). Significant selenium decreases are common in schizophrenia patients, and, interestingly, increased prevalence of the condition has been linked to soils containing low selenium levels, perhaps indicating environmental factors at play (Brown 1994). In fact, exposure to toxic metals has also been cited as one of the major factors in the onset of schizophrenia (Nawaz et al. 2014).

Although blood copper has been reported as significantly higher in many schizophrenia patients, ceruloplasmin levels are not indicated as a biomarker, since contradictory results have been observed, with both nonsignificant and increased plasma concentrations reported in schizophrenia patients (Virit et al. 2008).

Regarding oxidative stress, several studies have reported both increased and decreased activity of the metalloenzyme SOD in the serum of chronic schizophrenic patients (Gama et al. 2006; Wu et al. 2013; Kunz et al. 2008; Mukerjee et al. 1996; Ranjekar et al. 2003). A link between SOD activities in this biofluid and medication intake has also been observed, where patients put on antipsychotic medication showed significantly increased plasma SOD activities (Wu et al. 2013).

2.5 Conclusions

Metal dyshomeostasis is present in most neurodegenerative and mental diseases, especially regarding essential elements. However, there is still no consensus on which metals are altered, and if they are increased or decreased in these diseases, as many discrepancies concerning these aspects are reported in the literature. In addition, inconsistencies have also been reported for different biofluids, making these types of investigations of extreme importance in this regard. Oxidative stress probably caused by metals is also prevalent in most brain diseases. However, many inconsistencies are also reported regarding the levels and expression of endogenous antioxidant metalloproteins and metalloenzymes in different biofluids. In addition, it is also noteworthy that, until now, it is not known if it is the altered metal levels that affect protein levels, or vice-versa.

Furthermore, many of the discussed disorders are characterized by differential metal, metalloprotein, and metalloenzyme levels during different stages or conditions, and confounding factors such as specific population groups, age, or sex may also interfere in these investigations.

It is clear that applications of metallomic strategies in the study of brain diseases in general are still scarce, and, in most cases, only exploratory studies have been performed, which demonstrate that the paths for further quantitative and systematic research are still open.

References

Abdalla DS, Monteiro HP, Oliveira JA, Bechara E (1986) Activities of superoxide dismutase and glutathione peroxidase in schizophrenic and manic-depressive patients. Clin Chem 32(5):805–807

Ahmed SSSJ, Santosh W (2010) Metallomic profiling and linkage map analysis of early Parkinson's disease: a new insight to aluminum marker for the possible diagnosis. PLoS One 5(6):e11252

Arinola G, Idonije B, Akinlade K, Ihenyen O (2010) Essential trace metals and heavy metals in newly diagnosed schizophrenic patients and those on anti-psychotic medication. J Res Med Sci 15(5):245–249

Arnal N, Cristalli DO, de Alaniz MJT, Marra C (2010) Clinical utility of copper, ceruloplasmin, and metallothionein plasma determinations in human neurodegenerative patients and their first-degree relatives. Brain Res 1319:118–130

Aschner M (1996) The functional significance of brain metallothioneins. FASEB J 10(10):1129–1136

Ayton S, Lei P, Duce JA, Wong BX, Sedjahtera A, Adlard PA, Bush AI, Finkelstein DI (2013) Ceruloplasmin dysfunction and therapeutic potential for Parkinson disease. Ann Neurol 73:554–559

Barbariga M, Curnis F, Andolfo A, Zanardi A, Lazzaro M, Conti A, Magnani G, Volontè MA, Ferrari L, Comi G, Corti A, Alessio M (2015) Ceruloplasmin functional changes in Parkinson's disease-cerebrospinal fluid. Mol Neurodegener 10:59

Barnham KJ, Bush AI (2014) Biological metals and metal-targeting compounds in major neurodegenerative disorders. Chem Soc Rev 43:6727–6749

Basun H, Forssell LG, Wetterberg L, Winblad B (1991) Metals and trace elements in plasma and cerebrospinal fluid in normal aging and Alzheimer's disease. J Neural Transm Park Dis Dement Sect 3:231–258

Bharucha KJ, Friedman JK, Vincent AS, Ross ED (2009) Lower serum ceruloplasmin levels correlate with younger age of onset in Parkinson's disease. J Neurol 255(12):1957–1962

Blennow K, Zetterberg H (2009) Cerebrospinal fluid biomarkers for Alzheimer's disease. J Alzheimers Dis 18:413–417

Blom ES, Giedraitis V, Zetterberg H, Fukumoto H, Blennow K, Hyman BT, Irizarry MC, Wahlund LO, Lannfelt L, Ingelsson M (2009) Rapid progression from mild cognitive impairment to

Alzheimer's disease in subjects with elevated levels of tau in cerebrospinal fluid and the APOE epsilon4/epsilon4 genotype. Dement Geriatr Cogn Disord 27:458–464

Boaru SG, Merle U, Uerlings R, Zimmermann A, Weiskirchen S, Matusch A, Stremmel W, Weiskirchen R (2014) Simultaneous monitoring of cerebral metal accumulation in an experimental model of Wilson's disease by laser ablation inductively coupled plasma mass spectrometry. BMC Neurosci 15:98–111

Bocca B, Alimonti A, Petrucci F, Violante N, Sancesario G, Forte G, Senofonte O (2004) Quantification of trace elements by sector field inductively coupled plasma mass spectrometry in urine, serum, blood and cerebrospinal fluid of patients with Parkinson's disease. Spectrochim Acta B 59:559–566

Bolann BJ, Khazen-Rahil R, Henriksen H, Isrenn R, Ulvik RJ (2007) Evaluation of methods for trace-element determination with emphasis on their usability in the clinical routine laboratory. Scand J Clin Lab Invest 67(4):353–366

Boll MC, Alcaraz-Zubeldia M, Montes S, Rios C (2008) Free copper, ferroxidase and SOD1 activities, lipid peroxidation and NO(x) content in the CSF. A different marker profile in four neurodegenerative diseases. Neurochem Res 33:1717–1723

Bonilla E, Estevez J, Suarez H, Morales LM, Chacin de Bonilla L, Villalobos R, D'Avila JO (1991) Serum ferritin deficiency in Huntington's disease patients. Neurosci Lett 129:22–24

Braak H, Braak E (1991) Neuropathological stageing of Alzheimer-related changes. Acta Neuropathol 82:239–259

Brown JSJ (1994) Role of selenium and other trace elements in the geography of schizophrenia. Schizophr Bull 20:387–398

Chahine L, Stern M, Chen-Plotkin A (2014) Blood-based biomarkers for Parkinson's disease. Parkinsonism Relat Disord 20(1):S99–S103

Cristóvão JS, Santos R, Gomes CM (2016) Metals and neuronal metal binding proteins implicated in Alzheimer's disease. Oxidative Med Cell Longev 2016:1–13

Crouch PJ, Barnham KJ (2012) Therapeutic redistribution of metal ions to treat Alzheimer's disease. Acc Chem Res 45:1604–1611

Ebadi M, Sharma S (2006) Metallothioneins 1 and 2 attenuate peroxynitrite-induced oxidative stress in Parkinson disease. Exp Biol Med 231(9):1576–1583

El-Agnaf OM, Salem SA, Paleologou KE, Cooper LJ, Fullwood NJ, Gibson MJ, Curran MD, Court JA, Mann DM, Ikeda S, Cookson MR, Hardy J, Allsop D (2003) Alpha-synuclein implicated in Parkinson's disease is present in extracellular biological fluids, including human plasma. FASEB J 17:1945–1947

Fattal O, Budur K, Vaughan AJ, Franco K (2006) Review of the literature on major mental disorders in adult patients with mitochondrial diseases. Psychosomatics 47(1):1–7

Ferenci P (2004) Diagnosis and current therapy of Wilson's disease. Aliment Pharmacol Ther 19(2):157–165

Forte G, Bocca B, Senofonte O, Petrucci F, Brusa L, Stanzione P, Zannino S, Violante N, Alimonti A, Sancesario G (2004) Trace and major elements in whole blood, serum, cerebrospinal fluid and urine of patients with Parkinson's disease. J Neural Transm 111:1031–1040

Fukushima T, Tan X, Luo Y, Wang P, Song J, Kanda H, Hayakawa T, Kumagai T, Kakamu T, Tsuji M, Hidaka T, Mori Y (2013) Heavy metals in blood and urine and its relation to depressive symptoms in Parkinson's disease patients. Fukushima J Med Sci 59(2):76–80

Gama CS, Salvador M, Andreazza AC, Kapczinski F, Silva Belmonte-de-Abreu P (2006) Elevated serum superoxide dismutase and thiobarbituric acid reactive substances in schizophrenia: a study of patients treated with haloperidol or clozapine. Prog Neuro-Psychopharmacol Biol Psychiatry 30(3):512–515

Gerhardsson L, Lundh T, Minthon L, Londos E (2008) Metal concentrations in plasma and cerebrospinal fluid in patients with Alzheimer's disease. Dement Geriatr Cogn Disord 25:508–515

Goldfischer S, Sternlieb I (1968) Changes in the distribution of hepatic copper in relation to the progression of Wilson's disease (hepatolenticular degeneration). Am J Pathol 53:883–901

González-Domínguez R, García-Barrera T, Gómez-Ariza JL (2014) Characterization of metal profiles in serum during the progression of Alzheimer's disease. Metallomics 6:292–300

González-Estecha M, Trasobares EM, Tajima K, Cano S, Fernández C, López JL, Unzeta B, Arroyo M, Fuentenebro F (2011) Trace elements in bipolar disorder. J Trace Elem Med Biol 25(1):S78–S83

Halliwell B (2001) Role of free radicals in the neurodegenerative diseases. Drugs Aging 18:685–716

Halliwell B (2006) Phagocyte-derived reactive species: salvation or suicide? Trends Biochem Sci 31(9):509–515

Hampel H, Burger K, Teipel S, Bokde A, Zetterberg H, Blennow K (2008) Core candidate neurochemical and imaging biomarkers of Alzheimer's disease. Alzheimers Dement 4:38–48

Hands SL, Mason R, Sajjad MU, Giorgini F, Wyttenbach A (2010) Metallothioneins and copper metabolism are candidate therapeutic targets in Huntington's disease. Biochem Soc Trans 38(2):552–558

Hare DJ, Kysenius K, Paul B, Knauer B, Hutchinson RW, O'Connor C, … Doble PA (2017). Imaging Metals in Brain Tissue by Laser Ablation - Inductively Coupled Plasma - Mass Spectrometry (LA-ICP-MS). J Vis Exp 22(119) 55042. Advance online publication. http://doi.org/10.3791/55042

He J, Qiao JP, Zhu S, Xue M, Chen W, Wang X, Tempier A, Huang Q, Kong J, Li XM (2013) Serum β-amyloid peptide levels spike in the early stage of Alzheimer-like plaque pathology in an APP/PS1 double transgenic mouse model. Curr Alzheimer Res 10(9):979–986

Hutchinson RW, Cox AG, McLeod C, Marshall P, Harper A, Dawson EL, Howlett DR (2005) Imaging and spatial distribution of beta-amyloid peptide and metal ions in Alzheimer's plaques by laser ablation-inductively coupled plasma-mass spectrometry. Anal Biochem 346:225–233

Ihara Y, Chuda D, Kuroda S, Haybara T (1999) Hydroxyl radical and superoxide dismutase in blood of patients with Parkinson's disease: relationship to clinical data. J Neurol Sci 170(2):90–95

Jin L, Wang J, Zhao L, Jin H, Fei G, Zhang Y, Zeng M, Zhong C (2011) Decreased serum ceruloplasmin levels characteristically aggravate nigral iron deposition in Parkinson's disease. Brain 134:50–58

Kim H-J (2013) Alpha-Synuclein expression in patients with Parkinson's disease: a Clinician's perspective. Exp Neurol 22(2):77–83

Kosaka T, Imagawa M, Seki K, Arai H, Sasaki H, Tsuji S, Asami-Odaka A, Fukushima T, Imai K, Iwatsubo T (1997) The beta APP717 Alzheimer mutation increases the percentage of plasma amyloid-beta protein ending at A beta42(43). Neurology 48:741–745

Kunz M, Gama CS, Andreazza AC, Salvador M, Cereser KM, Gomes FA, Belmonte-de-Abreu PS, Berk M, Kapczinski F (2008) Elevated serum superoxide dismutase and thiobarbituric acid reactive substances in different phases of bipolar disorder and in schizophrenia. Prog Neuro-Psychopharmacol Biol Psychiatry 32(7):1677–1681

Lee JY, Kim JH, Choi DW, Lee DW, Park JH, Yoon HJ, Pyo HS, Kwon HJ, Park KS (2012) The association of heavy metal of blood and serum in the Alzheimer's diseases. Toxicol Res 28(2):93–98

Lee PH, Lee G, Park HJ, Bang OY, Joo IS, Huh K (2006) The plasma alpha-synuclein levels in patients with Parkinson's disease and multiple system atrophy. J Neural Transm 113:1435–1439

Lee Y-H, Wong DT (2009) Saliva: an emerging biofluid for early detection of diseases. Am J Dent 22(4):241–248

Li QX, Mok SS, Laughton KM, McLean CA, Cappai R, Masters CL, Culvenor JG, Horne MK (2007) Plasma alpha-synuclein is decreased in subjects with Parkinson's disease. Exp Neurol 204:583–588

Loeffier DA, Connor JR, Juneau PL, Snyder BS, Kanaley L, DeMaggio AJ, Nguyen H, Brickman CM, LeWitt PA (1995) Transferrin and iron in normal, Alzheimer's disease, and Parkinson's disease brain regions. J Neurochem 65:710–716

Maeda S, Sahara N, Saito Y, Murayama M, Yoshiike Y, Kim H, Miyasaka T, Murayama S, Ikai A, Takashima A (2007) Granular tau oligomers as intermediates of tau filaments. Biochemistry 46(12):3856–3861

Manto M (2014) Abnormal copper homeostasis: mechanisms and roles in neurodegeneration. Toxics 2:27–345

Marklund SL (1982) Human copper-containing superoxide dismutase of high molecular weight. Proc Natl Acad Sci U S A 79(24):7634–7638

Martínez-Hernández R, Montes S, Higuera-Calleja J, Yescas P, Boll MC, Diaz-Ruiz A, Rios C (2011) Plasma ceruloplasmin ferroxidase activity correlates with the nigral sonographic area in Parkinson's disease patients: a pilot study. Neurochem Res 36:2111–2115

Matusch A, Depboylu C, Palm C, Wu B, Höglinger GU, Schäfer KH, Becker JS (2010) Cerebral bioimaging of Cu, Fe, Zn, and Mn in the MPTP mouse model of Parkinson's disease using laser ablation inductively coupled plasma mass spectrometry (LA-ICP-MS). J Am Soc Mass Spectrom 21:161–171

Matusch A, Fenn LS, Depboylu C, Klietz M, Strohmer S, McLean JA, Becker JS (2012) Combined elemental and biomolecular mass spectrometry imaging (MSI) for probing the inventory of tissue at a micrometer scale. Anal Chem 84:3170–3178

Mehta PD, Pirttila T, Mehta SP, Sersen EA, Aisen PS, Wisniewski HM (2000) Plasma and cerebrospinal fluid levels of amyloid beta proteins 1-40 and 1-42 in Alzheimer disease. Arch Neurol 57:100–105

Michael GJ, Esmailzadeh S, Moran LB, Christian L, Pearce RKB, Graeber MB (2011) Up-regulation of metallothionein gene expression in Parkinsonian astrocytes. Neurogenetics 12(4):295–305

Molina J, Jiménez-Jiménez F, Aguilar M, Meseguer I, Mateos-Vega CJ, González-Muñoz MJ, de Bustos F, Porta J, Ortí-Pareja M, Zurdo M, Barrios E, Martínez-Para MC (1998) Cerebrospinal fluid levels of transition metals in patients with Alzheimer's disease. J Neural Transm 105:479–488

Morrison PJ, Nevin NC (1994) Serum iron, total iron binding capacity and ferritin in early Huntington disease patients. Ir J Med Sci 163:236–237

Mukerjee S, Mahadik SP, Scheffer R, Correnti EE, Kelkar H (1996) Impaired antioxidant defense at the onset of psychosis. Schizophr Res 19(1):19–26

Mulder TP, Janssens AR, Verspaget HW, van Hattum J, Lamers CB (1992) Metallothionein concentration in the liver of patients with Wilson's disease, primary biliary cirrhosis, and liver metastasis of colorectal cancer. J Hepatol 16(3):346–350

Muller M, Leavitt BR (2014) Iron dysregulation in Huntington's disease. J Neurochem 130:328–350

Murphy MP, LeVine H (2010) Alzheimer's disease and the β-amyloid peptide. J Alzheimers Dis 19(1):311–327

Nartey NO, Frei JV, Cherian MG (1987) Hepatic copper and metallothionein distribution in Wilson's disease (hepatolenticular degeneration). Lab Investig 57(4):397–401

Nawaz R, Zahir E, Siddiqui S, Usmani A, Shad KF (2014) The role of trace metals and environmental factors in the onset and progression of schizophrenia in Pakistani population. World J Neurosci 4:450–460

Ogihara H, Ogihara T, Miki M, Yasuda H, Mino M (1995) Plasma copper and antioxidant status in Wilson's disease. Pediatr Res 37(2):219–226

Olatunbosun DA, Akindele MO, Adadevoh BK, Asuni T (1975) Serum copper in schizophrenia in Nigerians. Br J Psychiatry 127:191–121

Orisakwe OE (2014) The role of lead and cadmium in psychiatry. N Am J Med Sci 6(8):370–376

Park JH, Lee DW, Park KS (2014) Elevated serum copper and ceruloplasmin levels in Alzheimer's disease. Asia-Pacific Psychiatry 6(1):38–45

Pomara N, Willoughby LM, Sidtis JJ, Mehta PD (2005) Selective reductions in plasma Abeta 1-42 in healthy elderly subjects during longitudinal follow-up: a preliminary report. Am J Geriatr Psychiatr 13:914–917

Prakash A, Dhaliwal GK, Kumar P, Majeed AB (2017) Brain biometals and Alzheimer's disease - boon or bane? Int J Neurosci 127:99–108

Ranjekar PK, Hinge A, Hegde MV, Ghate M, Kale A, Sitasawad S, Wagh UV, Debsikdar VB, Mahadik SP (2003) Decreased antioxidant enzymes and membrane essential poly- unsaturated fatty acids in schizophrenic and bipolar mood disorder patients. Psychiatry Res 121(2):109–122

Rasia RM, Bertoncini CW, Marsh D, Hpyser W, Cherny D, Zweckstetter M, Griesinger C, Jovin TM, Fernández CO (2005) Structural characterization of copper(II) binding to α-synuclein: insights into the bioinorganic chemistry of Parkinson's disease. Proc Natl Acad Sci U S A 102(12):4294–4299

Rodriguez-Castro KI, Hevia-Urrutia FJ, Sturniolo GC (2015) Wilson's disease: a review of what we have learned. World J Hepatol 7(29):2859–2870

Rohani M, Jalali N, Khademolhosseini M, Miri S, Shahidi GA, Ghourchian S (2013) Serum ceruloplasmin and ferritin levels in patients with Parkinson's disease. Mov Disord 28:893

Samgard K, Zetterberg H, Blennow K, Hansson O, Minthon L, Londos E (2010) Cerebrospinal fluid total tau as a marker of Alzheimer's disease intensity. Int J Geriatr Psychiatry 25:403–410

Sanyal J, Shiek SSJ, Ng HKT, Naiya T, Ghosh E, Banerjee TK, Lakshmi J, Guha G, Rao VR (2016) Metallomic biomarkers in cerebrospinal fluid and serum in patients with Parkinson's disease in Indian population. Sci Rep 6:35097

Sas K, Robotka H, Toldi J, Vécsei L (2007) Mitochondria, metabolic disturbances, oxidative stress and the kynurenine system, with focus on neurodegenerative disorders. J Neurol Sci 257:221–239

Savas HA, Gergerlioglu HS, Armutcu F, Herken H, Yilmaz HR, Kocoglu E, Selek S, Tutkun H, Zoroglu SS, Akyol O (2006) Elevated serum nitric oxide and superoxide dismutase in euthymic bipolar patients: impact of past episodes. World J Biol Psychiatry 7(1):51–55

Scheuner D, Eckman C, Jensen M, Song X, Citron M, Suzuki N, Bird T, Hardy J, Hutton M, Kukull W, Larson E, Levy-Lahad L, Viitanen M, Peskind E, Poorkaj P, Schellenberg G, Tanzi R, Wasco W, Lannfelt SD, Younkin S (1996) Secreted amyloid beta-protein similar to that in the senile plaques of Alzheimer's disease is increased in vivo by the presenilin 1 and 2 and APP mutations linked to familial Alzheimer's disease. Nat Med 2:864–870

Schneider P, Hampel H, Buerger K (2009) Biological marker candidates of Alzheimer's disease in blood, plasma, and serum. CNS Neurosci Ther 15:358–374

Schupf N, Tang MX, Fukuyama H, Manly J, Andrews H, Mehta P, Ravetch J, Mayeux R (2008) Peripheral Abeta subspecies as risk biomarkers of Alzheimer's disease. Proc Natl Acad Sci U S A 105:14052–14057

Selek S, Savas HA, Gergerlioglu HS, Bulbul F, Uz E, Yumru M (2008) The course of nitric oxide and superoxide dismutase during treatment of bipolar depressive episode. J Affect Disord 107(1–3):89–94

Sharma SK, Sood S, Sharma A, Gupta ID (2013) Estimation of serum zinc and copper levels patients with schizophrenia: a preliminary study. Sri Lanka J Psychiatry 5(1):14–17

Shekhar S, Kumar R, Rai N, Kumar V, Singh K, Upadhyay AD, Tripathi M, Dwivedi S, Dey AB, Dey S (2016) Estimation of tau and phosphorylated Tau181 in serum of Alzheimer's disease and mild cognitive impairment patients. PLoS One 11(7):e0159099

Singleton AB, Farrer M, Johnson J, Singleton A, Hague S, Kachergus J, Hulihan M, Peuralinna T, Dutra A, Nussbaum R, Lincoln S, Crawley A, Hanson M, Maraganore D, Adler C, Cookson MR, Muenter M, Baptista M, Miller D, Blancato J, Hardy J, Gwinn-Hardy K (2003) α-Synuclein locus triplication causes Parkinson's disease. Science 302(5646):841

Siwek M, Styczén K, Sowa-Kućma M, Dudek D, Reczyński W, Szewczyk B, Misztak P, Opoka W, Topór-Mądry R, Nowak G, Rybakowski JK (2017) The serum concentration of copper in bipolar disorder. Psychiatr Pol 51:469–481

Southwell AL, Smith EP, Davis TR, Caron NS, Villanueva EB, Xie Y, Collins JA, Ye ML, Sturrock A, Leavitt BR, Schrum AG, Hayden MR (2015) Ultrasensitive measurement of huntingtin protein in cerebrospinal fluid demonstrates increase with Huntington disease stage and decrease following brain huntingtin suppression. Sci Rep 5:12166

Squitti R, Lupoi D, Pasqualetti P, Dal Forno G, Vernieri F, Chiovenda P, Rossi L, Cortesi M, Cassetta E, Rossini PM (2002) Elevation of serum copper levels in Alzheimer's disease. Neurology 59(8):1153–1161

Squitti R, Polimanti R (2013) Copper phenotype in Alzheimer's disease: dissecting the pathway. Am J Neurodegener Dis 2:46–56

Steinacker P, Hawlik A, Lehnert S, Jahn O, Meier S, Görz E, Braunstein KE, Krzovska M, Schwalenstöker B, Jesse S, Pröpper C, Böckers T, Ludolph A, Otto M (2010) Neuroprotective function of cellular prion protein in a mouse model of amyotrophic lateral sclerosis. Am J Pathol 176:1409–1420

Subramanian KS (1996) Determination of metals in biofluids and tissues: sample preparation methods for atomic spectroscopic techniques. Spectrochim Acta B At Spectrosc 51(3):291–319

Sussulini A, Becker JS, Becker JS (2017) Laser ablation ICP-MS: application in biomedical research. Mass Spectrom Rev 36:47–57

Sussulini A, Kratzin H, Jahn O, Banzato CE, Arruda MA, Becker JS (2010) Metallomics studies of human blood serum from treated bipolar disorder patients. Anal Chem 82(13):5859–5864

Sussulini A, Matusch A, Klietz M, Bauer A, Depboylu C, Becker JS (2012) Quantitative imaging of Cu, Fe, Mn and Zn in the L-DOPA-treated unilateral 6-hydroxydopamine Parkinson's disease mouse model by LA-ICP-MS. Biomed Spectrosc Imaging 1:125–136

Tórsdóttir G, Kristinsson J, Sveinbjörnsdóttir S, Snaedal J, Jóhannesson T (1999) Copper, ceruloplasmin, superoxide dismutase and iron parameters in Parkinson's disease. Pharmacol Toxicol 85(5):239–243

Tyrer SP, Delves HT, Weller MP (1979) CSF copper in schizophrenia. Am J Psychiatry 136(7):937–939

van Oijen M, Hofman A, Soares HD, Koudstaal PJ, Breteler MM (2006) Plasma Abeta(1-40) and Abeta(1-42) and the risk of dementia: a prospective case-cohort study. Lancet Neurol 5:655–660

Ventriglia M, Bucossi S, Panetta V, Squitti R (2012) Copper in Alzheimer's disease: a meta-analysis of serum, plasma, and cerebrospinal fluid studies. J Alzheimers Dis 30:981–984

Virit O, Altindag A, Selek S, Yumru M, Bulut M, Erel O, Savas HA, Herken H (2008) Increased plasma ceruloplasmin levels in schizophrenia. Klinik Psikofarmakoloji Bülteni 18:282–287

Wisnieff C, Ramanan S, Olesik J, Gauthier S, Wang Y, Pitt D (2015) Quantitative susceptibility mapping (QSM) of white matter multiple sclerosis lesions: interpreting positive susceptibility and the presence of iron. Magn Reson Med 74:564–570

Wu JQ, Kosten TR, Zhang XY (2013) Free radicals antioxidant defense systems, and schizophrenia. Prog Neuro-Psychopharmacol Biol Psychiatry 46:200–206

Yanik M, Kocyigit A, Tutkun H, Vural H, Herken H (2004) Plasma manganese, selenium, zinc, copper, and iron concentrations in patients with schizophrenia. Biol Trace Elem Res 98:109–117

Zetterberg H, Wilson D, Andreasson U, Minthon L, Blennow K, Randall J, Hansson O (2013) Plasma tau levels in Alzheimer's disease. Alzheimers Res Ther 5(2):9

Chapter 3
Environmental Metallomics

Gema Rodríguez-Moro, Sara Ramírez-Acosta, Ana Arias-Borrego, Tamara García-Barrera, and José Luis Gómez-Ariza

Abstract Metallomics is the new paradigm about the metallobiomolecules related to living organisms, considering the interactions between toxic and essential metals, transport through biological fluids, passing across biological membranes and interfaces, synergic and antagonist actions among metal species, and alterations in metabolic pathways triggered by overexpression or inhibition of these metallobiomolecules. These challenging studies require the development of new analytical approaches in order to get suitable information of these species close to their native environment which has promoted the application of new tools based in mass spectrometry under the double focus of elemental (ICP-MS) and molecular (Qq-TOF-MS) mass spectrometry, generally arranged with chromatography in multidimensional platforms. The driving force for the design of these new analytical instrumental arrangements is the analyst imagination who adapts the new metallomic methodology to the new problems. In this work the most recent metallomic approaches proposed have been considered, deepening their application to the most frequent problems related to metal toxicity in environmental issues, such as exposure experiments of mice to toxic metals, interactions and homeostasis of metals, metal imaging, metabolic alterations caused by metallobiomolecules over- or down-expressed, and more interestingly real-life consequences of metal species expression in environmental field studies. In this way, the application of two-dimensional chromatographic approaches with ICP-MS detection, the use of multidimensional chromatography-column-switching-ICP-MS devices, metal imaging

G. Rodríguez-Moro · S. Ramírez-Acosta · A. Arias-Borrego · T. García-Barrera
J. L. Gómez-Ariza (✉)
Department of Chemistry, Faculty of Experimental Sciences, University of Huelva, Campus de El Carmen, Huelva, Spain

Research Center on Natural Resources, Health and the Environment (RENSMA), University of Huelva, Huelva, Spain

International Campus of Excellence on Agrofood (ceiA3), University of Huelva, Huelva, Spain
e-mail: gema.moro@dqcm.uhu.es; sara.ramirez@dqcm.uhu.es; ana.arias@dqcm.uhu.es; tamara.garcia@dqcm.uhu.es; ariza@uhu.es

M. A. Z. Arruda (ed.), *Metallomics*, Advances in Experimental Medicine and Biology 1055, https://doi.org/10.1007/978-3-319-90143-5_3

with LA-ICP-MS, combined application of metallomics and metabolomics for environmental toxicological appraisal, and the application of these metallomic techniques in environmental field studies have been reviewed.

Keywords Environmental metallomics · Two-dimensional chromatographic metallomics · Column-switching metallomics · Metal exposure · Laser ablation, Metabolomics

Abbreviations

AB	Arsenobetaine
AEC	Anion exchange chromatography
AE-SPE	Anion exchange solid-phase extraction
AF-HPLC	Affinity chromatography
AsC	Arsenocholine
BSA	Bovine serum albumin
CRC	Collision/reaction cell
CSVs	Column-switching valves
DIMS	Direct infusion mass spectrometry
DMASV	Dimethylmonothioarsinic acid
DMA^V	Dimethylarsinate
DNP	Doñana National Park
eGPx	Glutathione peroxidase
ESI	Electrospray ionization
GPAsC	Glycerylphosphorylarsenocholine
HPLC	High-performance liquid chromatography
IC-ICP-MS	Ion chromatography coupled with inductively coupled plasma mass spectrometry
ICP-MS	Inductively coupled plasma mass spectrometry
IDA	Isotope dilution analysis
IEF	Isoelectric focusing
LA-ICP-MS	Laser ablation inductively coupled plasma mass spectrometry
MALDI-MS	Matrix-assisted laser desorption/ionization mass spectrometry
MA^V	Monomethylarsonate
MT	Metallothionein
nESI-QqQ-TOF-MS	Nano-electrospray ionization-triple quadrupole-time-of-flight-mass spectrometry
PCs	Phosphatidylcholines
QqQ-TOF-MS	Triple quadrupole-time-of-flight-mass spectrometry

RBC	Red blood cells
ROS	Reactive oxygen species
RP-HPLC	Reversed-phase HPLC
SAM	S-adenosyl-methionine
SeAlb	Selenoalbumin
SEC	Size exclusion chromatography
SeP	Selenoprotein P
SOD	Superoxide dismutase
SUID	Species-unspecific isotope dilution
TCA	Tricarboxylic acid cycle
TETRA	Tetra-methyl-arsonium
TMAO	Trimethyl-arsine oxide

3.1 Introduction

The study of life and its environment is the paradigm of present in science especially since the epigenetics demonstrated the importance of the medium in the production of stable heritable traits (phenotypes) that cannot be explained by changes in DNA sequence (Wrobel et al. 2009). Although the term environment is very broad, we refer here to the air, water, soil, particulate matter, and other components that surround these organisms and contributed to their well-being. In this context, metals play an important role on living organisms, because many proteins require a metal cofactor (Cu, Fe, Zn, Mo, and others) to perform their functions (Tainer et al. 1991), which can affect up- and downregulation of protein expression. In addition, other metal-binding proteins (such as metallothioneins) are essential in homeostasis and detoxification processes (Maret 2004), being critical the knowledge about metal traffic and interactions in cells and tissues, considering their transport by biological fluids and the pass through biological barriers (Gómez-Ariza et al. 2011).

At the intracellular level, the concentration and distribution of metals among the different cell compartments and the incorporation in metalloproteins are carefully controlled (Outten and O'Halloran 2001), and metallochaperones nurse the metals through the cytoplasm protecting and, most importantly, directing these metals to their objectives in order to fulfill its function. In this context, it is necessary to understand the mechanisms by which a metal is sensed, stored, or incorporated to other biomolecules (Mounicou et al. 2009), and, in addition, we have to consider that metallobiomolecules have not to be considered in isolation, but the spatial, temporal, and physiological environment have also to be taken into account to go deeper into the role and consequences of an active metallobiomolecule in health. This means that we need to study not only the molecules themselves but also to consider the complexity of their in vivo worlds and the impact of their native environments in their functions. The in-cell environments are crowded and inhomogeneous, as well as most fluids, tissues, and organs. Physicochemical properties of intracellular and extracellular environments are full with macromolecules, salts, metabolites, and other biomolecules, which introduce important interactions with the key analytes and limit intracellular and extracellular space for molecules to roam, affecting correlatively their behavior (Gierasch and Gershenson 2009).

For all these reasons, it is necessary to develop powerful new analytical methodological approaches to decipher these hard questions, allowing a more real vision of the behavior of metallobiomolecules in living organisms and their consequences. The complexity of these biological systems has triggered the combined application of multidimensional analytical platforms that approach these problems from elemental and molecular focuses on the basis of ICP-MS and organic MS, generally coupled to liquid and/or gas chromatography (Mounicou et al. 2009; García-Sevillano et al. 2014a, 2015a). This results in multidimensional analytical platforms suitable for studying the changes of all of metal and metalloid species present in a cell, fluid, or tissue, considering their identity, quantity, and localization in the living organisms as consequence of environmental factors (Mounicou et al. 2009); for these analytical approaches, the term environmental metallomics has been coined, which has to be complemented with the comprehensive analysis of metabolic changes in the same context, associated to different metabolic processes in organisms exposed to these environmental issues (García-Sevillano et al. 2015a), the so-called environmental metabolomics.

In this work, the new analytical platforms proposed for current metallomic challenges have been commented emphasizing more in the problem to be solved than in the methodology or technique that is applied, although considering that both have to go together. Therefore, metallomic issues related to metal exposure of mammals (particularly mice), metallobiomolecule transport and interactions in living organisms, metal homeostasis, metabolic response to metal contamination, and metallomics in real environmental problems have been considered, especially in connection with the more recent analytical tools reported for this purpose, particularly multidimensional chromatography approaches coupled to ICP-MS, the use of chromatographic column-switching devices, LA-ICP-MS imaging, and the use of the methodological and conceptual binomial metallomics-metabolomics.

3.2 Experimental Approaches for Environmental Metallomics

Different analytical approaches have been proposed for environmental metallomics, considering that most of the studies are based on the use of samples obtained from in vivo specimen exposed to pollutants in laboratory experiments (García-Sevillano et al. 2013a, 2014b, c, d). Similar studies have also been carry out with cell lines (Wolters et al. 2012; Nischwitz et al. 2013; Wang et al. 2015; Ogra et al. 2016), but with this second type of organisms, the information about the changes and interactions between the contaminants during the transport and crossing through the different biological interfaces, digestive tract, tissues, blood-brain barrier, as well as metal homeostasis inside the organisms, beside their consequences at metabolic level would get lost. Finally, other works have been focused on free-living organisms in their natural habitat, which allows translating the laboratory methodology to the field. A critical point is to avoid changes in metallospecies identity during sample collection, storage, and treatment. For this reason, direct analysis procedures have been proposed, such as X-ray absorption spectroscopy (XAS), to get information about metal-coordination environment and element oxidation state. However, the requirements of high level of metal concentration and sample purity in this technique represent serious drawbacks for its application (Mounicou et al. 2009). Therefore, hyphenated techniques combining high-resolution separations by chromatographic or electrophoretic devices with sensitive detection by elemental or molecular MS are usually the approaches of choice.

3.3 Metallomic Approaches Based on Two-Dimensional Chromatography with ICP-MS Detection

Many studies for metal pollution assessment are based on nontargeted analysis of metal-tagged molecules in organs and fluids of living organisms exposed to the different elements, which provides a metal profiling of metallobiomolecules over-expressed or inhibited under the action of contamination. In Fig. 3.1, is shown a typical metallomic workflow to study metalloproteins in the liver of mice (*Mus spretus*) affected by metal pollution. A two-dimensional approach is used for this purpose involving a first separation of metallobiomolecules by the size (SEC-ICP-MS chromatogram), followed by the purification of the 32 kDa fraction, traced by Cu and Zn, by anion exchange chromatography (AEC) with ICP-MS detection (AEC-ICP-MS) and final identification of the Cu,Zn-peak by fraction collection from AEC column, desalting, lyophilization, and tryptic digestion for analysis by nano-ESI-triple quadrupole-TOF-MS, using MASCOT database for protein identification, which confirm the presence of Cu,Zn-superoxide dismutase associated to the peak of 32 kDa (García-Sevillano et al. 2012a). These two-dimensional chromatographic approaches allow to overcome the problems associated to the low resolution of SEC and are used frequently in metallomic studies.

A number of metallomic studies that are based on exposure experiments of different organisms in the laboratory, such as mice, clams, crustacean, and others, to toxic elements, mainly mercury, arsenic, or cadmium; toxic mixtures, such as As+Cd; or antagonist interactions that explore the protective action of selenium or zinc on Hg or As have been performed. Several examples of these researchers will be considered in the following headings.

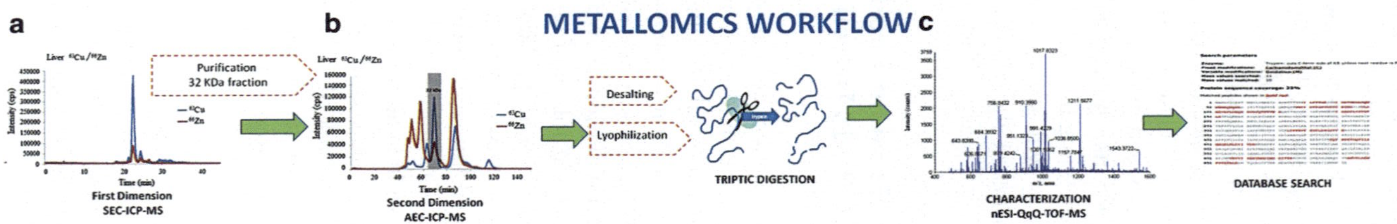

Fig. 3.1 Typical workflow in environmental metabolomic experiment to study metalloproteins in liver extract of mice from metal-contaminated area: (**a**) separation of metallobiomolecules by SEC-ICP-MS; (**b**) purification of 32 kDa fraction by AEC-ICP-MS; (**c**) identification Cu,Zn-peak (32 kDa Cu,Zn-SOD) by triptic digestion and analysis by nano-ESI-QqQ-TOF-MS using MASCOT database

3.3.1 Mercury Exposure Experiments

The intake of inorganic mercury adversely affects the metabolic pathways of organisms related to energy metabolism, amino acid metabolism, and gut microflora (Nicholson et al. 1985; Wei et al. 2008). However, it is difficult to have a comprehensive view of all biochemical changes associated with the exposure of mammals to inorganic mercury.

For this purpose exposure experiments to this element have been performed in mice *Mus musculus* using inorganic mercury ($HgCl_2$) for 14 days (García-Sevillano et al. 2014b). The intake of mercury reduced the intensity of a Cu-containing biomolecule in the liver that coeluted with a standard of Cu,Zn-SOD (32 KDa) using a coupling SEC-ICP-ORS-MS (Superdex-75 column with fractionation range from 3 to 70 kDa). The intensity of this peak decreased by about 60% during the exposure to mercury; in contrast, the intensity of mercury bound to this biomolecule (a putative Hg-SOD) increased by about 30%, although the intensity of the Hg peak and the corresponding mercury presence in liver was comparatively low with respect to this element intake during exposure. For this reason, the presence of Hg was also studied in the kidneys, in which the concentration of Cu and Zn increased markedly along the experiment, reaching values 100 times greater than that in the liver. Additionally, a peak traced by Hg at about 7 kDa coeluted with a Cu,Zn,Cd-metallothionein standard, which increased strongly along the exposure experiment. The peak corresponding to copper metallothionein (Cu-MT) presented a similar trend. These results about mercury and copper distribution between the liver and kidney during mercury exposure can be explained by the competition between both elements for the Cu-containing proteins, so that Cu-SOD peak intensity decreased as defense mechanism triggered by the presence of mercury. Nevertheless, the presence of mercury in the liver is scarce being transported to the kidney for excretion in the urine as Hg-MT (Zalups 2000; Zalups and Koropatnick 2000). The identity of this 7 kDa-peaks was previously confirmed in a similar exposure experiment of *M. musculus* to Cd, which was performed by preselecting the retention time of the peak by SEC-HPLC-ICP-MS, later collection of this peak for purification using a second chromatographic separation by RP-HPLC, and identification of putative metallothionein (Cd-MT) by ESI-Q-TOF-MS (Jara-Biedma et al. 2013).

3.3.2 Arsenic Exposure Experiments

It is well known the carcinogenic activity of arsenic is associated to oxidative stress (Chen et al. 1986; Brown et al. 1989; Warner et al. 1994; Szymańska-Chabowska et al. 2007), although other harmful effects related to liver damage and cardiovascular disease have also been reported (Carlson-Lynch et al. 1994), as well as its capacity to cross the blood-brain barrier (BBB) (García-Chávez et al. 2003; Piao et al. 2005). Considering that toxicity, accumulation, and metabolic involvement of

Table 3.1 Concentration of arsenic metabolites (mean ± sd/μg g^{-1}) in the liver of mice *Mus musculus* after 7 days exposure to this element

Exposure	AB	MMA^V	DMA^V	iAs^{III}	iAs^V	SUM As species	TOTAL As
Control	<LOD	<LOD	26.6 ± 0.6	< LOD	< LOD	26.6	27.02 ± 1.12
Day 2	<LOD	105.4 ± 1.8	511.4 ± 3.2	< LOD	< LOD	616.8	631.2 ± 2.1
Day 4	<LOD	101.9 ± 0.9	952.0 ± 4.1	3.02 ± 0.32	< LOD	1057	1210 ± 21
Day 7	<LOD	177.1 ± 1.1	665.7 ± 3.7	29.62 ± 0.81	< LOD	872.4	1345 ± 25

arsenic in living organisms depend on the chemical form of this element, being inorganic forms highly toxic, followed by methylated ones, other organic species such as arsenobetaine (AB), arsenocholine (AsC), and arsenosugars are regarded to be innocuous (Geiszinger et al. 2002; Fattorini et al. 2006). In mammals, highly toxic inorganic arsenic is mainly metabolized in the liver, after absorption from the gastrointestinal tract, to produce methylated species such as MA^V and DMA^V, which are excreted in the urine (Del Razo et al. 1997; Hughes et al. 2003).

The combination of SEC-HPLC-ICP-MS and AEC-HPLC-ICP-MS has been used in exposure experiments of mice to inorganic arsenic (As_2O_3) (García-Sevillano et al. 2013a), revealing the presence of low molecular mass As species (< 1.35 kDa) in liver cytosolic extracts when analyzed by SEC-HPLC-ICP-MS. The higher intensity of these signals was obtained from the liver of mice exposed for 7 days. On the other hand, the increase of levels of proteins traced by As with molecular mass about 32 kDa was also observed along the exposure; this bound of As to Cu,Zn-SOD or carbonic anhydrase (molecular mass 32 and 35 kDa, respectively) through cysteine residues by covalent arsenic-monothiol of arsenic-dithiol bonds (Kitchin and Wallace 2005) can lead to the inhibition of the enzymatic activity of these enzymes, which is regarded as one of the causes of the carcinogenic action of As (Kitchin and Wallace 2008). This fact was confirmed by the decreasing of SOD activity during the previously commented exposure experiment (García-Sevillano et al. 2013a).

In addition, anion exchange chromatography of liver extracts (García-Sevillano et al. 2013a) shows that methylated arsenicals (MA^V and DMA^V) are the predominant species in this organ. This fact agrees with previous reports, establishing that inorganic arsenic is biotransformed into methylated species in the liver of mammals (Suzuki 2005; Naranmandura and Suzuki 2008), which are finally excreted in the urine as pentavalent methylated arsenic forms (Suzuki et al. 2002). It is remarkable that concentration of DMA^V decreases along the arsenic exposure and correlatively increases the levels of MA^V and iAs^{III} (Table 3.1). This fact can be due to the saturation of methylation capacity of inorganic arsenic in the liver, which has been already observed by other authors in mice (Vahter and Norin 1980; Vahter 1981).

Another study considers the bioaccessibility of arsenic species in human after ingestion of the anemone *Anemonia sulcata*, a seafood very appreciated in the south of Spain, and its metabolization in the human body after ingestion (Contreras-Acuña et al. 2014). Analysis of As-traced biomolecules in anemone tissue extracts and human fluids was carried out by anion/cation exchange chromatography with inductively coupled plasma mass spectrometry ((IC)-ICP-MS) (Contreras-Acuña et al. 2013). Results obtained show that about 54% of the arsenic present in the anemones is lost after cooking, around 95% of total arsenic in cooked anemones is bioaccessible, and 85% of the amount ingested is eliminated from the body by urine in 90 h. The relative abundance of As in human urine, cooked anemone, and human serum varies in the following order:

Human urine (after 10 h of ingestion)	Arsenobetaine (AB) ≈ DMA^V (dimethylarsinate) > arsenocholine (AsC) > TETRA-methyl-arsonium (TETRA) > trimethyl-arsine oxide (TMAO) > MA^V (monomethylarsonate) > iAs^V > dimethylarsenoethanol (DMAE)
Cooked anemone	AB > DMA^V > dimethylmonothioarsinic acid (DMASV) > AsC > TETRA > TMAO > iAs^V > glycerylphosphorylarsenocholine (GPAsC) > MA^V > DMA
Human serum (after 30 min of ingestion)	DMA^V > AB > MA^V > AsC

Finally, organic mass spectrometry has been used for the identification of species for which no standards were available such as AsC, GPAsC, $DMAS^V$, and DMAE.

3.3.3 Cadmium Exposure Experiments

Cadmium is another toxic environmental pollutant that can be accumulated in living organisms with carcinogenic activity both in human (Kolonel 1976; Bako et al. 1982; Elinder et al. 1985) and mice (Gunn et al. 1963; Haddow et al. 1964). Cd has a long half-life in the body, with serious adverse health effects under chronic exposures, affecting especially the kidney and bone (Åkesson et al. 2005, 2006), as well as the endocrine system (Piasek et al. 2001; Åkesson et al. 2008). On the other hand, selenium has important antagonistic properties against the toxic action of cadmium (Early Jr. and Schnell 1981; Flora et al. 1982; Messaoudi et al. 2009), and the knowledge of them, using metallomic and metabolomic techniques, is of great interest.

The coupling SEC-ICP-MS has been used to characterize the biological response of mice *Mus musculus* to Cd exposure by subcutaneous administration during 12 days, considering the change of metallobiomolecules profiles. In addition, interactions of cadmium with selenium (orally administrated as selenite) have also been considered, and complementarily the presence of selenoproteins can be evaluated by in-series coupling of SEC with multiaffinity chromatography (AF) and ICP-MS detection using IDA for quantification (García-Sevillano et al. 2014c). Exposure to Cd causes change in the distribution of endogenous Zn and Cu in tissues and biological fluids, since both metals compete with Cd for S-containing proteins such as metallothioneins, therefore Zn and Cu protect organisms against Cd, due to their contribution in MTs induction (Shimoda et al. 2003; Martelli et al. 2006). The Cd exposure increases the intensity of Cu-traced peak at about 67 kDa in plasma that reflects the induction on copper transport proteins in the blood stream, such as BSA (67 kDa) and transferrin (79 kDa), which are used to move the Cd to kidney. In connection to this, the peak of Cd-metallothionein (Cd-MT) in the kidney increased markedly with the exposure, and correlatively the Cu-peak at 32 kDa (SOD) increased at the same time besides the 7 kDa Cu-MT peak. These facts reflect the elimination of Cd as Cd-MT via the kidney and urine as main route for excretion of this element (Suzuki and Yoshikawa 1981).

Additionally, the application of multidimensional chromatography 2D-SEC-AF-ICP-MS has been applied to achieve a better understanding of metal interactions caused by the simultaneous exposure to Cd and Se. The application of affinity column to plasma samples shows two remarkable peaks, one close to void volume and the other more intense at retention time matching with SeP peak that increases along the exposure time. Additionally, the chromatogram obtained by SEC-ICP-MS shows a Cd-peak related to high molecular mass proteins (about 55 kDa), corresponding to the molecular mass of SeP, and other peak eluting at 19 min that can be related to Cd-MT fraction at 19 min. Therefore, these results suggest a cadmium binding to the selenoprotein P (Srivastava et al. 1988).

3.3.4 Arsenic/Cadmium Exposure Experiments

The biochemical effects of independent exposure to As and Cd have been extensively studied in experimental animals (Gonzalez-Fernández et al. 2011; Jara-Biedma et al. 2013; García-Sevillano et al. 2013a, b). However, the biological response of mammals under simultaneous exposure to both toxics has been scarcely studied, and the toxic effects caused by the joint administration of the two metals remain unclear. In addition, changes in the relative concentration of metal-containing biomolecules and their general homeostasis should not only be expected as a consequence of simultaneous exposure to these two toxics but also significant changes in cellular metabolic functions (Dudley et al. 1985; García-Sevillano et al. 2013a) due to the interaction of As/Cd, more damaging that separate exposure to these element, which has been reported by Yañez et al. (Yáñez et al. 1991), which have checked the induction of lipid peroxidation and both glutathione and metallothionein upregulation in rats.

In this way, a multidimensional metallomic approach based on SEC-ICP-MS and 2D-SEC-AF-SUID-ICP-MS has been applied to a better understanding of detoxification mechanisms triggered by exposure of mice *Mus musculus* to As/Cd during 12 days (García-Sevillano et al. 2014d). The presence of low molecular mass As

Table 3.2 Quantification of selenium species in mice plasma (*Mus musculus*) under As/Cd exposure

	Concentration of Se-containing proteins (mean ± SD)			
Group of exposure	Control	As exposure	Cd exposure	As/Cd exposure
GPx (ng g^{-1})	5.61 ± 0.42	5.92 ± 0.51	6.12 ± 0.84	7.41 ± 1.1
SeP (ng g^{-1})	148 ± 5.2	132 ± 11	121 ± 9.2	165 ± 13
SeAlb (ng g^{-1})	16.10 ± 0.82	12.1 ± 1.4	8.24 ± 0.91	8.8 ± 0.82
Se-metabolites (ng g^{-1})	32.3 ± 6.2	3.24 ± 1.8	5.45 ± 2.1	<LOD

species (<300 Da) in liver cytosolic extracts has been observed. The higher intensity of signals was obtained from the liver of mice exposed to As/Cd during 6 days; in addition, a peak traced by As can be observed at about 32 kDa that can be related to the interaction of As with enzymes such as carbonic anhydrase (CA) and superoxide dismutase (Cu/Zn-SOD) with molecular masses of 35 kDa and 32 kDa, respectively. The increase of this peak is more pronounced when As is administered alone in comparison with the joint administration As/Cd. Another peak traced by As was observed at about 70 kDa, which increases with the exposure during 6 days; this can be related to the well-known affinity of arsenite to albumin and hemoglobin, with molecular masses of 67 kDa and 68 kDa, respectively (Bogdan et al. 1994; Lu et al. 2007). In this case, this peak presents higher intensity in mice exposed to the mixture As/Cd during 6 days. On the other hand, very intense signals from Cd associated to MT (7 kDa) are observed in liver cytosolic extracts that are more pronounced when Cd is administered alone; however, an important depletion of this peak occurs when As/Cd is ingested, which confirms the antagonist interactions between these two elements and with other essential metals such as Cu, Zn, and others.

Similarly, the application of 2D-SEC-AF-SUID-ICP-MS to mice serum reveals differences in the expression of Se-proteins depending of the toxic element administrated, since the levels of selenium and Se-proteins decrease under As or Cd exposure (Table 3.2), but co-administration of both elements reduces the toxicity of each other (Saïd et al. 2010), and levels of Se in As/Cd exposure increase, as well as Se-proteins, particularly SeP, decreasing the presence of Se-metabolites involved in the synthesis of SeP.

3.4 Metallomic Approaches Based on Multidimensional Chromatography-Column-Switching Valve-ICP-MS Devices

These experimental platforms are very suitable to study complex matrices, such as plasma, serum, amniotic fluid, and cytosolic and mitochondrial extracts (García-Sevillano et al. 2013c, 2014e, f). The approaches consider the potential of ICP-MS as detector for different elements, such as Zn, Cu, I, Se, and others, which have essential properties to assure living organism functions. Thereby, several critical metallobiomolecules containing Se in their moieties, such as extracellular glutathione peroxidase (eGPx), selenoprotein P, selenoalbumin (SeAlb), as well as selenometabolites (selenium amino acids and inorganic forms of selenium), have an important role in the antioxidative status of mammals plasma, although their biological action mechanisms are not completely elucidated, which have launched the development of suitable analytical methods for the simultaneous determination of SeP, eGPx, SeAlb, and low molecular mass selenium species free of interferences. Other metallobiomolecules, such as superoxide dismutase Cu,Zn-SOD, have antioxidant properties catalyzing the dismutation of superoxide radicals into oxygen and hydrogen peroxide, which represents one important defense mechanism against oxidative stress in different tissues, such as hepatic cell cytosol (Fridovich 1995). For this reason, it is necessary to provide accurate and precise analytical methods for the determination of this metalloprotein involved in the redox balance of living cells. On the other hand, the mitochondria, one of the most important cell organelles involved in the energy cycle, is also affected by the action of reactive oxygen species (ROS) that can damage mitochondrial membranes, proteins, and DNA (Girotti 1985; Ames 1989; Stadtman 1992; Kasai 1997), and this process can be counteracted by the presence of SOD.

Generally, the methods for Cu,Zn-SOD determination are based on enzymatic reactions and spectrophotometric quantification; nevertheless, these methods are affected by many interferences, so that analytical approaches based on elemental mass spectrometry, mainly ICP-ORS-MS, represent a good alternative due to its sensitivity and selectivity, as well as the capability for isotope ratio measurement that allows isotope dilution analysis (IDA), which increases the accuracy and precise quantification of multi-isotopic elements that is the case of Cu-containing proteins (Nuevo Ordoñez et al. 2010). In addition, when the composition of studied species is not known or reference isotopically labeled standards are not commercially available to be applied to the IDA approach, the species-unspecific dilution mode (SUID) could be alternatively used (Rodríguez-González et al. 2005).

Similar scenery presents the simultaneous analysis of selenium species in biological matrices, especially in serum, due to the importance of SeP as antioxidant, complemented by the action of GPx and the transporter activity of SeAlb, which carry Se until the liver for SeP synthesis for later releasing into the bloodstream (Suzuki and Ogra 2002). In addition, the presence of inorganic selenium and selenometabolites in plasma has also to be considered. In connection to this, the methods proposed for speciation of SeAlb, SelP, and GPx in serum are mainly based on the combination of affinity chromatography-HPLC (AF-HPLC) for protein separation with online selenium detection by ICP-MS with collision/reaction cell (CRC) (Hinojosa Reyes et al. 2003; Shigeta et al. 2007; Jitaru et al. 2008), which provide a selective species separation and accurate quantification by post-column isotope dilution (ID) that those based on size exclusion chromatography (SEC-HPLC) (Palacios et al. 2005; Palacios and Lobinski 2007). However, there is still remaining challenges, such as the elimination of interferences caused by Br^- and Cl^- (very abundant in serum), and the possible presence of selenoamino acids, such as selenomethionine, which may be derived from the hydrolysis of selenoalbumin, or selenocysteine from GPx or SeP. These selenoamino acids together with inorganic selenium (Se(IV) and Se(VI)) can interfere with GPx when affinity chromatography-HPLC is used. In this context, Br^- and Cl^- interferences can be alleviated by serum clean up using anion exchange solid-phase extraction (AE-SPE), whose capability to remove Cl^- and Br^- from human serum was demonstrated elsewhere (Jitaru et al. 2008). The main drawback of this off-line approach is the introduction of a laborious step into the analytical process and that they are prone to contamination. On the other hand, an approach based on enzymatic hydrolysis of the whole serum followed by the determination of selenium amino acids including SeMet by reversed phase-HPLC (RP-HPLC) coupled to ICP-QMS (Jitaru et al. 2010), using SSID analysis with isotopically marked SeMet. This method allows to check the presence of free SeMet together to intact SeAlb.

More recently, suitable speciation methods based on orthogonal chromatographic techniques, online coupled between each other and with the ICP-MS detector through a switching valve with a well-designed assembling, allow to address the complex speciation of selenium in biological fluids, especially in serum, overcoming the various problems previously discussed, particularly the interference of Cl^- and Br^-, and that of selenoamino acids and inorganic selenium in the quantification of GPx, which was an important drawback in most procedures previously proposed in the literature (García-Sevillano et al. 2013c). This study proposed a method for the quantification of selenoproteins (GPx and SelP), selenium-containing proteins (SeAlb), selenoamino acids, and inorganic selenium species in human serum using species-unspecific isotope dilution (SUID)-ICP-ORS-qMS online coupled to 3D/SE-AF-AEC-HPLC involving five online connected columns. Using this chromatographic arrangement, the spectral interferences produced by bromide and chloride are removed, and the total chromatographic runtime is less than 35 min. Figure. 3.2 shows the schematic diagram of 3D/SE-AF-AEC-HPLC-SUID-ICP-ORS-MS arrangement for the complete resolution of GPx SeP, SeAlb, selenometabolites, selenite, and selenate using a column-switching approach with an 8-port valve.

Further, it is possible to separate selenium-containing proteins and selenometabolites using a two-dimensional chromatographic separations based on double size exclusion prior to double affinity chromatography. The separation of analytes using two stacked 5 ml HiTrap®Desalting columns (based on size exclusion) in series connected with the dual affinity column arrangement (1 ml heparin-sepharose column (HEP-HP) and 1 ml blue-sepharose column (BLU-HP)) is truly novel and makes it possible for the first time the complete online separation of the various selenium species without the interference of Cl^- and Br^-, allowing the separation of GPx and selenoamino acids.

In addition, separation of selenium-containing proteins and inorganic selenium metabolites can be performed using three-dimensional chromatographic separations based on the use of double size exclusion separation prior to double affinity chromatography and anionic exchange chromatography, following the schematic arrangement previously described (Fig. 3.2). This approach includes the separation of the analytes in a first step with the two-dimensional chromatographic arrangement (2D/SE-AF-HPLC-SUID-ICP-ORS-MS), connected to an 8-port

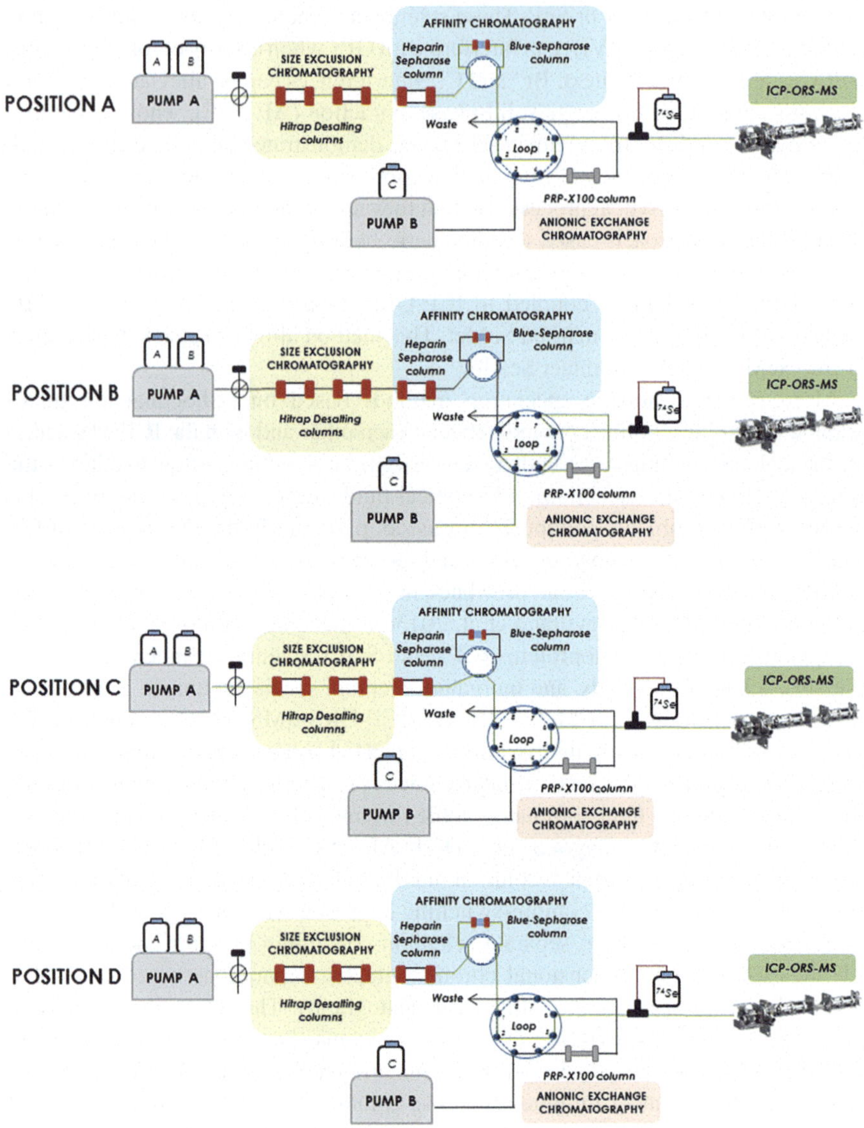

Fig. 3.2 Schematic diagram of 3D/SE-AF-AEC-HPLC-SUID-ICP-ORS-MS arrangement for selenium-containing proteins and inorganic selenium speciation in plasma and serum samples

column-switching valve that compensates the pressure gradient between low- and high-pressure pumps used in the arrangement (Fig. 3.2). When the fraction containing low molecular weight selenium species fills the loop of 3 ml, the 8-port valve is changed to other position, and all these selenium species are driven by the pump B through the anion exchange column to the ICP-MS. In this way, complete separation of selenium species is achieved, since the dual SEC unit accomplishes the separation of GPx, selenometabolites, and Br^- and the affinity chromatographic unit separates SelP and SeAlb. Finally, the anion chromatography column completes the analysis with the separation of selenium inorganic species.

Analogously, column-switching valve multidimensional chromatography devices have been used for the absolute quantification of Cu,Zn-SOD in cytosol and mitochondria of mice hepatic cells after mercury exposure, as previously mentioned (García-Sevillano et al. 2014g). Separation of metallobiomolecules from cytosolic and mitochondrial extracts was carried out with a glass analytical size exclusion column with an exclusion limit of 100 kDa. After SEC separation of Cu-containing fractions, they were collected into a loop of 3 mL, tracing the pass of Cu,Zn-SOD with ultraviolet detection. After that, the 8-port valve is changed to inject the content of the loop to the AEC column. As was mentioned in the separation of seleno-biomolecules, the use of an 8-port valve is required to equilibrate the different pressure of the two pumps. When the position of the 8-port valve is changed from position A to B, the fraction containing the 32 kDa Cu-traced peak (at retention time 9.5–12.5 min), which has been retained in the 3 mL loop, is driven by the pump B through the anion exchange column with ICP-ORS-MS detection. In this way, this SEC fraction is online chromatographically purified and resolved by the AEC column.

Therefore, analytical platforms based on column-switching valves (CSVs) provide great versatility for multidimensional chromatographic methods, serving as examples the speciation of selenoproteins and selenium species of low molecular weight in serum, or the online purification of metalloproteins, as is the case of SOD from liver extracts. In these fields the use of CSVs presents a great future.

3.5 Metallomic Approaches Based on Laser Ablation ICP-MS

Considering that metal-containing biomolecules, mainly metalloproteins, are in general inhomogeneously distributed in biological tissues and that their functions are depending of tissue microenvironment, imaging analytical techniques with both good spatial resolution and high signal/noise ratio (low detection limits) represent a good alternative in metallomics. Conventionally, native distributions of free metals in tissues are determined by specific chemical stains. Autoradiography of radioisotopes applied in vivo allows measurements of metal kinetics and binding sites. Binding sites are additionally accessible ex vivo. However, these methods do not provide multielement capability neither measures the absolute total metal content, including tightly bound or precipitated metal species. Therefore, advanced imaging mass spectrometric techniques combining a sampling processes from solid surfaces either using a focused laser beam such as laser ablation inductively coupled plasma mass spectrometry (LA-ICP-MS) or matrix-assisted laser desorption/ionization mass spectrometry (MALDI-MS), with sensitive measurement of atomic and molecular ions formed in the ion source, represent a good alternative (Becker 2010). LA-ICP-MS has been used under two main experimental strategies: (a) the study of metal and nonmetal distributions in tissues and (b) the separation of metallobiomolecules, in particular proteins, from selected regions of the organisms by one-dimensional or two-dimensional gel electrophoresis and later detection of metalloproteins in the gel by LA-ICP-MS after screening. The upregulation of Cu,Zn-SOD in the digestive gland of oyster (*Crassostrea gigas*) after Cu exposure (Xu et al. 2014a), the distribution of total Zn in a section of slag samples (used as environmental monitors for metal studies) and the detection of Zn-containing proteins after one-dimensional separation by gel electrophoresis (Becker et al. 2011), the identification of selenium-containing proteins in sunflower leave extracts by two-dimensional gel electrophoresis and detection by LA-ICP-MS (da Silva and Arruda 2012), and the evaluation of the selenium addition in transgenic and non-transgenic using the same procedure (Maciel et al. 2014) represent good examples of the application of this methodology.

Environmental studies consider the use of LA-ICP-MS to evaluate the accumulation and adverse effects of uranium that can be found in aquatic ecosystems because of its natural occurrence and/or anthropogenic activities; the U toxicity under exposure has been reported, which produces accumulation in living organism's organs (Paquet et al. 2006; Goulet et al. 2011). However, the mechanisms of uranium toxicity are not well known; to delve into it, is necessary the identification of U-biomolecules present in these tissues, especially U-proteins (Maret 2004), which is a big challenge because of the electrostatic nature of this binding that requires the preservation of coordinate and multi-subunits structure of these metalloproteins using denaturing experimental conditions (Xu et al. 2014b, c). These authors have optimized an approach for gel electrophoresis and elemental imaging by LA-ICP-MS to be applied to the cytosol of a crayfish (*Procambarus clarkii*) considered as U-bioaccumulating environmental model. Latterly, the spots from 2D-PAGE gels imaged by LA-ICP-MS were submitted to a proteolytic digestion followed by μRPC-ESI-MS/MS analysis of the produced peptides to identify the proteins. Several U-proteins were identified using this approach including U-ferritin involved in the detoxification of this element and U-glyceraldehyde-3-phosphate dehydrogenase. In addition, six other metals such as Mn, Fe, Co, Ni, Cu, and Zn were also imaged in order to investigate the potential interaction of metalloproteins with U. These metals can be regarded as potential in vivo competitors of U when their biocoordination chemistry is similar (Xu et al. 2014c).

Lobinski et al. (Bucher et al. 2016) have also studied the accumulation and adverse effects of uranium in zebra fish, which produces deleterious effects on reproduction, growth, and respiration under chronic exposure (Bucher et al. 2014). Although, in this case, in vivo *U* speciation was studied by two-dimensional (2D) nondenaturing mapping of the gill cytosolic proteins by isoelectric focusing (IEF) (isoelectric point [pI] discrimination) and by size exclusion chromatography (SEC) separation coupled to U, Fe, and P detection by (ICPMS), potential U target proteins were identified by complementary top-down and bottom-up metallomic approaches. In addition, metal imaging in nondenaturing 2D electrophoresis gels by LA-ICP-MS has been used for detection of metalloproteins in rat kidneys, which can be used for the study of metalloproteins distribution in model organisms exposed to pollutants (Becker et al. 2008, 2009).

Finally, LA-ICP-MS has been used for direct metal imaging in tissues generating qualitative and quantitative maps of elemental distribution in thin tissue sections, which can be used as methodological approaching to metal distribution and interactions in the tissues, normally associated to proteins (metalloproteins) (Sussulini et al. 2017), which can be used in the study of contamination processes (Becker et al. 2011), biomedical studies (Knauer et al. 2017; Portbury et al. 2017), and other issues.

3.6 Combination of Metallomics and Metabolomics in Environmental Issues

Metals and other contaminants present in the environment as consequence of pollution episodes not only input living organisms where they are transported and distributed through different organs, tissues, and fluids, producing important changes in metalloproteins and other metallo-metabolites present in tissues (Gómez-Ariza et al. 2011), but also they induce changes in cellular metabolic pathways whose knowledge is crucial for understanding the biological responses to these contaminants (García-Sevillano et al. 2015a). Therefore to have a deep insight into the global metabolic changes caused by pollution in exposed model organisms, metabolomics represents a powerful tool to understand as the complex biological machinery works and to identify suitable metabolic biomarkers associated to metal-tagged biomarkers that provides a broader knowledge about contaminants mode of action (MOA) (Gómez-Ariza et al. 2011).

It has been checked that there is a common pattern of metabolic response to exposure of living organisms to different toxic metals (García-Sevillano et al. 2013a, 2014b, g, d, 2015b, c). Metabolomics based on both direct infusion mass spectrometry (DIMS) and GC-MS has been applied to different tissues (liver, kidney), serum red blood cells (RBC), and liver mitochondria from mice *Mus musculus* exposed to mercury, arsenic, cadmium, and arsenic plus cadmium, whose results can be summarized as follows.

3.6.1 Alteration of Energy Metabolism

It has been demonstrated by decreased glucose levels in plasma, serum and RBCs, together with other Krebs cycle intermediates indicating an increased cellular energy metabolism under Hg exposure (García-Sevillano et al. 2014b, 2015b). In the case of As exposure, reduction of the levels of glucose and glyceraldehyde-3-phosphate was observed; this fact can be related to alteration of carbohydrate metabolism, which has been previously reported in relation to the toxic action of arsenic in rats (Szinicz and Forth 1988). The continuous decreasing pyruvate levels as end products of glycolysis cycle support this idea. In addition, low levels of this altered metabolite can be related with the formation of acetyl-CoA via pyruvate metabolism. Acetyl-CoA transports the carbon atoms within the acetyl group to the citric acid cycle (Krebs cycle), which is oxidized for energy production. In addition, high levels of citric acid (intermediate metabolite in the Krebs cycle) have been observed during the exposure.

In the case of Cd exposure, inhibition of tricarboxylic acid cycle (TCA) marked by decreasing of glucose and increasing concentration in lactic acid was also observed (García-Sevillano et al. 2014d) under Cd exposure. However, As exposure produces a decrease of lactic acid levels, and the combined exposure to As/Cd does not provoke any alteration in this metabolite. Further perturbations (increasing) of energy metabolism intermediates, such as citric acid, isocitric acid, α-ketoglutarate, and glutamic acid, were also observed under As exposure that has been related with the synthesis of glutathione (GSH) (García-Sevillano et al. 2013a). Under As/Cd exposure, similar effects were observed.

3.6.2 Cell Membrane Alteration

Exposure of mice to inorganic mercury triggers the presence of choline-containing metabolites in plasma, such as choline and phosphocholine, which are the final product of cell membrane alteration (García-Sevillano et al. 2014b). These metabolites may be regarded as degradation products of phosphatidylcholines (PCs), one of the most abundant phospholipids in the bulk of cell membranes, which are related to signal transduction, apoptosis, necrosis, and protein sorting (Bashir et al. 2006). Correlatively, increased levels of free fatty acids were detected, including pipecolic, arachidonic, linoleic, and docosahexanoic acids, as a consequence of the PC degradation. A similar behavior was observed in the case of As exposure (García-Sevillano et al. 2013a).

3.6.3 Alteration Glutathione Expression

Glutathione plays an important role in arsenic exposure due to the antioxidant properties of this metabolite that protects cells against free radicals induced by arsenic, contributing, in addition to arsenic methylation (Jin et al. 2010). These facts can explain the decrease of glutathione levels as a consequence of oxidative stress caused by arsenic and the correlative increase of L-cysteine and L-glutamic acid involved in the production of this metabolite (García-Sevillano et al. 2013a). Also in relation to As exposure, can be mentioned the increase of methionine content that is converted to S-adenosyl-methionine (SAM) by methionine adenosyltransferase, which donates a methyl group to arsenic to form methyl and dimethyl arsenic induced by arsenic exposure (Thomas et al. 2001), fact that was experimentally checked (García-Sevillano et al. 2013a).

3.6.4 Amino Acid Alteration

Alteration of amino acid metabolism was observed in Hg exposure (García-Sevillano et al. 2015b) by increased concentrations of valine and isoleucine, possibly caused by depleted metabolism of branched amino acids by inhibition of α-keto acid dehydrogenase complex (Platell et al. 2000). Decreased glutamine levels were also detected, which amino acid contributes to the production of NH_4^+ that reduces cellular acidosis (Hems 1972). In addition, increased arginine levels were observed in serum and red blood cells under mercury exposure that protects mice from oxidative stress (Dasgupta et al. 2006).

In addition, the study mice exposure to As reveals the alteration of several amino acids in the mitochondrial fraction of the liver (García-Sevillano et al. 2015c). The oxidative degradation of amino acids contributes to total metabolic energy by conversion in intermediates of TCA cycle, such as pyruvate, α-ketoglutarate, succinyl-CoA, fumarate, oxaloacetate, acetyl-CoA, or acetoacetate.

3.6.5 Increased Levels of Taurine

Metabolite involved in antioxidant activity, as well as osmoregulation, membrane stabilization, and modulation of calcium signaling functions. This metabolite exhibits an opposite trend in relation to GSH during As exposure whose biosynthesis of GSH is blocked by the toxic action of arsenic (García-Sevillano et al. 2013a).

3.6.6 Alteration of Arginine Levels

This metabolic perturbation was observed upon As but also Hg exposure. Arginine has a protective effect on oxidative stress (Dasgupta et al. 2006), and its levels decrease in mice plasma upon the exposure to toxic elements to compensate the oxidative stress.

3.6.7 Alteration in Creatine Levels

Increased levels of creatine upon As and Hg exposure and creatinine under the action of Hg were observed. In mammals, creatine is phosphorylated, but this reaction is reversed under oxidative stress, and creatine and phosphocreatine are nonenzymatically converted into creatinine (Almar et al. 2002; Wang et al. 2005). Laterly, creatinine is transferred to the kidneys by the bloodstream, whereupon it is eliminated from the body by glomerular filtration and partial tubular excretion by urine (Wang et al. 2005).

3.7 Environmental Metallomics to Study Terrestrial Areas Affected by Metal Pollution: A Case Study – Doñana National Park (Southwest Spain)

The possibilities of environmental metallomic and metabolomic approaches, previously described, to deal with the environmental status of an area affected by contamination have to be tested by applying them to a real case of study. For this purpose Doñana Natural Park (located at the southwest Spain) has been chosen.

Doñana National Park (DNP) is a 50,000 ha wildlife reserve located at the north of the Guadalquivir River Estuary, in the Huelva province, and it was declared as UNESCO World Heritage Site due to its ecological importance. The Park is an important area for migrating birds that each year land for feeding and breeding in the way from Africa to Europe. Nevertheless, this area suffers from contamination threats caused by adjacent agricultural, mining, and industrial activities giving rise to the presence of metal species (Grimalt et al. 1999; Bonilla-Valverde et al. 2004; Vioque-Fernández et al. 2009; Montes Nieto et al. 2010; Gonzalez-Fernández et al. 2011; García-Sevillano et al. 2012a, b).

The mouse *Mus spretus* was selected as bioindicator of contamination since it attains high population densities in the Park and it demonstrated a marked genetic homology with the already sequenced model laboratory mouse *Mus musculus* (Ruiz-laguna et al. 2006; Montes-Nieto et al. 2007), which has been used in all the exposure experiments previously described that allows translate the results from laboratory study to the wild mice.

Free-living mice (*Mus spretus*) were collected from two areas of the Park: (i) "Lucio del Palacio" (LDP) located in the center of the Park, which was considered as control area with low contamination, and (ii) "El Matochal" (MAT) next to Guadiamar river that is affected by agricultural and mining contamination caused by metals. Twenty mice specimens were sampled in each area (García-Sevillano et al. 2014a), and the metallomic approach previously described was based on SEC-ICP-MS (Section 2.1.1), using two-dimensional chromatography (SEC and AEC) with ICP-MS detection (Section 2.1.2) in the case of arsenic.

The biological response to metals in plasma from free-living mice *Mus spretus* in both contaminated (MAT) and non-contaminated (LDP) sites was studied. The over- or down-expression of metallobiomolecules caused by pollutants was traced by SEC-ICP-MS. In the case of Cu, a peak of high intensity at about 67 kDa was observed that can be related with transport proteins such as Cu/Zn-BSA (bovine serum albumin) and Cu-Tf (transferrin), of 67 kDa and 79 kDa, respectively. The results for Zn were analogous. Higher intensities of both metal species were detected in plasma from mice captured in the most contaminated area (MAT). The serum profile traced by Cd shows a peak at 7 kDa (putative Cd-metallothionein/Cd-MT), which is only present in MAT, denoting the contamination of this area against the non-polluted LDP. This response of wild mice with respect to Cd was similar to that obtained in the exposure experiment of *Mus musculus* to this element during 12 days (Section 2.1.3).

The response of *Mus spretus* to the As presence in the contaminated area MAT reveals the presence of low molecular species of this element, iAs and especially DMA, which is in agreement with the results of exposure experiments of *Mus musculus* to this element previously described (Section 2.1.2). The high presence of arsenic in the surrounding of DNP has been reported before (García-Sevillano et al. 2012b).

In addition, the influence of metal pollution in the antioxidant mechanisms of *Mus spretus* was tested by the quantification of selenium-containing proteins in plasma of free-living mice *Mus spretus*, using the two-dimensional chromatography-column-switching valve-ICP-MS device described in Section 2.2, (2D/SEC-AF-HPLC-SUID-ICP-ORS-qMS). The concentrations of eGPx, SeP, SeAlb, and low molecular weight selenium species were analyzed in plasma from mice plasma captured in contaminated (MAT) and non-contaminated (LDP) areas. It was observed that concentration of SeP, SeAlb, and Se-metabolites decreased in mice from MAT in comparison with LDP, correlatively the concentration of eGPx increased. Similar metabolic effects were verified when laboratory mice *Mus musculus* were exposed to As or Cd (Table 3.2).

Finally, metabolites from plasma of *Mus spretus* were extracted and analyzed to get the metabolomic profile using the DI-ESI-QqQ-TOF-MS platform. Results show an increase in the levels of creatinine, LPCs, choline, and glutamic acid in plasma from mice captured in MAT in comparison with the control area (LDP). In addition, a decrease in the levels of GSH, PCs, arginine, and glucose was observed. These results are very similar to those obtained from plasma of *Mus musculus* exposed to As (Section 3).

3.8 Gaps and Future Trends

Metallomics considered as the science of living organisms is a very young discipline pivoting between the apparent *chance* associated to the behavior and functioning of life and the *necessity* of chemical processes regulated by inviolable laws. This approach makes metallomics a science of frontiers in which diverse disciplines converge, such as analytical chemistry, bioinorganic chemistry, metabolism, environmental and medical chemistry, and others. It could also be considered as a science of interfaces because most of the interactions between metals and biomolecules are very dynamic and usually take place between organs and biofluids, or involve several organs at the same time producing systemic alterations, or inside the cell, at the cytoplasm, membrane, or organelles. Therefore, one of the gaps associated to metallomic studies is related to the comprehensive character of this discipline and the need of complementary methodologies that allows a global approach for a global problem. Several drawbacks associated to the application of metallomics can be mentioned; sampling and sample preservation are very critical in order to assure the precise replication of the biological system under study; in addition, efforts have to be made to get unequivocal identification of metallobiomolecules as well as their identification and localization in the organisms, which opens the way to the collaborative use of elemental and molecular mass spectrometry. Other problems related to the spatial, temporal, and physiological environment of the metallobiomolecules have to be taken into account especially in environmental, toxicological, and health issues; therefore, in vivo analysis has to be enhanced, and new life science paradigms, such as epigenetic, have to be interphased with metallomics. Finally, recent methodological alternatives related to single-cell analysis, nanoparticles in metallomics, and metallometabolomics have to be considered to achieve progress and future in this discipline.

Acknowledgments This work was supported by the projects CTM2015-67902-C2-1-P and CTM2012-38720-C03-03 from the Spanish Ministry of Economy and Competitiveness and P12-FQM- 0442 from the Regional Ministry of Economy, Innovation, Science and Employment (Andalusian, Government, Spain). Gema Rodríguez Moro and Sara Ramírez Acosta thank the Ministry of Economy and Competitiveness for PhD scholarships BES-2013-064501 and BES-2016-076354, respectively. Finally, the authors are grateful to FEDER (European Community) for the financial support, grant number UNHU13-1E-1611 and UNHU15-CE-3140.

References

Åkesson A, Bjellerup P, Lundh T, Lidfeldt J, Nerbrand C, Samsioe G, Skerfving S, Marie Vahter M (2006) Cadmium-induced effects on bone in a population-based study of women. Environ Health Perspect 114:830–834

Åkesson A, Julin B, Wolk A (2008) Long-term dietary cadmium intake and postmenopausal endometrial cancer incidence: a population-based prospective cohort study. Cancer Res 68:6435–6441

Åkesson A, Lundh T, Vahter M, Bjellerup P, Lidfeldt J, Nerbrand C, Samsioe G, Strömberg U, Skerfving S (2005) Tubular and glomerular kidney effects in Swedish women with low environmental cadmium exposure. Environ Health Perspect 113:1627–1631

Almar M, Villa JG, Cuevas MJ, Rodríguez-Marroyo JA, Avila C, González-Gallego J (2002) Urinary levels of 8-hydroxydeoxyguanosine as a marker of oxidative damage in road cycling. Free Radic Res 36:247–253

Ames BN (1989) Endogenous oxidative DNA damage, aging, and cancer. Free Radic Res 7:121–128

Bako G, Smith ES, Hanson J, Dewar R (1982) The geographical distribution of high cadmium concentrations in the environment and prostate cancer in Alberta. Can J Public Health 73:92–94

Bashir S, Sharma Y, Irshad M, Gupta S, Dogra TD (2006) Arsenic-induced cell death in liver and brain of experimental rats. Basic Clin Pharmacol Toxicol 98:38–43

Becker JS (2010) Bioimaging of metals in brain tissue from micrometre to nanometre scale by laser ablation inductively coupled plasma mass spectrometry: state of the art and perspectives. Int J Mass Spectrom 289:65–75

Becker JS, Mounicou S, Zoriy MV, Becker JS, Lobinski R (2008) Analysis of metal-binding proteins separated by non-denaturating gel electrophoresis using matrix-assisted laser desorption/ionization mass spectrometry (MALDI-MS) and laser ablation inductively coupled plasma mass spectrometry (LA-ICP-MS). Talanta 76:1183–1188

Becker JS, Lobinski R, Sabine Becker J (2009) Metal imaging in non-denaturating 2D electrophoresis gels by laser ablation inductively coupled plasma mass spectrometry (LA-ICP-MS) for the detection of metalloproteins. Metallomics 1:312–316

Becker JS, Pozebon D, Matusch A, Dressler VL, Becker JS (2011) Detection of Zn-containing proteins in slug (genus Arion) tissue using laser ablation ICP-MS after separation by gel electrophoresis. Int J Mass Spectrom 307:66–69

Bogdan GM, Sampayo-Reyes A, Vasken Aposhian H (1994) Arsenic binding proteins of mammalian systems: I. Isolation of three arsenite-binding proteins of rabbit liver. Toxicology 93:175–193

Bonilla-Valverde D, Ruiz-Laguna J, Muñoz A, Ballesteros J, Lorenzo F, Gómez-Ariza JL, López-Barea J (2004) Evolution of biological effects of Aznalcóllar mining spill in the Algerian mouse (Mus spretus) using biochemical biomarkers. Toxicology 197:123–138

Brown KG, Boyle KE, Chen CW, Gibb HJ (1989) A dose-response analysis of skin Cancer from inorganic arsenic in drinking water. Risk Anal 9:519–528

Bucher G, Mounicou S, Simon O, Floriani M, Lobinski R, Frelon S (2016) Insights into the nature of uranium target proteins within zebrafish gills after chronic and acute waterborne exposures. Environ Toxicol Chem 35:736–741

Bucher G, Mounicou S, Simon O, Floriani M, Lobinski R, Frelon S (2014) Different uranium distribution patterns in cytosolic protein pool of zebrafish gills after chronic and acute waterborne exposures. Chemosphere 111:412–417

Carlson-Lynch H, Beck BD, Boardman PD (1994) Arsenic risk assessment. Environ Health Perspect 102:354–356

Chen C-J, Chuang Y-C, You S-L, Lin T-M, Wu H-Y (1986) A retrospective study on malignant neoplasms of bladder, lung and liver in Blackfoot disease endemic area in Taiwan. Br J Cancer 53:399–405

Contreras-Acuña M, García-Barrera T, García-Sevillano MA, Gómez-Ariza JL (2014) Arsenic metabolites in human serum and urine after seafood (Anemonia sulcata) consumption and bioaccessibility assessment using liquid chromatography coupled to inorganic and organic mass spectrometry. Microchem J 112:56–64

Contreras-Acuña M, García-Barrera T, García-Sevillano MA, Gómez-Ariza JL (2013) Speciation of arsenic in marine food (Anemonia sulcata) by liquid chromatography coupled to inductively coupled plasma mass spectrometry and organic mass spectrometry. J Chromatogr A 1282:133–141

da Silva MAO, Arruda MAZ (2012) Identification of selenium in the leaf protein of sunflowers by a combination of 2D-PAGE and laser ablation ICP-MS. Microchim Acta 176:131–136

Dasgupta T, Hebbel RP, Kaul DK (2006) Protective effect of arginine on oxidative stress in transgenic sickle mouse models. Free Radic Biol Med 41:1771–1780

Del Razo LM, García-Vargas GG, Vargas H, Albores A, Gonsebatt ME, Montero R, Ostrosky-Wegman P, Kelsh M, Cebrián ME (1997) Altered profile of urinary arsenic metabolites in adults with chronic arsenicism. A pilot study. Arch Toxicol 71:211–217

Dudley RE, Gammal LM, Klaassen CD (1985) Cadmium-induced hepatic and renal injury in chronically exposed rats: likely role of hepatic cadmium-metallothionein in nephrotoxicity. Toxicol Appl Pharmacol 77:414–426

Early JL Jr, Schnell RC (1981) Selenium antagonism of cadmium-induced inhibition of hepatic drug metabolism in the male rat. Toxicol Appl Pharmacol 58:57–66

Elinder C-G, Kjellstrom T, Hogstedt C, Andersson K, Spång G (1985) Cancer mortality of cadmium workers. Br J Ind Med 42:651–655

Fattorini D, Notti A, Regoli F (2006) Characterization of arsenic content in marine organisms from temperate, tropical, and polar environments. Chem Ecol 22:405–414

Flora SJS, Behari JR, Ashquin M, Tandon SK (1982) Time-dependent protective effect of selenium against cadmium-induced nephrotoxicity and hepatotoxicity. Chem Biol Interact 42:345–351

Fridovich I (1995) Superoxide radical and superoxide dismutases. Annu Rev Biochem 64:97–112

García-Chávez E, Santamaría A, Díaz-Barriga F, Mandeville P, Juárez BI, Jiménez-Capdeville ME (2003) Arsenite-induced formation of hydroxyl radical in the striatum of awake rats. Brain Res 976:82–89

García-Sevillano MA, González-Fernández M, Jara-Biedma R, García-Barrera T, López-Barea J, Pueyo C, Gómez-Ariza JL (2012a) Biological response of free-living mouse Mus spretus from Doñana National Park under environmental stress based on assessment of metal-binding biomolecules by SEC-ICP-MS. Anal Bioanal Chem 404:1967–1981

García-Sevillano MA, González-Fernández M, Jara-Biedma R, García-Barrera T, Vioque-Fernández A, López-Barea J, Pueyo C, Gómez-Ariza JL (2012b) Speciation of arsenic metabolites in the free-living mouse Mus spretus from Doñana National Park used as a bio-indicator for environmental pollution monitoring. Chem Pap 66:914

García-Sevillano MA, García-Barrera T, Navarro F, Gómez-Ariza JL (2013a) Analysis of the biological response of mouse liver (Mus musculus) exposed to As2O3 based on integrated -omics approaches. Metallomics 5:1644–1655

García-Sevillano MA, Jara-Biedma R, González-Fernández M, García-Barrera T, Gómez-Ariza JL (2013b) Metal interactions in mice under environmental stress. Biometals 26:651–666

García-Sevillano MA, García-Barrera T, Gómez-Ariza JL (2013c) Development of a new column switching method for simultaneous speciation of selenometabolites and selenoproteins in human serum. J Chromatogr A 1318:171–179

García-Sevillano MA, García-Barrera T, Gómez-Ariza JL (2014a) Application of metallomic and metabolomic approaches in exposure experiments on laboratory mice for environmental metal toxicity assessment. Metallomics 6:237–248

García-Sevillano MA, García-Barrera T, Navarro F, Gailer J, Gómez-Ariza JL (2014b) Use of elemental and molecular-mass spectrometry to assess the toxicological effects of inorganic mercury in the mouse Mus musculus. Anal Bioanal Chem 406:5853–5865

García-Sevillano MA, García-Barrera T, Navarro F, Gómez-Ariza JL (2014c) Cadmium toxicity in Mus musculus mice based on a metallomic study. Antagonistic interaction between se and cd in the bloodstream. Metallomics 6:672–681

García-Sevillano MT, García-Barrera T, Navarro-Roldán F, Montero-Lobato Z, Gómez-Ariza JL (2014d) A combination of metallomics and metabolomics studies to evaluate the effects of metal interactions in mammals. Application to Mus musculus mice under arsenic/cadmium exposure. J Proteome 104:66–79

García-Sevillano MA, García-Barrera T, Gómez-Ariza JL (2014e) Simultaneous speciation of selenoproteins and selenometabolites in plasma and serum by dual size exclusion-affinity chromatography with online isotope dilution inductively coupled plasma mass spectrometry. Anal Bioanal Chem 406:2719–2725

García-Sevillano MA, García-Barrera T, Navarro F, Gómez-Ariza JL (2014f) Absolute quantification of superoxide dismutase in cytosol and mitochondria of mice hepatic cells exposed to mercury by a novel metallomic approach. Anal Chim Acta 842:42–50

García-Sevillano MA, Contreras-Acuña M, García-Barrera T, Navarro F, Gómez-Ariza JL (2014g) Metabolomic study in plasma, liver and kidney of mice exposed to inorganic arsenic based on mass spectrometry. Anal Bioanal Chem 406:1455–1469

García-Sevillano MÁ, García-Barrera T, Gómez-Ariza JL (2015a) Environmental metabolomics: biological markers for metal toxicity. Electrophoresis 36:2348–2365
García-Sevillano MA, García-Barrera T, Navarro F, Abril N, Pueyo C, López-Barea J, Gómez-Ariza JL (2015b) Combination of direct infusion mass spectrometry and gas chromatography mass spectrometry for toxicometabolomic study of red blood cells and serum of mice Mus musculus after mercury exposure. J Chromatogr B Analyt Technol Biomed Life Sci 985:75–84
García-Sevillano MA, García-Barrera T, Navarro F, Montero-Lobato Z, Gómez-Ariza JL (2015c) Shotgun metabolomic approach based on mass spectrometry for hepatic mitochondria of mice under arsenic exposure. Biometals 28:341–351
Geiszinger AE, Goessler W, Francesconi KA (2002) The marine polychaete Arenicola marina: its unusual arsenic compound pattern and its uptake of arsenate from seawater. Mar Environ Res 53:37–50
Gierasch LM, Gershenson A (2009) Post-reductionist protein science, or putting humpty dumpty back together again. Nat Chem Biol 5:774–777
Girotti AW (1985) Mechanisms of lipid peroxidation. J Free Radic Biol Med 1:87–95
Gómez-Ariza JL, Jahromi EZ, González-Fernández M, García-Barrera T, Gailer J (2011) Liquid chromatography-inductively coupled plasma-based metallomic approaches to probe health-relevant interactions between xenobiotics and mammalian organisms. Metallomics 3:566–577
Gonzalez-Fernández M, García-Sevillano MA, Jara-Biedma R, García-Barrera T, Vioque A, López-Barea J, Pueyo C, Gómez-Ariza JL (2011) Size characterization of metal species in liver and brain from free-living (Mus spretus) and laboratory (Mus Musculus) mice by SEC-ICP-MS: application to environmental contamination assessment. J Anal At Spectrom 26:141–149
Goulet RR, Fortin C, Spry DJ (2011) Uranium. Fish Physiol 31:391–428
Grimalt JO, Ferrer M, MacPherson E (1999) The mine tailing accident in Aznalcollar. Sci Total Environ 242:3–11
Gunn SA, Gould TC, Anderson WAD (1963) Cadmium-induced interstitial cell tumors in rats and mice and their prevention by zinc. J Natl Cancer Inst 31:745–749
Haddow A, Roe FJ, Dukes CE, Mitchley AC (1964) Cadmium neoplasia: sarcomata at the site of injection of cadmium sulphate in rats and mice. Br J Cancer 18:667–673
Hems DA (1972) Metabolism of glutamine and glutamic acid by isolated perfused kidneys of normal and acidotic rats. Biochem J 130:671–680
Hinojosa Reyes L, Marchante-Gayón JM, García Alonso JI, Sanz-Medel A (2003) Quantitative speciation of selenium in human serum by affinity chromatography coupled to post-column isotope dilution analysis ICP-MS. J Anal At Spectrom 18:1210–1216
Hughes MF, Kenyon EM, Edwards BC, Mitchell CT, Del Razo LM, Thomas DJ (2003) Accumulation and metabolism of arsenic in mice after repeated oral administration of arsenate. Toxicol Appl Pharmacol 191:202–210
Jara-Biedma R, González-Dominguez R, García-Barrera T, Lopez-Barea J, Pueyo C, Gómez-Ariza JL (2013) Evolution of metallothionein isoforms complexes in hepatic cells of Mus musculus along cadmium exposure. Biometals 26:639–650
Jin Y, Zhao F, Zhong Y, Yu X, Sun D, Liao Y, Lv X, Li G, Sun G (2010) Effects of exogenous GSH and methionine on methylation of inorganic arsenic in mice exposed to arsenite through drinking water. Environ Toxicol 25:361–366
Jitaru P, Goenaga-Infante H, Vaslin-Reimann S, Fisicaro P (2010) A systematic approach to the accurate quantification of selenium in serum selenoalbumin by HPLC-ICP-MS. Anal Chim Acta 657:100–107
Jitaru P, Prete M, Cozzi G, Turetta C, Cairns W, Seraglia R, Traldi P, Cescon P, Barbante C (2008) Speciation analysis of selenoproteins in human serum by solid-phase extraction and affinity HPLC hyphenated to ICP-quadrupole MS. J Anal At Spectrom 23:402–406
Kasai H (1997) Analysis of a form of oxidative DNA damage, 8-hydroxy-2′-deoxyguanosine, as a marker of cellular oxidative stress during carcinogenesis. Mutat Res 387:147–163

Kitchin KT, Wallace K (2005) Arsenite binding to synthetic peptides based on the Zn finger region and the estrogen binding region of the human estrogen receptor-α. Toxicol Appl Pharmacol 206:66–72
Kitchin KT, Wallace K (2008) The role of protein binding of trivalent arsenicals in arsenic carcinogenesis and toxicity. J Inorg Biochem 102:532–539
Knauer B, Majka P, Watkins KJ, Taylor AWR, Malamanova D, Paul B, Yu H-H, Bush AI, Hare DJ, Reser DH (2017) Whole-brain metallomic analysis of the common marmoset (: Callithrix jacchus). Metallomics 9:411–423
Kolonel LN (1976) Association of cadmium with renal cancer. Cancer 37:1782–1787
Lu M, Wang H, Li X-F, Arnold LL, Cohen SM, Le XC (2007) Binding of dimethylarsinous acid to Cys-13α of rat hemoglobin is responsible for the retention of arsenic in rat blood. Chem Res Toxicol 20:27–37
Maciel BCM, Barbosa HS, Pessôa GS, Salazar MM, Pereira GAG, Gonçalves DC, Ramos CHI, Arruda MAZ (2014) Comparative proteomics and metallomics studies in Arabidopsis thaliana leaf tissues: evaluation of the selenium addition in transgenic and nontransgenic plants using two-dimensional difference gel electrophoresis and laser ablation imaging. Proteomics 14:904–912
Maret W (2004) Exploring the zinc proteome. J Anal At Spectrom 19:15–19
Martelli A, Rousselet E, Dycke C, Bouron A, Moulis J-M (2006) Cadmium toxicity in animal cells by interference with essential metals. Biochimie 88:1807–1814
Messaoudi I, El Heni J, Hammouda F, Saïd K, Kerkeni A (2009) Protective effects of selenium, zinc, or their combination on cadmium-induced oxidative stress in rat kidney. Biol Trace Elem Res 130:152–161
Montes-Nieto R, Fuentes-Almagro CA, Bonilla-Valverde D, Prieto-Alamo MJ, Jurado J, Carrascal M, Gómez-Ariza JL, López-Barea J, Pueyo C (2007) Proteomics in free-living Mus spretus to monitor terrestrial ecosystems. Proteomics 7:4376–4387
Montes Nieto R, García-Barrera T, Gómez-Ariza JL, López-Barea J (2010) Environmental monitoring of Domingo Rubio stream (Huelva estuary, SW Spain) by combining conventional biomarkers and proteomic analysis in Carcinus maenas. Environ Pollut 158:401–408
Mounicou S, Szpunar J, Lobinski R (2009) Metallomics: the concept and methodology. Chem Soc Rev 38:1119–1138
Naranmandura H, Suzuki KT (2008) Formation of dimethylthioarsenicals in red blood cells. Toxicol Appl Pharmacol 227:390–399
Nicholson JK, Timbrell JA, Sadler PJ (1985) Proton NMR spectra of urine as indicators of renal damage. Mercury-induced nephrotoxicity in rats. Mol Pharmacol 27:644–651
Nischwitz V, Davies JT, Marshall D, González M, Gómez Ariza JL, Goenaga-Infante H (2013) Speciation studies of vanadium in human liver (HepG2) cells after in vitro exposure to bis(maltolato)oxovanadium(iv) using HPLC online with elemental and molecular mass spectrometry. Metallomics 5:1685–1697
Nuevo Ordoñez Y, Montes-Bayón M, Blanco-González E, Sanz-Medel A (2010) Quantitative analysis and simultaneous activity measurements of cu, Zn-superoxide dismutase in red blood cells by HPLC-ICPMS. Anal Chem 82:2387–2394
Ogra Y, Nagasaki S, Yawata A, Anan Y, Hamada K, Mizutani A (2016) Metallomics approach to changes in element concentration during differentiation from fibroblasts into adipocytes by element array analysis. J Toxicol Sci 41:241–244
Outten CE, O'Halloran TV (2001) Femtomolar sensitivity of metalloregulatory proteins controlling zinc homeostasis. Science 292:2488–2492
Palacios Ò, Encinar JR, Bertin G, Lobinski R (2005) Analysis of the selenium species distribution in cow blood by size exclusion liquid chromatography-inductively coupled plasma collision cell mass spectrometry (SEC-ICPccMS). Anal Bioanal Chem 383:516–522
Palacios O, Lobinski R (2007) Investigation of the stability of selenoproteins during storage of human serum by size-exclusion LC-ICP-MS. Talanta 71:1813–1816

Paquet F, Houpert P, Blanchardon E, Delissen O, Maubert C, Dhieux B, Moreels AM, Frelon S, Voisin P, Gourmelon P (2006) Accumulation and distribution of uranium in rats after chronic exposure by ingestion. Health Phys 90:139–147

Piao F, Ma N, Hiraku Y, Murata M, Oikawa S, Cheng F, Zhong L, Yamauchi T, Kawanishi S, Yokoyama K (2005) Oxidative DNA damage in relation to neurotoxicity in the brain of mice exposed to arsenic at environmentally relevant levels. J Occup Health 47:445–449

Piasek M, Blanuša M, Kostial K, Laskey JW (2001) Placental cadmium and progesterone concentrations in cigarette smokers. Reprod Toxicol 15:673–681

Platell C, Kong S-E, McCauley R, Hall JC (2000) Branched-chain amino acids. J Gastroenterol Hepatol 15:706–717

Portbury SD, Hare DJ, Sgambelloni CJ, Bishop DP, Finkelstein DI, Doble PA, Adlard PA (2017) Age modulates the injury-induced metallomic profile in the brain. Metallomics 9:402–410

Rodríguez-González P, Marchante-Gayón JM, García Alonso JI, Sanz-Medel A (2005) Isotope dilution analysis for elemental speciation: a tutorial review. Spectrochim Acta Part B: Spectrosc 60:151–207

Ruiz-Laguna J, Abril N, García-Barrera T, Gómez-Ariza JL, López-Barea J, Pueyo C (2006) Absolute transcript expression signatures of Cyp and Gst genes in Mus spretus to detect environmental contamination. Environ Sci Technol 40:3646–3652

Saïd L, Banni M, Kerkeni A, Saïd K, Messaoudi I (2010) Influence of combined treatment with zinc and selenium on cadmium induced testicular pathophysiology in rat. Food Chem Toxicol 48:2759–2765

Shigeta K, Sato K, Furuta N (2007) Determination of selenoprotein P in submicrolitre samples of human plasma using micro-affinity chromatography coupled with low flow ICP-MS. J Anal At Spectrom 22:911–916

Shimoda R, Achanzar WE, Qu W, Nagamine T, Takagi H, Mori M, Waalkes MP (2003) Metallothionein is a potential negative regulator of apoptosis. Toxicol Sci 73:294–300

Srivastava RC, Ahmad I, Kaur G, Hasan SK (1988) Alterations in the metabolism of endogenous trace metals due to cadmium, manganese and nickel-effect of partial hepatectomy. J Environ Sci Health Part A: Environ Sci Eng 23:95–101

Stadtman ER (1992) Protein oxidation and aging. Science 257:1220–1224

Sussulini A, Becker JS, Becker JS (2017) Laser ablation ICP-MS: application in biomedical research. Mass Spectrom Rev 36:47–57

Suzuki KT (2005) Metabolomics of arsenic based on speciation studies. Anal Chim Acta 540:71–76

Suzuki KT, Mandal BK, Ogra Y (2002) Speciation of arsenic in body fluids. Talanta 58:111–119

Suzuki KT, Ogra Y (2002) Metabolic pathway for selenium in the body: speciation by HPLC-ICP MS with enriched se. Food Addit Contam 19:974–983

Suzuki Y, Yoshikawa H (1981) Cadmium, copper, and zinc excretion and their binding to metallothionein in urine of cadmium exposed rats. J Toxicol Environ Health 8:479–487

Szinicz L, Forth W (1988) Effect of As2O3 on gluconeogenesis. Arch Toxicol 61:444–449

Szymańska-Chabowska A, Antonowicz-Juchniewicz J, Andrzejak R (2007) The concentration of selected cancer markers (TPA, TPS, CYFRA 21-1, CEA) in workers occupationally exposed to arsenic (as) and some heavy metals (Pb, cd) during a two-year observation study. Int J Occup Med Environ Health 20:229–239

Tainer JA, Roberts VA, Getzoff ED (1991) Metal-binding sites in proteins. Curr Opin Biotechnol 2:582–591

Thomas DJ, Styblo M, Lin S (2001) The cellular metabolism and systemic toxicity of arsenic. Toxicol Appl Pharmacol 176:127–144

Vahter M (1981) Biotransformation of trivalent and pentavalent inorganic arsenic in mice and rats. Environ Res 25:286–293

Vahter M, Norin H (1980) Metabolism of 74As-labeled trivalent and pentavalent inorganic arsenic in mice. Environ Res 21:446–457

Vioque-Fernández A, Alves de Almeida E, López-Barea J (2009) Assessment of Doñana National Park contamination in Procambarus clarkii: integration of conventional biomarkers and proteomic approaches. Sci Total Environ 407:1784–1797

Wang H, Wu Z, Chen B, He M, Hu B (2015) Chip-based array magnetic solid phase microextraction on-line coupled with inductively coupled plasma mass spectrometry for the determination of trace heavy metals in cells. Analyst 140:5619–5626

Wang Y, Tang H, Nicholson JK, Hylands PJ, Sampson J, Holmes E (2005) A metabonomic strategy for the detection of the metabolic effects of chamomile (Matricaria recutita L.) ingestion. J Agric Food Chem 53:191–196

Warner ML, Moore LE, Smith MT, Fanning E, Smith AH, Kalman DA (1994) Increased micronuclei in exfoliated bladder cells of individuals who chronically ingest arsenic-contaminated water in Nevada. Cancer Epidemiol Biomark Prev 3:583–590

Wei L, Liao P, Wu H, Li X, Pei F, Li W, Wu Y (2008) Toxicological effects of cinnabar in rats by NMR-based metabolic profiling of urine and serum. Toxicol Appl Pharmacol 227:417–429

Wolters DA, Stefanopoulou M, Dyson PJ, Groessl M (2012) Combination of metallomics and proteomics to study the effects of the metallodrug RAPTA-T on human cancer cells. Metallomics 4:1185–1196

Wrobel K, Wrobel K, Caruso JA (2009) Epigenetics: an important challenge for ICP-MS in metallomics studies. Anal Bioanal Chem 393:481–486

Xu M, Bijoux H, Gonzalez P, Mounicou S (2014a) Investigating the response of cuproproteins from oysters (Crassostrea gigas) after waterborne copper exposure by metallomic and proteomic approaches. Metallomics 6:338–346

Xu M, Frelon S, Simon O, Lobinski R, Mounicou S (2014b) Non-denaturating isoelectric focusing gel electrophoresis for uranium-protein complexes quantitative analysis with LA-ICP MS. Anal Bioanal Chem 406:1063–1072

Xu M, Frelon S, Simon O, Lobinski R, Mounicou S (2014c) Development of a non-denaturing 2D gel electrophoresis protocol for screening in vivo uranium-protein targets in Procambarus clarkii with laser ablation ICP MS followed by protein identification by HPLC-Orbitrap MS. Talanta 128:187–195

Yáñez L, Carrizales L, Zanatta MT, de Jesús Mejía J, Batres L, Díaz-Barriga F (1991) Arsenic-cadmium interaction in rats: toxic effects in the heart and tissue metal shifts. Toxicology 67:227–234

Zalups RK (2000) Molecular interactions with mercury in the kidney. Pharmacol Rev 52:113–143

Zalups RK, Koropatnick J (2000) Temporal changes in metallothionein gene transcription in rat kidney and liver: relationship to content of mercury and metallothionein protein. J Pharmacol Exp Ther 295:74–82

Chapter 4
Metallomics Study in Plants Exposed to Arsenic, Mercury, Selenium and Sulphur

Jörg Feldmann, Katharina Bluemlein, Eva Maria Krupp, Martin Mueller, and Barry Alan Wood

Abstract This chapter is focussing on the interaction of arsenic, mercury and selenium with plans. Aspects of biotransformations are discussed, before the analytical methodologies are listed and critically appraised in the second part. A holistic view is given, starting from the soil environment and continuing to the plant roots and the translocations into the upper part of the plants. Under different soil conditions, different kinds of elemental species are identified, which have an impact on how the elemental species are taken up by the plant. The uptake mechanisms of these elemental species are explained and compared before the biotransformation reactions of all elemental species in the plant root; their transport into the vacuoles and translocation to the leaves and grains are discussed. Here in particular the interaction with sulphur-rich phytochelatins is described for all three elemental species. Since the sulphur chemistry is so important for the uptake, bioaccumulation and translocation of the metals and metalloids, a subchapter about sulphur chemistry in plants has been added. All aspects of biotransformation dealt with in this chapter is finally rounded up by a thorough description of the analytical methodology given with a focus on the use of HPLC-ICPMS/ESI-MS for both quantitative and molecular analysis.

Keywords Arsenic · Phytochelatins · Mercury · Selenium · Sulphur · ICP-MS · ESI-MS · Hyphenated techniques · HPLC-ICPMS/ESI-MS

J. Feldmann · E. M. Krupp (✉) · M. Mueller
TESLA (Trace Element Speciation Laboratory), Department of Chemistry, University of Aberdeen, Scotland, UK
e-mail: e.krupp@abdn.ac.uk

K. Bluemlein
Fraunhofer Institute for Toxicology and Experimental Medicine, Hannover, Germany

B. A. Wood
Reading Scientific Services Ltd, The Reading Science Centre, Reading, Berkshire, UK

M. A. Z. Arruda (ed.), *Metallomics*, Advances in Experimental Medicine and Biology 1055, https://doi.org/10.1007/978-3-319-90143-5_4

Abbreviations

As(III)	Arsenite
As(V)	Arsenate
DMA(V)	Dimethylarsinic acid
ESI-MS	Electrospray ionisation mass spectrometry
GSH	Glutathione
HPLC	High performance liquid chromatography
ICPMS	Inductively coupled plasma mass spectrometry
MeHg	Methylmercury
MMA(III)	Monomethylarsonious acid
MMA(V)	Monomethylarsonic acid
PC	Phytochelatins
Se(IV)	Selenite
Se(VI)	Selenate
Se^o	Elemental selenium
TETRA	Tetramethylarsonium
TMA	Trimethylarsine
TMAO	Trimethylarsine oxide

4.1 Introduction

Metallomics, the molecular forms of metals and metalloids in plants, can be viewed from different angles with different degrees of complexity. In this chapter, we have focussed on the exposure to mainly toxic elements such as arsenic and mercury, whilst selenium and sulphur were covered as well. Selenium is mainly considered toxic, but there is evidence that selenium might be essential to plants in a similar way as in animals and humans. There are a lot of similarities on how plants react to direct exposure to these elements so that common schemes can be developed.

In this book chapter, we do not cover other toxic elements such as cadmium or essential elements such as zinc or copper. Furthermore, we exclude the formation of essential metalloproteins, since other analytical methods and platforms are necessary for the analysis of those elements, which goes beyond the merit of this chapter.

The focus of our chapter is on arsenic, mercury, selenium and sulphur; thus we cover the most studied elements with regard to speciation analysis in plants. We will describe the speciation in which these elements occur in soil and soil porewater and the form in which they are taken up by the roots. The process of accumulation in vacuoles including long-range transport (roots/shot translocation) will be covered by identifying their molecular forms including their biotransformations. Here we have added this subchapter on sulphur since this element seems important for biotransformation of thiophilic elements arsenic, mercury and selenium.

4.2 Biotransformation of Arsenic

Arsenic is a metalloid with ubiquitous occurrence in our environment – of either natural or anthropogenic origin. Thomas and Troncy (2009), referring to the ambivalent character associated with arsenic, called it "a beneficial therapeutic poison" in their review about arsenical-based medicine, recent and past. Its "fame" as poison, however, is undisputable. Marie Lafarge (Griffin 2015) and Madelaine Smith (MacGowan 2007; Scotsman 2005) were probably not the only women who rid themselves of their husband or lover using arsenic. With such a reputation, it is not surprising that arsenic even made it on stage. In "Arsenic and Old Lace" (Kesselring 1939), two elderly ladies take it in their skilled hands to remedy old bachelors from their suffering, with home-made organic elderberry wine doing the trick. Not all arsenic-related incidents were as intentional, to just name the Bradford sweets poisoning in 1858 (Jones 2000) as one. In recent days, the widespread arsenic poisoning in Southeast Asia has been of great concern, with arsenic-contaminated drinking water being one of the culprits (McCarty et al. 2011). The considerable daily consumption of rice poses another significant route of arsenic uptake (Meharg and Rahman 2003; Ohno et al. 2007), which on a second thought raises the question of how plants defend themselves against arsenic exposure.

4.2.1 Arsenic Species in Bulk Soil and Rhizosphere

The arsenic uptake by terrestrial plants mainly takes place via the roots. The speciation and bioavailability of arsenic in any given soil depend among other things on the soil type, pH, redox potential and the content of metal oxides/hydroxides (e.g. Fe, Al). The most prominent arsenic species found in the soil environment are the two inorganic forms As(III) and As(V). Under aerobic conditions, arsenate is the major arsenic species, whereas under anaerobic conditions (e.g. flooded rice paddy fields), arsenite prevails (Zhao et al. 2008; Masscheleyn et al. 1991). Both forms are readily adsorbed by metal oxides such as Fe oxides/hydroxides, an often-found soil constituent, as was demonstrated by a study undertaken by Pierce and Moore (Pierce and Moore 1982). In general, arsenate showed a stronger sorption behaviour than arsenite. At a pH of 4, the adsorption of arsenate to Fe oxides/hydroxides was the highest and decreased with increasing pH. The adsorption of arsenite to Fe oxides/hydroxides was most efficient at a pH of 7. Having said that, Pierce and Moore also revealed that the sorption behaviour of the two inorganic arsenic species is more complex than that. The findings of the study showed that the adsorption behaviour of both species is influenced by their concentration: the inorganic arsenic – adsorbent ratio as well as the presence of ions competing for the sorption sites on Fe oxides/hydroxides, such as sulphate.

Adsorption of inorganic arsenic to metal oxides/hydroxides puts plants out of harm's way – a false sense of safety. Adsorbed arsenate no longer being bioavailable can be mobilised again by reduction to arsenite due to either reducing conditions or microbial activities. Reductive dissolution of Fe oxides/hydroxides is another process that increases the bioavailability of inorganic arsenic in soil (Masscheleyn et al. 1991).

Apart from inorganic arsenic, also organic species are found in soils. Here, monomethylarsonic acid (MMA(V)) and dimethylarsinic acid (DMA(V)) are to be named as the two prominent methylated arsenic species found in arsenic-contaminated soils. If not a result of arsenical treatment of the soil (e.g. biocides), MMA(V) and DMA(V) are formed from arsenite by microorganisms (Abedin et al. 2002). These two organo-arsenic species, as the inorganic ones, can be further metabolised to volatile (methylated) arsines, removing arsenic from the soil (Cheng and Focht 1979). Mestrot et al. (2013) have reviewed biovolatilisation of arsenic from soils and sediments, concluding the amount of volatile arsenic species being very low in comparison to the total arsenic content of the bulk soil.

Rhizosphere is the soil in the immediate proximity of plant roots and should be regarded as an independent soil entity. The rhizosphere serves among other things as nutrient pool for plants. Root exudates, such as organic acids, are means used by plants to increase the bioavailability of nutrients by changing pH and redox potential of the rhizosphere. These two soil parameters affect the adsorption and desorption characteristics and thereby the bioavailability, of not only nutrients but also contaminants such as arsenic. In their pursuit to optimise their nutrient uptake, plants often team up with fungi and microorganisms, increasing the complexity of arsenic speciation and bioavailability even further. Mechanistic studies regarding a. arsenic speciation in soil and/or plants and b. arsenic uptake and translocation by plants require a well-designed experimental set-up considering above-mentioned information, to ensure that no false conclusions are drawn. A study ran by Lomax and co-workers (Lomax et al. 2012) demonstrated that when looking at methylated arsenic species in plants, axenic exposure media should be chosen to rule out any microorganism- or fungi-derived contamination of the growth media with methylated arsenic species, such as MMA(V) or DMA(V). This work and the findings by Zhu and co-workers (Jia et al. 2012) indicate that plants are not able of arsenic methylation, thereby refuting previously made assumptions (Raab et al. 2005).

Methodical investigations regarding the effect of selected root exudates and microbial siderophores on the dissolution of arsenic from As-goethite were conducted by Liu et al. (2017). Dissolution testing in the presence of selected root exudates and/or siderophores demonstrated their beneficial effect on the solubilisation of inorganic arsenic. Analogous experiments with the arsenic hyperaccumulator *P. vittata* indicated continuous dissolution of As from goethite due to its depletion from the solution by *P. vittata*. The investigated ligands also affected the As uptake by *P. vittata* as well as its translocation in the plant with differing extent.

Gonzaga and co-workers (Gonzaga et al. 2006) studied the influence of the As hyperaccumulator *P. vittata* and the non-accumulator *N. exaltata* on bulk soil and rhizosphere parameters as well as on arsenic soil distribution. After 8 weeks of growth on As-contaminated soil in a greenhouse, the water-soluble arsenic concentration, pH and dissolved organic carbon (a measure for root exudates) were determined in bulk soil and rhizosphere and compared to soil with no plants growing. For the bulk soil, no changes were observed. In the rhizosphere, however, plant species-dependent changes were monitored. Noticeable was also that both plant species depleted the most abundant arsenic pool, the amorphous Al and Fe hydrous-oxide bound, rather than the most available one, an observation that could be explained by the release of root exudates.

Having discussed the increased bioavailability of inorganic arsenic by means of root exudates, there has got to be an example demonstrating the opposite. Liu et al. (2004) suggest that the release of O_2 by the roots of rice plant can further the formation of an iron plaque in the presence of Fe(II) due to oxidation processes. The formation of an iron plaque on the root surface could facilitate the trapping of arsenate, thereby reducing its bioavailability and uptake by rice plants.

More detailed information on the subject of "arsenic species in bulk soil and rhizosphere" can be found in review articles, for example, by Zhao et al. (2008) and Fitz and Wenzel (2002).

4.2.2 Arsenic Speciation in Terrestrial Plants

It might not come as a surprise that different plant species have developed different mechanism to cope with soil-derived arsenic exposure – ranging from low arsenic uptake to hyperaccumulation (Zhao et al. 2010a). Inorganic arsenic, when taken up as As(V), is readily reduced, enzymatically and non-enzymatically, to As(III) inside the plant. Both forms are toxic, by different modes of action though, as explained in more detail in a review paper by Finnegan and Chen (2012). As(III) as the major arsenic species is generally found in plant roots (Xu et al. 2017 b) after exposure to inorganic arsenic is dealt with in various ways by plants. One option is its efflux from the roots by, for example, aquaporin channels that also facilitate arsenite uptake. Zhao et al. (2010b) investigated the efflux capacity of rice aquaporin Lsi1, concluding bidirectional permeability of this passive, but major, arsenite transport route. Their findings also indicated that arsenite efflux will depend on the arsenite concentration in the external medium – the higher the concentration gradient the higher the efflux rate. Also, the existence of further efflux mechanisms was brought forward by the researchers. The review by Chen et al. gives a more comprehensive account on arsenite efflux mechanisms in plants, also covering the findings from transgenic plants (Chen et al. 2017). The work by Liu et al. (2010) addressed the impact of phytochelatins on arsenite efflux and root-to-shoot translocation in *Arabidopsis* using wild-type as well as GSH-deficient *cad2-1* and PC-deficient *cad3-1* mutants. The experiments meddling with the GSH and PC synthesis showed that the arsenite efflux is more pronounced in *Arabidopsis* plant with insufficient GSH/PC supply, suggesting the trapping of arsenite by complexation with GSH and PCs.

Glutathione (GSH) and phytochelatins (PCs) are often discussed cysteine-rich peptides, contributing to arsenic detoxification in plants, non-tolerant, tolerant and hyperaccumulating ones. Their contribution to arsenic detoxification, however, probably differs depending on the plant species, as the varying contribution of As(III)-PC complexes to the total arsenic content suggests (Raab et al. 2004, 2005). The complexation of arsenite by GSH and/or PCs and their subsequent sequestering from the cytosol into vacuoles, whether in roots or shoots, are another arsenic detoxification pathway found in plants. Song and co-workers identified the two vascular transporters AtABCC1 and AtABCC2 in *Arabidopsis thaliana*, supporting the hypothesis of As(III)-PC vacuole sequestering (Song et al. 2010). The complexation of As(III) by PCs does not only impact on the arsenite efflux, but it also reduces its mobility in terms of root-to-shoot translocation- an aspect of particular interest for crop plants such as rice, where the above-ground tissues (e.g. rice grains) are of vital importance for the nutrition of millions of people. Batista et al. concluded from their study on six rice cultivars that the complexation of arsenite by phytochelatins suppresses the arsenite root-to-shoot translocation (Batista et al. 2014).

The number of possible phytochelatins is vast. The core of each phytochelatin is the repetitive y-glutamylcysteine unit whereas the terminal amino acid can vary, depending on the plant and metal(loid) species. Apart from the most frequently discussed PCs ($[y\text{-Glu-Cys}]_n$-Gly), iso-PCs ($[y\text{-Glu-Cys}]_n$-Glu; ($[y\text{-Glu-Cys}]_n$-Gln), desglycine-PCs ($[y\text{-Glu-Cys}]_n$) and hydroxymethyl-PCs ($[y\text{-Glu-Cys}]_n$-Ser) have been associated with arsenic exposure in higher plants (Batista et al. 2014; Mishra et al. 2017). Arsenite requires the formation of three As-S bonds to either PCs or glutathione to ensure that its effectiveness to interact with enzymes and proteins is diminished. With plants having GSH, PCs, iso-PCs, desglycine-PCs and hydroxymethyl-PCs at their "hand", the number of conceivable As(III)-"PC" complexes is considerable. Raab et al. (2005) detected not only As(III)-PC complexes in root and above-ground tissue of *Helianthus annuus* but also a mixed complex with arsenite being chelated by GSH and PC_2. They also detected for the first time the coordination of MMA(III) with phytochelatins – namely, the MMA(III)-PC_2 complex. Mishra et al. (2017) studied arsenic speciation in *Oryza sativa L.* after exposure to As(V), DMA(V) and MMA(V). Their findings confirmed the capability of (some) plants to reduce MMA(V) to MMA(III) and subsequent complexation by thiols (including, e.g. GSH, PCs, hydroxymethyl-PCs) of the latter. DMA(V) taken up by the rice plants was mainly detected as such in the plant tissue extracts, with one unidentified compound being the exception. Could it have been dimethylthioarsinoyl glutathione (DMA(V)S-GS), which was detected in *Brassica oleracea* by Raab et al. (2007a).

Despite the arsenic species found in roots and above-ground tissues being quite diverse, the two shuttles, xylem and phloem exudates, are only entered by As(V), As(III), MMA (V) and DMA(V) (Ye et al. 2010). Even though free oxidised glutathione and phytochelatins were found, they did not detect any arsenic-thiol complexes in phloem and xylem exudates of castor beans. The authors discuss that the relative fractions of As(III) and As(V) found in xylem sap are most likely determined by several factors such as:

- As(V)-to-As(III) reduction capacity in roots
- As(III) vs As(V) xylem-loading transporters
- Availability of As(III) in roots, determined by thiol complexation in roots
- Competing ions

MMA(V) and DMA(V) are two arsenic species produced by soil microorganisms. They are taken up more slowly than inorganic arsenic, the efficiency with which they are transported via xylem and phloem, however, is higher (Raab et al. 2007b).

Arsenic hyperaccumulating plants translocate most of the arsenic entering the roots into above-ground tissues. Thiol complexation of arsenic can be considered as secondary in hyperaccumulators, as the low fraction (1%) of As(III)-PCs detected in *Pteris cretica* suggests (Raab et al. 2004). Lombi et al. (2002) looked into the arsenic distribution in the fronds of another arsenic hyperaccumulator, *Pteris vittata*. The majority (75%) of inorganic arsenic found in the fronds was As(III), with the remaining being As(V). A large fraction (96%) of total arsenic in the fronds was located in the pinnae. There, most of the inorganic arsenic was detected in upper and lower epidermal cells. X-ray microanalysis results suggest that the arsenic in these cells is compartmentalised in the vacuole. It has been suggested that the ACR3 found in duplicate in the vacuolar membrane of *Pteris vittata* is the key to arsenite translocation from cytosol into the vacuole in hyperaccumulating plants (Indriolo et al. 2010).

In summary, inorganic arsenic is as ubiquitous in plants as it is in our environment. Thiol complexation occurs in roots and above-ground tissues, but not in xylem and phloem. The two methylated arsenic species MMA(V) and DMA(V) are less abundant in plants but were detected in all tissues, including xylem and phloem. MMA(V) and DMA(V) are not produced by plants. They are taken up from the soil environment. MMA(V) can be reduced to MMA(III) which in turn is available for thiol complexation. After thiol activation DMA(V) can also bind to thiols. Diversity, abundance and distribution of the various arsenic species are determined among other things by plant species, soil parameters, other nutrients, available arsenic species, and its concentration, in soil environment as well as exposure time.

Arsenic species found in plants also comprise less spoken of species such as trimethylarsine oxide (TMAO), tetramethylarsonium ion (TETRA), arsenobetaine and arsenocholine (Geiszinger et al. 2002). Investigation on arsenic volatilisation from rice plants revealed that when TMAO was taken up by rice plants, it was reduced to the volatile species trimethylarsine (TMA) (Jia et al. 2012). The researchers also discovered that neither inorganic arsenic nor MMA or DMA was further methylated to TMA. Due to their absence in the scientific limelight, up to now little is known about their compartmentalisation and fate, once they are taken up by terrestrial plants. These compounds are, however, not the result of plants metabolising other arsenicals but originate from the soil surrounding the root system, where they are most likely produced and released by microorganisms.

4.3 Biotransformations of Mercury

Mercury (Hg) occurs in a variety of stable chemical species, with the main species being elemental mercury Hg^o, inorganic salts (e.g $HgCl_2$) and organic Hg compounds in which Hg can be bound to one or two carbon chains. Organic mercury is mainly present as methylmercury ($MeHg^+$, here defined as HgMe) and is well known to bioaccumulate through the aquatic food chain. In the environment, MeHg can be bound to a variety of anions, e.g. with chloride anions, whilst Hg^{2+} can bind with a variety of anions, often as well with Cl anions to form a stable $HgCl_4^{2-}$ complex in water. Mercury is also thiophile and forms strong complexes and/or chemical bonds with sulphur or selenium in the environment and in biota. The bioaccumulation of MeHg in biota is driven by binding through sulphur-containing proteins, as the cysteine moiety of amino acids was shown to bind with MeHg, e.g. in protein-rich rice seeds (Clarkson and Magos 2006).

Elemental mercury is highly volatile; Hg^0 volatilises in ambient temperatures; however $HgCl_2$ is similarly volatile and can be found in the air. Coal-fired power plants release several tonnes of Hg per year, and Hg does not remain in the ash but forms volatile Hg^0, $HgCl_2$ and other reactive Hg species (Pavageau et al. 2002).

Mercury species in general undergo a most complicated cycle in the environment: volatile Hg, e.g. from coal-fired power plants, is oxidised by the sun or other radicals and is removed from the atmosphere with rain as wet deposit but can also be deposited dry; as Hg and its species are volatile or can be volatilised, the Hg species can deposit into the seas or lands; inorganic Hg is prone to methylation in water and sediment, forming MeHg which bioaccumulates in the food chain. Crucially, airborne Hg is very mobile and undergoes a depositing-volatilising cycle, in which Hg is "distilled" to the poles, evidenced by Hg analysis of polar ice sheets and long-lived animals, e.g. polar bears or whales (Swain 2007).

In general, any Hg compound is toxic, with MeHg the most critical Hg compound as it is the most toxic, and known to accumulate. MeHg can also pass through the blood-brain barrier and can damage brain functions in the foetus. This has been evidenced tragically when the Minamata bay was contaminated with MeHg, taken up into the staple food, fish, and caused death or severe brain damage for hundreds of people. However, biota have developed detoxification strategies, e.g. the formation of HgSe nanoparticles or binding in the form of phytochelatin or other sulphur or selenium compounds. For example, HgSe nanoparticles were evidenced in the liver and brain of a pod of long-finned pilot whales, which suggests that MeHg is transported to the brain where it is detoxified through the formation of biochemical inert HgSe (Gajdosechova et al. 2016).

4.3.1 Hg in Soil and Biotransformation

Hg in soil mainly occurs in inorganic form of Hg^{2+} and can bind to anions, form stable complexes or form salts. In anaerobic conditions, Hg^{2+} can be methylated, and it is thought that the methylation process is microorganism induced, probably in an attempt to detoxify the inorganic mercury which is more toxic for the plant than MeHg (Xu et al. 2016). Rice cultivation practices are important. In wet conditions, MeHg is formed and can be taken up throughout, whilst it was found that alternative dry-wet conditions can mitigate MeHg formation however spikes upon re-establishing wet conditions (Rothenberg et al. 2014). Recently, it was suggested that S in soils may control Hg speciation in rice (Xu et al. 2017a, b).

4.3.2 Hg Species Accumulation in Rice

Rice is a staple food for most Asian countries, which makes it the most important food commodity worldwide. This underpins the importance of mercury analysis in rice and the research on uptake and accumulation processes of inorganic and methylmercury.

In a pot experiment using flooded conditions, rice plants were challenged with either inorganic Hg^{2+} or $MeHg^{+}$, which evidenced a stark difference in terms of uptake into different plant parts of the rice plant and showed a selective accumulation for MeHg into the grain, whilst Hg^{2+} was rather accumulated in roots. It was shown that rice plants challenged with Hg^{2+} was efficiently bound with a variety of phytochelatins, using two cysteine functional groups to bind Hg in the oxidised form, Hg^{2+}, which is deposited in the roots. In contrast, $MeHg^{+}$ does not form any complexes with phytochelatins but is transported and deposited mainly in the rice grains (Krupp et al. 2009; Krupp et al. 2008). This has important implications for food security in rice.

In this exposure experiment, it was evidenced that Hg^{2+} is bound to phytochelatins in the root but that MeHg was transported into the grains, probably as self-defence. This clearly demonstrates that the knowledge of molecular binding and speciation is of utmost importance to understand detoxification pathways.

MeHg in rice grain is accumulated to the highest concentrations and is uniformly distributed; some Hg2+ is also accumulated in bran and husk, similar to the As accumulation in bran and husk, but very different from grain where As is mainly found in husk and bran as discussed in Chap. 2 in this book. Whilst it is easy to remove As by husking and bran removal, MeHg cannot be removed as it is bound to proteins in the grain.

The way how the uptake and accumulation of Hg in roots and MeHg in rice grains works is shown in Fig. 4.1.

The picture is further complicated by the fact that Hg is taken up from the atmosphere: in a stable isotope experiment, it was evidenced that airborne Hg is taken up through aerial parts of the rice plant, showing that Hg^{2+} in rice grains is almost entirely from atmospheric Hg sources and is taken up through the leaves (Strickman and Mitchell 2017). This adds to the Hg burden in the rice plant, especially where coal-fired plants are exhaling mainly Hg^{0}, e.g. in the Wuhan District in China. Recently, Nature reported the uptake of atmospheric elemental Hg into vegetation in the tundra is approximately 70%, explaining high burdens of Hg in soil and water run-off in this part of the world (Obrist et al. 2017).

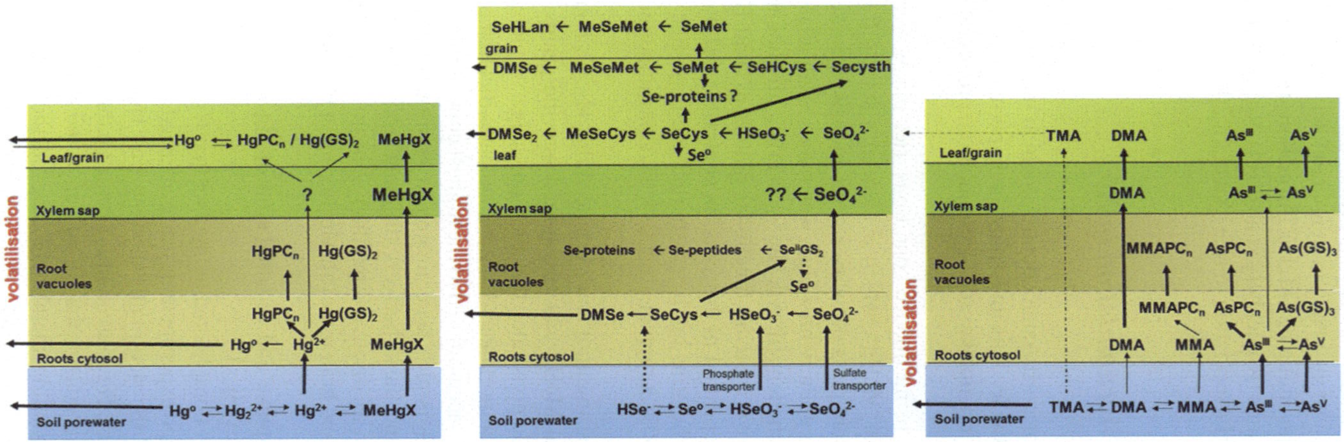

Fig. 4.1 Uptake, accumulation in roots, translocation into shoots and seeds for mercury, selenium and arsenic

4.3.3 Hg -Accumulating Plants in Terrestrial Plants and Phytoremediation

A variety of terrestrial plants are able to grow on highly Hg-contaminated soil; e.g. *Arabidopsis*, *Brassica*, *Datura*, *Marrubium* and *Sesbania*. For example, *Sesbania drummondii* was shown to accumulate up to 1000 mg/kg of Hg in their shoots but 40 times more in its roots. The search for hyperaccumulator plants for Hg is driven by the interest to reclaim contaminated lands. *S. drummondii* is a potential Hg bio-accumulator; however there are no confirmed hyperaccumulator plants for Hg (Venkatachalam et al. 2009).

There are attempts to detoxify Hg in plants using engineered transgenic plants which are bacterial genes *merA* and *merB* which are known to volatilize Hg^0, with the argument that Hg0 is less toxic than inorganic Hg or MeHg (Meagher 2000). Whilst this strategy may help to decontaminate a certain area, the volatile Hg^0 will be far more mobile and only dilutes the contamination, whilst Hg contamination in the atmosphere rises.

4.4 Selenium Biotransformations

Selenium is an important micronutrient and an essential element for animals including humans. It has been established almost 60 years ago (Schwarz and Foltz 1957) that the selenium is an integral part of glutathione peroxidase, an enzyme which is important to deter oxidative stress. However, so far selenium has not been established to be essential for plants. Although there are plants which hyperaccumulate selenium, as far as the authors know, there is neither a specific transporter known for the uptake of selenium and their species into the plant, nor a selenium-specific transporter known for the translocation of selenium inside the plant. Selenium is a very redox-active element, and it exist in many different oxidation stages from –II (as Se^{2-}), 0 (as Se^o), +IV ($HSeO_3^-$) and + VI (SeO_4^{2-}). In addition +II exists when selenium binds to sulphur ($SeSO_3^{2-}$). But it forms also a rather stable Se-C bond; hence we can expect a plethora of organoselenium species in environmental and biological samples (Wallschlaeger and Feldmann 2010).

4.4.1 Selenium Speciation in Soil and Soil Porewater

As can be seen from the Pourbaix diagram (Fig. 4.2), selenium can be thermodynamically stable as selenide (HSe^-) in water-logged soils which display strongly reducing conditions such as paddy soils. Under slightly reducing conditions, also elemental selenium is the most stable species, whilst selenite and selenate are the most common selenium species in aerobic soils. In acid soils selenite as $HSeO_3^-$ is the most prevalent form, whilst in alkaline soils selenate is much more stable.

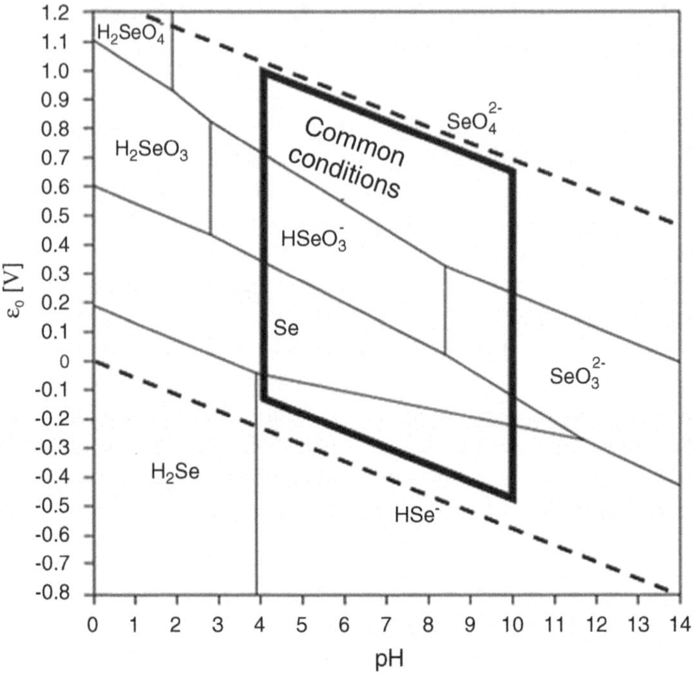

Fig. 4.2 Pourbaix diagram of selenium, which illustrates the thermodynamic stable species in soil porewaters in dependence of the redox potential and the pH

4.4.2 Selenium in Terrestrial Plants

Since selenium has not shown so far that it is an essential micronutrients for plants, no specific transporter of selenite or selenate has been identified. However, selenite is known to be taken up by phosphate transporters. In rice it has been shown using a rice mutant overexpressing OsPT2 to have enhanced selenite uptake (Zhang et al. 2014), and overexpression in OsPT8 in tobacco enhances also the selenite uptake as recently been reported (Song et al. 2017). Even the marine coccolithophore *E. huxleyi* uses a phosphate transporter, and selenite uptake is not influenced by selenate, sulphate or sulphite (Araie et al. 2011). Selenate on the other hand is taken up by the efficient sulphate transporter Sultr1;2 as established for *Arabidopsis thaliana* (Shibagaki et al. 2002).

Once in the roots, the different selenium species have also different long-range transport in the plants. Selenite is not well transported in the xylem to the shoots whilst selenate is efficiently transported (Li et al. 2008). However, inside the roots already biotransformation of the selenium species may take place. Selenate is reduced to selenite and further to elemental selenium in the root cells.

Elemental selenium can make up almost 20% in plant cells (e.g. *Thunbergia alata)* (Aborode et al. 2015). This reduction is not very efficient; hence there is the tendency that selenate is transported into the shoots whilst selenite is biotransformed into organoselenium species. This transformation encompasses a further reduction to selenide, which however is too reactive and consequently forms other organoselenium compounds and has therefore not been identified in plants.

Several organoselenium compounds can be identified in the roots, for example, selenium diglutathione ($Se(GS)_2$) and selenium phytochelatin2 ($SePC_2$), but additional selenium could replace sulphur in cysteine and form Se-containing glutathione (Aborode et al. 2016; Bluemlein et al. 2009b). This compound has however never been isolated on its own but in its oxidised form, when bound via the Se-S bond to another peptide chain. This finding illustrates that Se potentially be incorporated as selenocysteine (SeCys) or as selenomethionine (SeMet) in plant proteins. This however has not been intensively researched in contrast to selenium speciation in yeast (Bierla et al. 2012). What we know is that selenate (SeO_4^{2-}) gets reduced inside the roots to selenite ($HSeO_3^-$); this can be transferred via $Se(GS)_2$ and very reactive selenide (HSe^-) to form SeCys. SeCys can also be methylated to MeSeCys and to volatile dimethyldiselenide ($DMSe_2$) (Pilon-Smits et al. 1999). SeCys could however transfer to selenohomocysteine (SeHCys) which could convert to SeMet and eventually also to the volatile dimethylselenide (DMSe). This volatilisation is considered as a detoxification. Whilst selenium excluder usually generates only DMSe, hyperaccumulator tend to generate $DMSe_2$, which is more efficient to excrete selenium.

One aspect of selenium biotransformation and translocation is to introduce bioavailable forms of selenium into the edible parts of the plants. For example, a recent study showed the influence of nitrogen and sulphur fertilisers on the speciation of selenium compounds in wheat grains (Duncan et al. 2017). In the grains, up to 70 mg/kg selenium could be accumulated, and besides the expected SeMet and Se-methylselenocysteine (MeSeCys), selenohomolanthionine (SeHLan) was characterised to occur in the grain. Usually, in plants with lower selenium concentration, the main selenium species in the grain or seed is SeMet.

The transformation of SeCys into selenoproteins in plants is still controversially discussed, although it has been shown that in the algae *Chlamydomonas reinhardtii* glutathione peroxidase has been characterised (Fu et al. 2002). One obstacle is the reactivity of the selenium in the selenoproteins, and their small concentrations make it difficult to establish novel proteins. Using multidimensional HPLC coupled to ESI-qTOF and ICPMS established a selenium-containing peptide in which SeMet replaces methionine in soybean (Chan and Caruso 2012). There have been reports that glutathione peroxidase exists in *Aloe vera* (Sabeh et al. 1993), but later reports have concluded that this GPx protein does not contain SeCys. Hence, so far no specific selenoprotein in which selenium is incorporated specifically as SeCys has been established for terrestrial plants, and that is why the essentiality of selenium in plants is still controversially discussed.

4.5 Sulphur Biotransformation

Sulphur is important for plants as an essential, often growth-limiting, macronutrient and has indirect effects on the use efficiency of other plant nutrients. Plants are, contrary to animals and humans, capable of incorporating sulphur in organic compounds like the amino acids cysteine and methionine, as well as in essential vitamins and cofactors (Lewandowska and Sirko 2008; Blum et al. 2013).

In *Brassicaceae* (e.g. cabbage, broccoli, mustard) and *Allium* (e.g. onion, garlic, leek) species, sulphur compounds are additionally relevant as secondary metabolites. Glucosinolates and alliins are both enzymatically hydrolysed after tissue rupture to form a wide array of mostly volatile products. These are not only responsible for giving these vegetables their flavour but are also commonly associated with health benefits like reducing the risk of cancer (Hirai and Saito 2008; Kopriva et al. 2015).

The occurrence of sulphur in the environment ranges from the oxidation states −II in sulphides to +VI in sulphates. The biogeochemical cycle of sulphur as described by Takahashi et al. (2011) shows the free mobility between lithosphere, hydrosphere and atmosphere. Inputs of sulphur through atmospheric deposition that are less than the requirements of the crops lead to a sulphur deficiency in soils. The main causes of this deficit are the reduction of anthropogenic SO_2 emissions, the use of highly concentrated fertilisers with low sulphur content and the increased sulphur removal due to high-yielding crops (Scherer 2009; Solomon et al. 2009; Blum et al. 2013).

4.5.1 Sulphur Species in Soil

Of the total sulphur content in soil that ranges between 0.01% and 0.1%, only less than 5% is usually present as inorganic S, whereas organic sulphur compounds make up more than 95% (Balík et al. 2009; Solomon et al. 2009).

Sulphate constitutes the most important S source for the direct uptake by plants. Therefore, the process of mineralisation is pivotal to make organic S plant-available. Mineralisation can take place biologically as the breakdown of compounds with C-bonded S by microorganisms or biochemically as the hydrolysis of sulphate esters by sulphatase enzymes in the soil (Assefa et al. 2014).

The organic S compounds can be divided in the already mentioned two main groups of ester sulphates (C-O-S) and C-bonded S (C-S) (Scherer 2009). The traditionally used method for the soil sulphur speciation analysis is a wet-chemical sequential extraction. Prietzel et al. (2003) listed the commonly used extractions to determine sub-fractions of inorganic as well as organic S. Since not all important fractions are covered by these, indirect quantification by calculation from other fractions is necessary. Furthermore, the fractions are operationally defined and do not quantify distinct S species (Prietzel et al. 2009).

Table 4.1 Sulphur speciation in soils with different oxygen availability (Cambisol > Stagnosol > Histosol): (a) reduced inorganic S and reduced organic S, (b) intermediate S and oxidised S (Prietzel et al. 2009)

(a)	Inorganic sulphide S^{2-}	Inorganic sulphide S_2^{2-}	Organic disulphide S	Organic monosulphide/thiol S
Oxidation state	−2	−1	0.2	0.5
	mg S kg^{-1}			
Cambisol	0 ± 0	695 ± 49	842 ± 60	0 ± 0
Stagnosol	434 ± 15	954 ± 34	607 ± 21	0 ± 0
Histosol	0 ± 0	2347 ± 83	1369 ± 48	0 ± 0

(b)	Sulphoxide	Sulphite	Sulphone	Sulphonate	Ester sulphate	Sulphate
Oxidation state	2	3.7	4	5	6	6
	mg S kg^{-1}					
Cambisol	476 ± 34	439 ± 31	146 ± 10	329 ± 23	73 ± 5	659 ± 47
Stagnosol	520 ± 18	390 ± 14	260 ± 9	564 ± 20	87 ± 6	564 ± 20
Histosol	652 ± 23	652 ± 16	0 ± 0	587 ± 21	196 ± 14	717 ± 29

X-ray absorption near-edge spectroscopy (XANES) can be used for the direct speciation of soil S (Jokic et al. 2003; Prietzel et al. 2003) or the analysis of extracts of the humic substances in soil (Solomon et al. 2003). The energy of the maximal absorption of irradiated X-rays gives information on the oxidation state of S atoms. A quantitative analysis of samples of different soil types in Fichtelgebirge, Germany, shows the variability of S compounds in soil with varying oxygen availability. The results for the upper sections of the organic layer of the soil are depicted in Table 4.1. With decreasing levels of oxygen availability and higher influence of the groundwater in the sequence Cambisol-Stagnosol-Histosol, the contribution of reduced S species to the total S content increases (Prietzel et al. 2009).

4.5.2 Sulphur Metabolism and Speciation in Plants

Uptake of S takes place mainly in the form of sulphate through the roots of the plants. It is distributed throughout plant tissues by dedicated sulphate transporters. Excess sulphate is stored in vacuoles, but the majority is reduced in the shoot tissues (Maathuis 2009; Kopriva et al. 2015).

Reduction of sulphate takes place in a multistep pathway to form sulphide, which is used to form cysteine out of *O*-acetylserine. Cysteine is the starting material for the biosynthesis of all other reduced sulphur-containing metabolites like the amino acid methionine or the redox buffer glutathione. An alternative way of sulphur assimilation is sulphation, the addition of sulphate to a hydroxyl group. Sulphation plays a minor role in higher plants compared to reduction but is important in the biosynthesis of a range of metabolites including the glucosinolates in *Brassicaceae* (Leustek and Saito 1999; Kopriva et al. 2015).

Glucosinolates are secondary sulphur metabolites of importance to humans because of their presence in *Brassicaceae* crops like cabbage, broccoli and mustard. Their molecular structure consists of a thioglucoside group bound to an *N*-hydroximinosulphate ester plus a variable R group derived from different amino acids. Based on the amino acid, the compounds are classified as aliphatic, aromatic or indole glucosinolates. Tissue damage leads to the hydrolysis of glucosinolates by myrosinase to unstable intermediates that rearrange to isothiocyanates. The isothiocyanates are primarily responsible for flavour and taste of these plants (Prestera et al. 1996; Halkier and Gershenzon 2006).

Aires et al. (2006) analysed leaves and roots of broccoli sprouts (*Brassica oleracea*) and the effect of nitrogen and sulphur fertilisation on glucosinolates. The fertilisation showed a detrimental effect on the glucosinolate levels, and broccoli sprouts react sensitively to higher salt concentrations due to the fertilisation. The analysis of leaves and roots showed that aliphatic glucosinolates predominate in aerial parts of the plants, whereas aromatic and indole glucosinolates prevail in the roots. The total glucosinolate concentrations found in leaves and roots ranged from 21.8 to 56.1 μmol/g DW. The seven identified glucosinolates with their proportion of the total glucosinolate content in the broccoli sprouts are shown in Table 4.2.

Allium vegetables like garlic and onion are universally consumed because of their flavour but also as traditional medicine. Both flavour and beneficial health effects are attributed to *S*-alk(en)yl-$_{\text{L}}$-cysteine sulphoxides, another group of secondary sulphur metabolites. Their flavour and lachrymatory characteristics are connected with the transformation to thiosulphinates (e.g. allicin) and other volatile organosulphur compounds (e.g. diallylpolysulphides) by the enzyme alliinase when the plant tissue is damaged (Jansen et al. 1989; Ichikawa et al. 2006; Yamazaki et al. 2010).

The most common *S*-alk(en)yl-$_{\text{L}}$-cysteine sulphoxides in *Allium* plants are alliin, isoalliin and methiin. The three best known γ-glutamyl peptides occurring in *Alliums* γ-glutamyl-*S*-allyl-$_{\text{L}}$-cysteine (GSAC), γ-glutamyl-*S*-1-propenyl-$_{\text{L}}$-cysteine (GSPC) and γ-glutamyl-*S*-methyl-$_{\text{L}}$-cysteine (GSMC) are thought to be precursors for the corresponding sulphoxides and/or sulphur storage peptides (for structures, see Fig. 4.3) (Ichikawa et al. 2006; Yamazaki et al. 2010; Raab et al. 2017).

The common analysis methods involved the use of species-specific standards for quantification and identification (Ichikawa et al. 2006; Yamazaki et al. 2010). A recently developed method by Raab et al. (2017) uses a parallel ICP-MS/MS and ESI-MS system to gain elemental and molecular information simultaneously. Some of the newly identified sulphur compounds found in *Allium sativum* (garlic) are depicted in Fig. 4.3 together with the structures of the most commonly found sulphur compounds in *Allium* plants. The experiments with a changing sulphur fertilisation under hydroponic growth conditions show that the sulphur content in the bulbs was hardly affected, whereas the sulphur content in the roots was positively influenced (Raab et al. 2017).

The heavy metals arsenic, selenium and mercury are influenced by sulphur regarding the uptake and accumulation in plants. Thiols like glutathione play an important role in the detoxification process, and an increase in S supply improved the As tolerance in *Hydrilla verticillata* by enhancing the thiol metabolism (Srivastava and D'Souza 2009; Srivastava and D'Souza 2010). Reversely arsenic

Table 4.2 Basic structure of glucosinolates and different R groups occurring in aerial part and leaves of broccoli sprouts including the proportion of the total glucosinolate content (Aires et al. 2006)

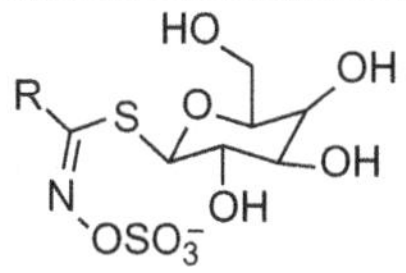

Structure of R	Trivial name	Aerial part/%	Roots/%
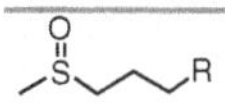	Glucoiberin	20.4	2.7
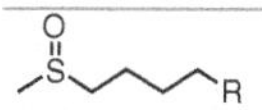	Glucoraphanin	48.2	6.8
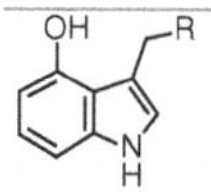	4-Hydroxy-glucobrassicin	4.8	4.9
	Glucoerucin	11.2	25.5
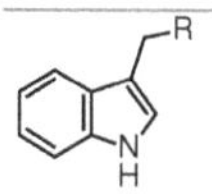	Glucobrassicin	5.7	5.5
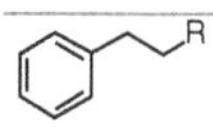	Gluconasturtiin	8.0	38.5
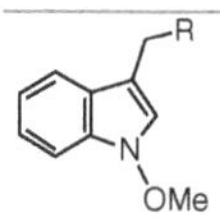	Neoglucobrassicin	1.7	16.1

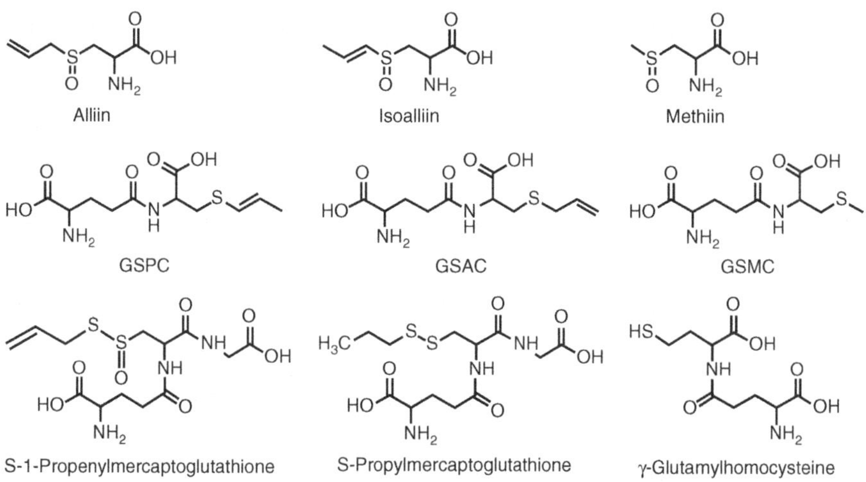

Fig. 4.3 Molecular structures of alliin, isoalliin and methiin, their precursor γ-glutamyl peptides and proposed structures for three newly identified sulphur compounds in garlic (Raab et al. 2017)

also affects the uptake of S. Watanabe et al. (2014) found an enhanced uptake of sulphur in the As hyperaccumulator *Pteris vittata* with As in the growth medium. Selenium reacts in a different way on higher sulphur fertilisation levels. Selenium concentrations in rapeseed (Liu et al. 2017a) as well as wheat (Liu et al. 2014; Yang et al. 2017) decreased significantly through the application of S fertilisers. Liu et al. (2014) explain these findings with changes in the soil resulting in the binding of Se to Fe-Mn oxide and organic matter. The effects of sulphur treatment on the accumulation of mercury in rice show a similar outcome. Both total mercury and methylmercury concentrations in the plants decrease due to the transformation of mercury to the form of RS-Hg-SR. Additionally mercury is inactivated in the soil to HgS and therefore reducing the health risk from these crops (Li et al. 2017).

4.6 Overview of the Analytical Approaches

4.6.1 Introduction

From an analytical perspective, metal(loid)-thiol complexes are species of sharp contrast depending on the metal(loid) being bound. With varying levels of stability in vivo and throughout the analytical procedure, the identification and quantification of such complexes can provide the analyst with a number of issues. This is exemplified by metal(loid)-phytochelatin complexes, formed by plants in the response to cellular influx of metal(loid) species.

The "traditional" complexes of phytochelatins, with metals such as cadmium and lead, are stable at high pH (>7), with little to no stability at low pH (<7) (Rauser 1990; Steffens 1990). Alkaline plant extracts can be stored at room temperature over a period of days with insignificant levels of PC complex degradation (Chassaigne et al. 2001; Johannig and Strasdeit 1998). The inherent stability of such complexes allows for 2D chromatography: size-exclusion chromatography (SEC) as the initial clean up and fraction collection step, followed by reverse phase LC-ESI-MS and LC-ICP-MS analysis as the identification and quantification techniques, respectively (de Knecht et al. 1994; Barałkiewicz et al. 2009).

In contrast, this is not the case for phytochelatin complexes of arsenic, mercury or selenium. Extracts containing such complexes show no stability at high pH (>7) and limited stability at low pH (<7) and must be stored chilled (<5 °C) prior to analysis to preserve the speciation profile (Meharg and Hartley-Whitaker 2002; Raab et al. 2004). However, the combination of low pH and temperature only extends sample stability to 24 h, after which time PC complex speciation is all but lost and only free unbound PC species are observed (Sneller et al. 1999). Sample storage at ambient temperatures leads only to faster sample speciation degradation. Attempts to extend the period of sample stability, such as freeze-drying of samples or extracts or their immediate storage at −78 °C, yield no appreciable extension in their shelf lives (Bluemlein et al. 2009a).

The instability of arsenic and selenium phytochelatin complexes is linked to the stability of their various oxidation states. In the natural environment, which would be primarily oxidising and where pH would be close to 7, the predominant forms of arsenic and selenium are the arsenate ($[AsO_4]^{3-}$, As(V)) and selenite ($[SeO_3]^{2-}$, Se(IV)) anions, respectively (Spuches et al. 2005; Zhu et al. 2008). However, the species bound by phytochelatins are arsenite ($As(OH)_3$, As(III)) and selenium (Se(II)), respectively, both showing better stability under reducing conditions (low pH and anoxic environments) (Aborode et al. 2016). The instability of PC-As and PC-Se complexes is linked to the driving force towards oxidised species: for arsenic, transformation of As(III) $\rightarrow$ As(V), and for Se, transformation of Se(II) $\rightarrow$ Se(IV). This can be mitigated against, but not removed completely, through the use of acidic diluents and low temperatures during extraction.

The peculiarity of the Hg-PC complex instability versus the inherent stability of Zn-PC or Cd-PC complexes could be due to the difference in their redox behaviours. Whereas Zn and Cd will displace H^+ from acidic solution, forming the Zn^{2+} and Cd^{2+} ions, respectively, Hg does not unless in the presence of oxidising acids. Furthermore, whilst Zn and Cd are amphoteric and can dissolve at low and high pH, mercury is not soluble at high pH, producing insoluble mercury oxides.

Whilst low pH conditions stabilise PC complexes with arsenic, selenium and mercury, there is competition for the binding sulphur atoms between hydrogen ions and the bound metal(loid). This destabilises the PC complex, leading to its eventual disintegration, and explains their transient stability at low pH.

Arsenic-PC complexes were the first identified to show inherent instability during isolation and extraction (Raab et al. 2004). In contrast to metal(loid)s such as lead and cadmium, which exist as the cations Cd^{2+} and Pb^{2+}, respectively, arsenic exists in the environment predominantly as the oxoanion arsenate, As(V), and smaller quantities of the neutral species arsenite, As(III) (Quaghebeur and Rengel 2005). Complexation requires the in vivo reduction of As(V) to As(III), which is then bound by thiols in the ratio of 1:3 with regard to As/S, with the loss of water as by-products (Scott et al. 1993). However, the driving force in most non-reducing environments is for the oxidation of As(III) to As(V), explaining As-PC complex instability at high pH and in oxygen-rich environments (Mandal and Suzuki 2002).

For arsenic and mercury, the case becomes complicated further, in that there is the potential for methylated analogues to exist within the environment. Monomethylarsenate ($CH_3AsO(OH)_2$, MMA(V)) and dimethylarsenate ($(CH_3)_2AsO(OH)$, DMA(V)) can exist in the environment as a result of natural methylation of arsenate in soils or from anthropogenic sources (e.g. historical herbicide use) (Le et al. 2000; Mandal and Suzuki 2002). Monomethylmercury (CH_3Hg^+) is almost exclusively produced via natural methylation of mercury in the environment but can be produced via anthropogenic actions such as burning of mercury wastes (Krupp et al. 2008). PC complexes can form with methylated arsenic and mercury species; however their stability is drastically reduced versus their inorganic analogues, as the number of M-S bonds decreases (Scott et al.

1993; Delnomdedieu et al. 1993). For arsenic, complex stability can be summarised as follows:

$$As(III)-PC > MMA(III)-PC > DMA(III)-PC$$

4.6.2 General Overview

Rapid sample preparation and analytical protocols are required to deal with the transient stability of PC complexes with arsenic, selenium or mercury. Low pH media are required to stabilise any complexes that may occur but not so low that they are detrimental to either the speciation profile or the analytical instrumentation used.

Current research suggests that the timeframe in which samples must be analysed is 24 h, and this makes the choice of analytical techniques crucial (Raab et al. 2004). The chromatographic method must resolve a number of nonbinding PC species and PC-metal(loid) complexes within a short period of time. Additionally, the detection step(s) must identify and quantify species in the extract simultaneously.

4.6.3 Sample Preparation

Given that the sample preparation step can impact heavily upon the subsequent chromatography, there must be compatibility between both the extraction and analytical steps, whilst maintaining conditions suitable for the isolation and stabilisation of PC-metal(loid) species.

Sample extraction is typically performed with chilled 1% formic acid solution (Bluemlein et al. 2008a; Wood et al. 2011). Plant material is frozen with liquid nitrogen and ground to a fine powder with a mortar and pestle. Extraction time is typically between 40 and 60 min, with intermittent shaking of the sample by hand. The extraction solution is around pH 2, which is tolerable by most LC columns, LC systems and analytical instrumentation. Furthermore, the sample injection volumes used are typically between 10 and 50 μL and are rapidly diluted by the eluent during chromatography, which is of a slightly higher pH. Finally, the use of a low pH extraction solution helps to minimise the uptake of large peptides and proteins into solution, which would typically be achieved via a clean-up step such as SEC, which could precipitate within the analytical system during measurement.

Samples can be filtered prior to analysis; however this may raise the concern of loss of species via adsorption to the filter discs used. Centrifugation is more commonly used, with samples spun at typically 3000–4000 rpm for 5 min. There is expected to be minimal change to the speciation, given the brief period of time that samples will be at ambient. However, this has not been proven.

4.6.4 Chromatographic Method

The instability of PC-As, PC-Se and PC-Hg complexes precludes the use of any additional chromatography techniques, such as SEC, prior to analysis. The chromatographic method must resolve a variety of free PC species and PC-M complexes in a single run, under conditions that not only preserve the speciation profile of the extract but are compatible with the subsequent analytical instrumentation.

For the chromatographic method, reverse phase analysis using a C_{18} column is typically employed (Raab et al. 2005). A gradient elution system, utilising 0.1% formic acid and methanol as eluents, can resolve a number of free PC species and PC-metal(loid) complexes in around 30–60 min based on a 1 mL/min flow rate, depending on the metal(loid) under investigation. The pH of the eluent system is typically around 4, which is tolerable by the C_{18} column, the LC system and the subsequent detection systems. The eluent system also helps mitigate the higher pH of the extraction solution (50 μL of pH 2 solution) via its rapid dilution once in the LC system.

UV data is generally not gathered nor required during analysis of plant extracts, as neither the poisoning metal(loid), free PC species, nor PC-M complexes are significantly UV active. Detection is garnered on both an elemental and molecular basis.

4.6.5 Analytical Detection

PC-metal(loid) complexes with arsenic, mercury or selenium in plant extracts are typically identified via ESI-MS and quantified on the metal(loid) via ICP-MS. Unlike PC-Cd complexes, which can be analysed sequentially by HPLC-ICP-MS and HPLC-ESI-MS (Barałkiewicz et al. 2009), the inherent instability of As, Se and Hg complexes means that both elemental and molecular data must to be collected simultaneously. The chromatography established by Raab et al., as described in the previous section, permits such analyses (Raab et al. 2004). Using LC-ESI-MS-ICP-MS (hyphenated mass spectrometry), both molecular and elemental information can be collected in a single sample run. The eluents used are tolerable by both systems, and the addition of formic acid helps species identification through enhanced protonation. However, the ICP-MS has to be configured to handle organic eluents. To achieve simultaneous identification and quantification, the eluent flow is split post-column, diverting portions of the eluent to both MS systems. Given the greater sensitivity of the ICP-MS versus the ESI-MS, eluent flow is typically spilt 2:8, respectively, between the systems.

Inevitably, there are drawbacks. The first is sensitivity. Given that the sample flow is split between two systems, there is a detriment to the signal intensities recorded by both the ESI-MS and ICP-MS, which disproportionately affects the ICP-MS more. Secondly, the use of methanol as an eluent leads to the formation of soot particles in the ICP-MS plasma, which could block the sampler and skimmer cones. This can be negated through the addition of oxygen into the plasma during analysis but does require that the ICP-MS hardware is capable of such a function. Additionally, as the sampler and skimmer cones are traditionally made of nickel which can be corroded when oxygen is added into the plasma, there is a requirement to change them to more expensive and delicate platinum cones for speciation analysis.

4.7 Analytical Strategy in Action

To put the analytical strategy into context, the precise requirements for arsenic, selenium and mercury-PC complex analysis are discussed.

4.7.1 ESI-MS Analysis: General Considerations

The predominant role of ESI-MS is to provide species identification for the array of glutathione, unbound PC and metal(loid)-bound PC species present in the plant extracts under investigation. Independent of the metal(loid)s to which the plant are being exposed, the ESI-MS should be optimised to detect unbound glutathione and PC species. Analysis is conducted in positive mode with species identification predominantly made on the protonated molecular ion $[M + H]^+$, due to the combination of formic acid in the eluent system with the multiple amide group nitrogen atoms present within GSH and PC structures (e.g. Raab et al. 2007c; Batista et al. 2014). Parameters applicable to the aforementioned species would apply equally to PC-metal(loid) complexes.

No direct quantification of species can be realistically made. Since ESI-MS is a relatively gentle ionisation technique and there is a substantial level of unrelated species present in the formic acid extracts, matrix suppression of the target PC and GSH species is expected. As no standards of PCs are readily available, it is not possible to counteract this by means of internal standard addition. Whilst bound PC and GSH can be quantified by ICP-MS on the central metal(loid), unbound species are largely unquantifiable due to the lack of a suitable heteroatom for analysis.

4.7.2 ICP-MS Analysis: General Considerations

The role of ICP-MS is to provide quantification for the various PC-metal(loid) complexes found in the plant extracts. The high temperature plasma efficiently desolvates, atomises and ionises the sample stream passed to it by the LC system (Gray 1989; Hill et al. 2005). Quantification is thus conducted on an elemental basis and can be applied across a range of PC-metal(loid) species provided a suitable elemental standard is used. The sensitivity of the ICP-MS can reveal PC-metal(loid) species that may not be evident initially from the ESI-MS data and is an important confirmatory tool in the analysis of such species.

There are, however, a number of factors which must be assessed and addressed during ICP-MS analysis: the isotope upon which to perform elemental identification and quantification, the possibility of ion suppression during analysis and the effect of methanol on plasma behaviour during analysis.

Phytochelatins and glutathione contain predominantly carbon, hydrogen, oxygen and hydrogen, none of which are readily quantified due to their relative ubiquity. The exception is sulphur; however, quantification is not a simple task. The main isotope is ^{32}S with a natural abundance of 95%, which is indistinguishable from $^{16}O_2$ when a single quadrupole instrument is used (Bluemlein et al. 2008b). Quantification is thus only routinely possible on PC-metal(loid) species with standard ICP-MS hardware, with identification and quantification being performed on the bound metal(loid).

As with ESI-MS analysis, there is the potential for ion suppression due to the sheer number of components that can be extracted during sample preparation. However, ICP-MS has two distinct advantages over ESI-MS that help to circumvent this problem:

1. The argon plasma is a considerably more powerful ionisation source, with a much higher degree of ionisation efficiency over ESI-MS.
2. Analysis is conducted on an elemental basis and is species-independent.

With ICP-MS, there should be less ion suppression effects due to the ferocity of the argon plasma, which operates with temperatures of up to 10,000 K. Where suppression becomes an issue, the application of an appropriate internal standard can correct for this. Typically, the internal standard is an element that is unlikely to be found in the sample matrix but has similar ionisation properties inside the plasma to that of the metal(loid) under investigation and is added post-column, as a separate continual stream throughout the duration of analysis. For arsenic, selenium and mercury, a common internal standard is rhodium with measurement on the only stable isotope, ^{103}Rh (Krupp et al. 2008; Bluemlein et al. 2009b).

Chromatographic separation of the PC complexes of arsenic, selenium and mercury utilises a gradient elution system composed of 0.1% formic acid and methanol as eluents, the latter of which has significant impacts on the plasma: eluent droplet size, degree of nebulisation and addition of fuel. Methanol levels of up to 10% in the eluent generally enhanced the signal area for arsenic when running equivalent standards. The effect was not linear, and gains in signal area intensity were lost with increasing methanol levels up to 20% (Bluemlein et al. 2008b). With mercury-PC complex analysis, increasing methanol levels up to 50% actually resulted in significantly diminished signal areas for standards compared to those at 0% methanol (Krupp et al. 2008). Initially, the addition of methanol to the plasma acts as a fuel, increasing the temperature of the plasma and enhancing the atomisation and ionisation processes (Forsgard et al. 2006). At higher levels, methanol removes energy from the plasma through combustion, diminishing the ionisation capability of the plasma. As described previously, increasing methanol levels requires ICP-MS hardware capable of allowing the addition of oxygen flows into the plasma to remove carbon and the use of expensive and delicate platinum sampler and skimmer cones for analysis.

4.7.3 ICP-MS Analysis: Arsenic-Phytochelatin Complexes

Arsenic is a monoisotopic element, being quantified on ^{75}As, m/z 75. There is the potential for interference of the m/z 75 signal by the argon chloride complex, $[^{40}Ar^{35}Cl]^+$, in systems where a substantial amount of chloride could be present (May and Wiedmeyer 1998). Interference of m/z 75 can readily be determined during analysis by comparison of the background isotope ratios for selenium between ^{77}Se, ^{78}Se and ^{82}Se. As chlorine has a significant component of ^{37}Cl, formation of $[^{40}Ar^{37}Cl]^+$ with mass m/z 77 would be expected and would impact on the observed signal for ^{77}Se. Whilst not routinely observed during analysis of terrestrial plants, where such argon chloride complex formation is observed, this can be dealt with through the use of a hydrogen collision cell.

The chromatographic method developed by Raab et al. (2004) utilises a gradient composed of 0.1% formic acid and methanol running from 0 to 20% methanol over the first 20 min and is able to resolve a number of free PC species, PC-As complexes and glutathione. A variety of plant species, including *Helianthus annuus* (sunflower), *Thunbergia alata* (black-eyed Susan) and *Arabidopsis thaliana* (thale cress), have shown a wide array of complexes, such as GS-As-PC_2, As-$(PC_2)_2$, As-PC_3 and As-PC_4, as determined by ESI-MS data (Raab et al. 2005; Bluemlein et al. 2008a; Liu et al. 2010).

The arsenic triglutathione complex As-SG_3 was hypothesised to exist in vivo but had never been observed equivocally. Liu et al (2010) demonstrated that the complex was present in extracts from the roots of the *cad2-1* mutant of *Arabidopsis thaliana*, a species deficient in PC synthase such that there was little to no PC production by the plant when exposed to arsenic (Liu et al. 2010). This result highlighted that only when there were no PCs available to bind arsenic would the As-SG_3 complex form, presumably as a result of its instability relative to other As-PC complexes.

Research to date has identified As-PC complexes where only 1 atom of arsenic is bound per complex, i.e. poly-arsenic complexes have not been identified. This simplifies quantification of the overall As-PC, given the 1:1 ratio between arsenic and complex. Where unknown As-containing species are identified, quantification by ICP-MS can still be achieved, and it is assumed that the complex contains only 1 central arsenic atom which may help with species identification.

4.7.4 ICP-MS Analysis: Selenium-Phytochelatin Complexes

The major isotope of selenium is ^{80}Se with a normal abundance of 49.6%. However, this is heavily interfered by $[^{40}Ar_2]^+$ during ICP-MS analysis (May and Wiedmeyer 1998). Analysis is typically conducted on ^{77}Se or ^{78}Se, with abundances of 7.6% and 23.7%, respectively. The latter can be interfered by $[^{40}Ar^{38}Ar]^+$, even though the abundances of both argon isotopes are 99.6 and 0.6%, respectively. This is due simply to the sheer volume of argon used to maintain the plasma, making ^{77}Se the preferred isotope for quantification by ICP-MS.

Bluemlein et al. investigated the co-exposure of *Thunbergia alata* to selenite and arsenate, using the same methodology as described in the arsenic case study (Bluemlein et al. 2009b). The Se(II)-PC_2 complex and Se-cysteinylserine glutathione were found in root extracts of *T. alata*, the latter of which highlights a situation specific to selenium. As sulphur and selenium belong to the same group of the periodic table, there is a close synergy in their chemical behaviours, so much so that selenium can replace sulphur in a number of biothiols in vivo, rather than forming Se-SG or Se-PC complexes.

Selenium is not only bound by biothiols, it can replace sulphur in the same molecules. However, selenium does not replicate the same metal(loid)-binding function as sulphur. Work by Blümlein et al. observed no complexes or species containing Se and As simultaneously, which would suggest the formation of Se-As bonds in vivo as part of the plants overall detoxification strategy does not occur.

Further work by Aborode and co-workers investigated the exposure of *T. alata* to selenium as sodium selenite (Aborode et al. 2015). A number of selenium species were observed: selenopeptides, incorporating Se-S bonds; Se(II)-PC complexes, including selenocysteinyl-2,3-dihydroxypropionyl-glutathione, Se(II)-PC_2 and Se(II)-$(SG)_2$; and a significant proportion of elemental selenium (approx. 20%) which was not recovered by the 0.1% formic acid extraction procedure.

In common with arsenic, research to date has suggested the formation of Se-PC complexes where only 1 atom of Se is bound per complex, i.e. a 1:1 ratio between selenium and the complex. Quantification of known and unknown Se-containing species can thus be performed on a relatively simple basis.

4.7.5 ICP-MS Analysis: Mercury-Phytochelatin Complexes

Mercury belongs to the same group of the periodic table as cadmium and lead, PC complexes of which are stable under alkaline conditions. However, two separate studies demonstrated that Hg-biothiol complexes are stable under acidic conditions (Chen et al. 2009; Krupp et al. 2008). Consequently, whilst mercury does not exhibit the degree of oxoanion formation observed for arsenic and selenium, the biothiol complexes of $Hg^{2+}$and $MeHg^{+}$ can be stabilised and analysed in a similar fashion as is employed for As- and Se-biothiol complexes.

Mercury has seven stable isotopes, of which ^{200}Hg and ^{202}Hg are the most abundant at 23.13% and 29.8%, respectively (Chen et al. 2009). The higher atomic masses of mercury isotopes mean that there is minimal risk from interferences from other elements or polyatomic species and ICP-MS analysis can be conducted on either ^{200}Hg or ^{202}Hg without the need for a collision cell (May and Wiedmeyer 1998). Other common isotopes are ^{198}Hg (10.02%), ^{199}Hg (16.84%) and ^{201}Hg (13.22%). Quantitation is not routinely performed on these isotopes, but the spread of isotopes and their high relative abundances generates unique isotope patterns for mercury compounds, which can be useful when identifying mercury-containing compounds.

Whilst Hg-PC_2, Hg-PC_3 and Hg-PC_4 complexes had previously been demonstrated to exist by ESI-MS/MS (Chen et al. 2009), no work had been done to analyse samples by liquid chromatography and tandem mass spectrometry, allowing for the identification and quantification of a number of Hg-PC complexes. Krupp et al. first examined the synthetic Hg-biothiol compounds Hg-$(Cys)_2$, Hg-$(GS)_2$, MeHg-Cys and MeHg-GS by ESI-MS/MS (Krupp et al. 2008). Then, a chromatographic method was established to allow for their separation and detection. Similar to the method utilised by Raab et al., a gradient run-up to 50% methanol was developed and was able to resolve the aforementioned species for detection and quantitation (Krupp et al. 2009). This method was utilised to identify a number of Hg- and MeHg-PC in vivo complexes in *Oryza sativa* (rice) and *Marrubium vulgare* (horehound) by tandem HPLC-ESI-MS/ICPMS.

In common with arsenic and selenium, research to date has suggested the formation of Hg-PC complexes where only 1 atom of Hg is bound per complex, i.e. a 1:1 ratio between mercury and the biothiol. Quantification of known and unknown Hg-containing species can thus be performed on a relatively simple basis.

4.7.6 Quantification of Free PC Species

Bound PC and GSH species can be quantified by ICP-MS upon the central metal(loid). However, unbound thiol species are not readily amenable for quantification, given that hydrogen, carbon, nitrogen and oxygen are relatively ubiquitous in the eluent and sample stream. The only other suitable heteroatom upon which to quantify is sulphur; however, this has a number of issues (Bluemlein et al. 2008b).

Sulphur has poor ionisation characteristics, with a first ionisation potential higher than argon. The main isotope, ^{32}S, is heavily interfered by molecular oxygen, $^{16}O_2$. Finally, as PCs and similar biothiols contain no chromophore, quantification by LC-UV is not possible.

A significant number of attempts have been made to tag unbound thiol species to permit their quantification by LC-UV or LC-ICP-MS. Wood and Feldmann have reviewed and assessed a number of techniques available for biothiol quantification, each of which has their advantages and limitations (Wood and Feldmann 2012). Pre-chromatographic derivatisation of biothiols is one technique. One class of reagents for this purpose is bimanes, highly UV-active reagents which have been used in a wide array of analyses, permitting quantification of free thiol species down to picomolar levels in extracts (Delnomdedieu et al. 1993; Sneller et al. 1999; Steffens 1990). However, bimanes are expensive reagents. Additionally, the use of pre-chromatographic derivatisation may require conditions incompatible with PC-metal(loid) complex stability, which would drastically alter the speciation profile of the extract.

Post-chromatographic derivatisation is another route to quantify free biothiols. Reagents such as 5,5′-dithiobis-(2-nitrobenzoic acid) (DTNB) permit rapid derivatisation of biothiol species (~10 s) by rapid disulphide exchange between the reagent and the reduced biothiol (Bräutigam et al. 2010; Zhang et al. 2005). During this reaction, the mixed disulphide species 2-nitro-5-thiobenzoate – biothiol species – is formed, along with the anion 2-nitro-5-thiobenzoate. The latter has a yellow colour in solution and absorbs strongly at 412 nm, permitting reduced biothiol quantitation by HPLC-UV (Riddles et al. 1983). Whilst this method is rapid and sensitive, it can only be used to quantify reduced biothiol species and requires a continual stream of reagent mix for the entire analytical procedure.

Whilst the aforementioned techniques permit biothiol quantitation by HPLC-UV, the analysis is in indirect with respect to the biothiol being measured – quantitation is based upon the spectrophotometric properties of the reagent. Additionally, the conditions with which to achieve analysis may impact upon the speciation profile of the sample extract under investigation. But crucially, the reagents are specific to reduced biothiols. Any oxidised species, such as oxidised glutathione GSSG, are not bound by the reagents without the addition of a reducing agent, which again alters the speciation profile of the extract under investigation (Hansen et al. 2009)

Single quadrupole ICP-MS cannot resolve the main sulphur isotope ^{32}S from molecular oxygen $^{16}O_2$ during analysis, both of which are measured as m/z 32. Distinction between both species can be achieved through the use of high-resolution ICP-MS systems, which utilise a high strength electromagnetic field to resolve ^{32}S (m/z 31.9271) from $^{16}O_2$ (m/z 31.9898) (Bluemlein et al. 2008a). With HR-ICP-MS, it is possible to quantify sulphur in extracts down to concentrations of 5 μg/L.(Liu et al. 2010; Ye et al. 2010). As quantification is performed directly on sulphur, reduced and oxidised species can be quantified in the same run. Additionally, provided the eluent conditions are compatible with the ICP-MS, no modification of the sample extract is required.

There are, however, significant drawbacks with HR-ICP-MS (Wood and Feldmann 2012). The cost of an HR-ICP-MS is typically two to three times that of a standard quadrupole ICP-MS. Secondly, high resolution is achieved in part via an electromagnet which bends the ion beam. The speed at which the magnetic field can be altered, defined by the settling time required by the magnet when jumping between masses, thus selecting a specific m/z ratio, is slower than that of a standard quadrupole by a factor of 4–5. The mass range 2–260 amu can be scanned in 100 ms by a standard ICP-MS but takes around 300–600 ms with a HR-ICP-MS.

Whilst a number of methods exist to perform quantification of unbound thiol species, each has their own distinct advantages and limitations, making their analysis a far from routine task.

4.8 Concluding Remarks

Looking at the different elements, it becomes obvious that there are a lot of different compounds formed within plants. It is therefore necessary to have a nontargeted analysis for any metal(loid) biomolecule (Feldmann et al. 2017). The most versatile method is the use of ICPMS as an element-specific detector and electrospray MS as a molecular-specific detector in HPLC-ICPMS/ESIMS as explained above.

There has however been a complete lack of studies which focus on element/element interaction, with some exceptions, in which co-exposure of arsenic and selenium has been studied (Aborode et al. 2016, Bluemlein et al. 2009b). In terms of selenium-sulphur interaction, *Brassica* have been studied in detail (Tian et al. 2017).

Furthermore, it can be foreseen that genes will be identified for the biotransformations of arsenic seen in the different plants. The transport and accumulation of mercury and especially methylmercury will see more investigations in the future. Whilst the motivation to study mercury and arsenic in plants is the toxicity of the element species accumulating in the different parts of the plants, the focus for sulphur and selenium is more about essentiality. Especially it is expected that the determination of selenoproteins will be a focus of many studies in the near future in order to establish finally whether selenium is essential or beneficial for plants or not.

Acknowledgements The authors and especially M.M. thank the TESLA research fund for financial support.

References

Abedin MJ, Feldmann J, Meharg AA (2002) Uptake kinetics of arsenic species in rice plants. Plant Physiol 128:1120–1128

Aborode FA, Raab A, Foster S, Lombi E, Maher W, Krupp EM, Feldmann J (2015) Selenopeptides and elemental selenium in Thunbergia alata after exposure to selenite: quantification method for elemental selenium. Metallomics 7:1056–1066

Aborode FA, Raab A, Voigt M, Costa LM, Krupp EM, Feldmann J (2016) The importance of glutathione and phytochelatins on the selenite and arsenate detoxification in *Arabidopsis thaliana*. J Environ Sci 49:150–161

Aires A, Rosa E, Carvalho R (2006) Effect of nitrogen and sulfur fertilization on glucosinolates in the leaves and roots of broccoli sprouts (*Brassica oleracea var. italica*). J Sci Food Agric 86(10):1512–1516

Araie H, Sakamoto K, Suzuki I, Shiraiwa Y (2011) Characterization of the selenite uptake mechanism in the coccolithophore *Emiliania huxleyi* (Haptophyta). Plant Cell Physiol 52:1204–1210

Assefa MK, von Tucher S, Schmidhalter U (2014) Soil sulfur availability due to mineralization: soil amended with biogas residues. J Soil Sci Environ Manag 5(1):13–19

Balík J, Kulhánek M, Černý J, Száková J, Pavlíková D, Čermák P (2009) Differences in soil sulfur fractions due to limitation of atmospheric deposition. Plant Soil Environ 55(8):344–352

Barałkiewicz D, Kózka M, Piechalak A, Tomaszewska B, Sobczak P (2009) Determination of cadmium and lead species and phytochelatins in pea (*Pisum sativum*) by HPLC–ICP-MS and HPLC–ESI-MSn. Talanta 79:493–498

Batista BL, Migar M, Mestrot A, Rocha BA, Júnior FB, Price AH, Raab A, Feldmann J (2014) Identification and quantification of phytochelatins in roots of rice to long-term exposure: evidence of individual role on arsenic accumulation and translocation. J Exp Bot 65:1467–1479

Bierla K, Szupunar J, Yiannikouris A, Lobinski R (2012) Comprehensive speciation of selenium in selenium-rich yeast. Trends Anal Chem 41:122–132

Bluemlein K, Raab A, Meharg AA, Charnock JM, Feldmann J (2008a) Can we trust mass spectrometry for determination of arsenicpeptides in plants: comparison of LC–ICP–MS and LC–ES-MS/ICP–MS with XANES/EXAFS in analysis of *Thunbergia alata*. Anal Bioanal Chem 390:1739–1751

Bluemlein K, Krupp EM, Feldmann J (2008b) Advantages and limitations of a desolvation system coupled online to HPLC-ICPqMS/ES-MS for the quantitative determination of sulfur and arsenic in arseno-peptide complexes. J Anal At Spectrom 24:108–113

Bluemlein K, Raab A, Feldmann J (2009a) Stability of arsenic peptides in plant extracts: off-line versus on-line parallel elemental and molecular mass spectrometric detection for liquid chromatographic separation. Anal Bioanal Chem 393:357–366

Bluemlein K, Klimm E, Raab A, Feldmann J (2009b) Selenite enhances arsenate toxicity in *Thunbergia alata*. Environ Chem 6:486–494

Blum SC, Lehmann J, Solomon D, Caires EF, Alleoni LRF (2013) Sulfur forms in organic substrates affecting S mineralization in soil. Geoderma 200:156–164

Bräutigam A, Wesenberg D, Preud'Homme H, Schaumloeffel D (2010) Rapid and simple UPLC-MS/MS method for precise phytochelatin quantification in alga extracts. Anal Bioanal Chem 398:877–883

Chan QL, Caruso JA (2012) A metallomics approach discovers selenium-containing proteins in selenium-enriched soybean. Anal Bioanal Chem 403:1311–1321

Chassaigne H, Vacchina V, Kutchan TM, Zenk MH (2001) Identification of phytochelatin-related peptides in maize seedlings exposed to cadmium and obtained enzymatically in vitro. Phytochemistry 56:657–668

Chen L, Yang L, Wang Q (2009) In vivo phytochelatins and Hg–phytochelatin complexes in Hg-stressed *Brassica chinensis L.* Metallomics 1:101–106

Chen Y, Han Y-H, Cao Y, Zhu Y-G, Rathinasabapthi B, Ma LQ (2017) Arsenic transport in rice and biological solution to reduce arsenic risk from rice. Front Plant Sci 8(268):1–11

Cheng C-N, Focht DD (1979) Production of arsine and methylarsines in soil and in culture. Appl Environ Microbiol 38:494–498

Clarkson TW, Magos L (2006) The toxicology of mercury and its chemical compounds. Crit Rev Toxicol 36:609–662

de Knecht JA, van Dillen M, Koevoets PLM, Schat H, Verkleij JAC, Ernst WHO (1994) Phytochelatins in cadmium-sensitive and cadmium-tolerant Silene vulgaris. Plant Physiol 104:255–261

Delnomdedieu M, Basti MM, Otvos JD, Thomas DJ (1993) Transfer of arsenite from glutathione to dithiols: a model of interaction. Chem Res Toxicol 6:598–602

Duncan EG, Maher WA, Jagtap R, Krikowa F, Roper MM, O'Sulivan CA (2017) Selenium speciation in wheat grain varies in the presence of nitrogen and Sulphur fertilisers. Environ Geochem Health 39:955–966

Feldmann J, Raab A, Krupp EM (2017) Importance of ICPMS for speciation analysis is changing: future trends for targeted and non-targeted element speciation analysis. Anal Bioanal Chem. https://doi.org/10.1007/s00216-017-0502-8

Finnegan PM, Chen W (2012) Arsenic toxicity: the effects on plant metabolism. Frontiers in physiology. Plant Physiol 3:182.1–18

Fitz WJ, Wenzel WW (2002) Arsenic transformation in the soil-rhizosphere-plant system: fundamentals and potential application to phytoremediation. J Biotechnol 99:259–278

Forsgard N, Nilsson E, Andersson M, Pettersson J (2006) Investigation of matrix effects in boron determination using organic solvents as modifiers for liquid chromatography coupled to ICP-MS. J Anal At Spectrom 21:305–310

Fu LH, Wang XF, Eyal Y, She YM, Donald LJ, Standing KG, Ben-Hayyim G (2002) A selenoprotein in the plant kingdom. J Biol Chem 277:25983–25991

Gajdosechova Z, Lawan MM, Urgast DS, Raab A, Scheckel KG, Lombi E, Kopittke PM, Loeschner K, Larsen EH, Woods G, Brownlow A, Read FL, Feldmann J, Krupp EM (2016) In vivo formation of natural HgSe nanoparticles in the liver and brain of pilot whales. Sci Rep 6:34361. https://doi.org/10.1038/srep34361

Geiszinger A, Goessler W, Kosmus W (2002) Organoarsenic compounds in plants and soil on top an ore vein. Appl Organomet Chem 16:245–249

Gonzaga MI, Santos JAG, Ma LQ (2006) Arsenic chemistry in the rhizosphere of *Pteris vittata* L. and *Nephrolepis exaltata* L. Environ Pollut 143:254–260

Gray AL (1989) Inductively coupled plasma mass spectrometry, Edited by Date AR, Gray AL. Blackie and Son Ltd., Bishopbriggs, Glasgow, pp 1–42

Griffin K (2015) Sex, lies and arsenic: how the 'king of poisons' lost its crown. https://www.theguardian.com/science/blog/2015/dec/22/sex-lies-and-arsenic-how-the-king-of-poisons-lost-its-crown. Accessed Aug 2017

Halkier BA, Gershenzon J (2006) Biology and biochemistry of glucosinolates. Annu Rev Plant Biol 57(1):303–333

Hansen R, Roth D, Winther JR (2009) Quantifying the global cellular thiol-disulfide status. Proc Natl Acad Sci U S A 106:422–427

Hill SJ, Fisher A, Liezers M (2005) ICP mass spectrometry handbook. Blackwell Publishing Ltd., Oxford, UK, pp 1–25

Hirai MY, Saito K (2008) Analysis of systemic sulfur metabolism in plants using integrated "-omics" strategies. Mol BioSyst 4(10):967–973

Ichikawa M, Ide N, Yoshida J, Yamaguchi H, Ono K (2006) Determination of seven organosulfur compounds in garlic by high-performance liquid chromatography. J Agric Food Chem 54(5):1535–1540

Indriolo E, Na GN, Ellis D, Salt DE, Banks JA (2010) A vacuolar arsenite transporter necessary for arsenic tolerance in the arsenic hyperaccumulating Fern *Pteris vittata* is missing in flowering plants. Plant Cell 22:2045–2057

Jansen H, Müller B, Knobloch K (1989) Characterization of an alliin lyase preparation from garlic (*Allium sativum*). Planta Med 55(5):434–439

Jia Y, Huang H, Sun G-X, Zhao F-J, Zhu Y-G (2012) Pathways and relative contributions to arsenic volatilization from rice plants and paddy soils. Environ Sci Technol 46:8090–8096

Johannig J, Strasdeit H (1998) A coordination-chemical basis for the biological function of the phytochelatins. Angew Chem Int Ed 37:2464–2466

Jokic A, Cutler JN, Ponomarenko E, van der Kamp G, Anderson DW (2003) Organic carbon and sulphur compounds in wetland soils: insights on structure and transformation processes using K-edge XANES and NMR spectroscopy. Geochim Cosmochim Acta 67(14):2585–2597

Jones IF (2000) Arsenic and the Bradford poisonings of 1858. Pharm J 265:938–939

Kesselring J (1939) "Arsenic and old lace" stage play, Fulton Theatre, Broadway

Kopriva S, Calderwood A, Weckopp SC, Koprivova A (2015) Plant sulfur and big data. Plant Sci 241:1–10

Krupp EM, Milne BF, Mestrot A, Meharg AA, Feldmann J (2008) Investigation into mercury bound to biothiols: structural identification using ESI–ion-trap MS and introduction of a method for their HPLC separation with simultaneous detection by ICP-MS and ESI-MS. Anal Bioanal Chem 390:1753–1764

Krupp EM, Mestrot A, Wielgus J, Meharg AA, Feldmann J (2009) The molecular form of mercury in biota: identification of novel mercury peptide complexes in plants. Chem Commun 28:4257–4259

Le XC, Ma M, Lu X, Cullen WR, Aposhian HV, Zheng B (2000) Determination of monomethylarsonous acid, a key arsenic methylation intermediate, in human urine. Environ Health Perspect 108:1015–1018

Leustek T, Saito K (1999) Sulfate transport and assimilation in plants. Plant Physiol 120(3):637–644

Lewandowska M, Sirko A (2008) Recent advances in understanding plant response to sulfur-deficiency stress. Acta Biochim Pol 55(3):457–471

Li HF, McGrath SP, Zhao FJ (2008) Arsenic uptake and metabolism in plants. New Phytol 178:92–102

Li Y, Zhao J, Guo J, Liu M, Xu Q, Li H, Li YF, Zheng L, Zhang Z, Gao Y (2017) Influence of sulfur on the accumulation of mercury in rice plant (*Oryza sativa* L.) growing in mercury contaminated soils. Chemosphere 182:293–300

Liu WJ, Zhu YG, Smith FA, Smith SE (2004) Do phosphorous nutrition and iron plaque alter arsenate (As) uptake by rice seedlings in hydroponic culture? New Phytol 162:481–488

Liu W-J, Wood BA, Raab A, McGrath SP, Zhao F-J, Feldmann J (2010) Complexation of arsenite with phytochelatins reduces Arsenite efflux and translocation from roots to shoots in arabidopsis. Plant Physiol 152:2211–2221

Liu X, Zhao Z, Duan B, Hu C, Zhao X, Guo Z (2014) Effect of applied sulphur on the uptake by wheat of selenium applied as selenite. Plant Soil 386(1–2):35–45

Liu X, Yang Y, Deng X, Li M, Zhang W, Zhao Z (2017a) Effects of sulfur and sulfate on selenium uptake and quality of seeds in rapeseed (*Brassica napus* L.) treated with selenite and selenate. Environ Exp Bot 135:13–20

Liu X, J-W F, Da Silva E, Shi X-X, Cao Y, Rathinasabapthi B, Chen Y, Ma LQ (2017b) Microbial siderophores and root exudates enhanced goethite dissolution and Fe/As uptake by As-hyperaccumulator Pteris vittata. Environ Pollut 223:230–237

Lomax C, Liu W-J, Wu L, Xue K, Hiong J, Zhou J, McGrath SP, Meharg AA, Miller AJ, Zhao F-J (2012) Methylated arsenic species in plants originate from soil microorganisms. New Phytol 193:665–672

Lombi E, Zhao F-J, Fuhrmann M, Ma QL, McGrath SP (2002) Arsenic distribution and speciation in the fronds of the hyperaccumulator *Pteris vittata*. New Phytol 156:195–203

Maathuis FJM (2009) Physiological functions of mineral macronutrients. Curr Opin Plant Biol 12(3):250–258

MacGowan D (2007) The strange affair of Madeleine Smith. In: Victorian Scotland's trial of the century. Mercat Press, Edinburgh. ISBN 978-1-8183-113-8

Mandal BK, Suzuki KT (2002) Arsenic round the world: a review. Talanta 58:201–235

Masscheleyn PH, Delaune RD, Patrick WH Jr (1991) Effect of redox potential and pH on arsenic speciation and solubility in a contaminated soil. Environ Sci Technol 25:1414–1419

May TW, Wiedmeyer RH (1998) A table of polyatomic interferences in ICP-MS. At Spectrosc 19:150–155

McCarty KM, Hanh HT, Kim K-W (2011) Arsenic geochemistry and human health in South East Asia. Rev Environ Health 26:71–78

Meagher BR (2000) Phytoremediation of toxic elemental and organic pollutants. Curr Opin Plant Biol 3:153–152

Meharg AA, Hartley-Whitaker J (2002) Arsenic uptake and metabolism in arsenic resistant and nonresistant plant species. New Phytol 154:29–43

Meharg AA, Rahman M (2003) Arsenic contamination of Bangladesh paddy field soils: implications for rice contribution to arsenic consumption. Environ Sci Technol 42:1051–1057
Mestrot A, Planer-Friedrich B, Feldmann J (2013) Biovolatilisation: a poorly studies pathway of the arsenic biogeochemical cycle. Environ Sci: Processes Impacts 15:1635–1651
Mishra S, Mattusch J, Wennrich R (2017) Accumulation and transformation of inorganic and organic arsenic in rice and role of thiol-complexation to restrict their translocation to shoot. Sci Rep 7:40522
Obrist D, Agnan Y, Jiskra M, Olson LC, Colegrove DP, Hueber J, Moore CW, Sonke JE, Helmig D (2017) Tundra uptake of atmospheric elemental mercury drives Arctic mercury pollution. Nature 547:201–204. https://doi.org/10.1038/nature22997
Ohno K, Yanase T, Matsuo Y, Kimura T, Rahman MH, Magara Y, Matsui Y (2007) Arsenic intake via water and food by a population living in an arsenic-affected area of Bangladesh. Sci Total Environ 381:68–76
Pavageau MP, Pecheyran C, Krupp EM, Morin A, Donard OFX (2002) Volatile metal species in coal combustion flue gas. Environ Sci Techol 36:1561–1573
Pierce ML, Moore CB (1982) Adsorption of arsenite and arsenate on amorphous iron hydroxide. Water Res 16:1247–1253
Pilon-Smits EAH, Hwang S, Lyle CM, Zhu Y, Tai JC, Bravo RC, Chen Y, Leustek T, Terry N (1999) Overexpression of ATP sulfurylase in Indian mustard leads to increased selenate uptake, reduction, and tolerance. Plant Physiol 119:123–132
Prestera T, Fahey JW, Holtzclaw WD, Abeygunawardana C, Kachinski JL, Talalay P (1996) Comprehensive chromatographic and spectroscopic methods for the separation and identification of intact glucosinolates. Anal Biochem 239:168–179
Prietzel J, Thieme J, Neuhäusler U, Susini J, Kögel-Knabner I (2003) Speciation of sulphur in soils and soil particles by X-ray spectromicroscopy. Eur J Soil Sci 54(2):423–433
Prietzel J, Thieme J, Tyufekchieva N, Paterson D, McNulty I, Kögel-Knabner I (2009) Sulfur speciation in well-aerated and wetland soils in a forested catchment assessed by sulfur K-edge X-ray absorption near-edge spectroscopy (XANES). J Plant Nutr Soil Sci 172(3):393–403
Quaghebeur M, Rengel Z (2005) Arsenic speciation governs arsenic uptake and transport in terrestrial plants. Microchim Acta 151:141–152
Raab A, Feldman J, Meharg AA (2004) The nature of arsenic-phytochelatin complexes in Holcus lanatus and Pteris cretica. Plant Physiol 134:1113–1122
Raab A, Schat H, Meharg AA, Feldmann J (2005) Uptake, translocation and transformation of arsenate and arsenite in sunflower (Helianthus annuus): formation of arsenic-phytochelatin complexes during exposure to high arsenic concentrations. New Phytol 168:551–558
Raab A, Wright SH, Jaspars M, Meharg AA, Feldmann J (2007a) Pentavalent arsenic can bind to biomolecules. Angew Chem Int Ed 46:2594–2597
Raab A, Ferreira K, Meharg AA, Feldmann J (2007b) Can arsenic–phytochelatin complex formation be used as an indicator for toxicity in Helianthus annuus? J Exp Bot 58:1333–1338
Raab A, Williams PN, Meharg A, Feldmann J (2007c) Uptake and translocation of inorganic and methylated arsenic species in plants. Environ Chem 4:197–203
Raab A, Ronzan M, Feldmann J (2017) Sulphur fertilization influences the sulphur species composition in Allium sativum: sulphomics using HPLC-ICPMS/MS-ESI-MS/MS. Metallomics. https://doi.org/10.1039/C7MT00098G
Rauser WE (1990) Phytochelatins. Annu Rev Biochem 59:61–86
Riddles PW, Blakeley RL, Zerner B (1983) Reassessment of Ellman's reagent. Methods Enzymol 91:49–60
Rothenberg SE, Windham-Myers L, Creswell JE (2014) Rice methylmercury exposure and mitigation: a comprehensive review. Environ Res 133:407–423
Sabeh F, Wright T, Norton SJ (1993) Purification and characterization of a glutathione peroxidase from Aloe vera plant. Enzyme Protoc 47:92–98
Scherer HW (2009) Sulfur in soils. J Plant Nutr Soil Sci 172(3):326–335
Schwarz K, Foltz CM (1957) Selenium as an integral part of factor 3 against dietary necrotic liver degeneration. J Am Chem Soc 70:3292–3293

Scotsman (2005) Madeleine Smith and her poisonous tale. http://www.scotsman.com/lifestyle/madeleine-smith-and-her-poisonous-tale-1-465286. Last accessed Aug 2017

Scott N, Hatlelid KM, MacKenzie NE, Carter DE (1993) Reactions of arsenic(III) and arsenic(V) species with glutathione. Chem Res Toxicol 6:102–106

Shibagaki N, Rose A, McDermott JP, Fujiwara T, Hayashi H, Yoneyama T, Davies JP (2002) Selenate-resistant mutants of Arabidopsis thaliana identify Sultr1;2, a sulfate transporter required for efficient transport of sulfate into roots. Plant J 29:475–486

Sneller FEC, van Heerwarden LM, Kraaijeveld-Smit FJL, Ten Bookum WM, Koevoets PLM, Schat H, Verkleij JAC (1999) Toxicity of arsenate in Silene vulgaris, accumulation and degradation of arsenate-induced phytochelatins. New Phytol 144:223–232

Solomon D, Lehmann J, Martínez CE (2003) Sulfur K-edge XANES spectroscopy as a tool for understanding sulfur dynamics in soil organic matter. Soil Sci Soc Am J 67:1721–1731

Solomon D, Lehmann J, Kinyangi J, Pell A, Theis J, Riha S, Ngoze S, Amelung W, Preez C, Machado S, Ellert B, Janzen H (2009) Anthropogenic and climate influences on biogeochemical dynamics and molecular-level speciation of soil sulfur. Ecol Appl 19(4):989–1002

Song WY, Park J, Mendoza-Cózatl DG, Suter-Grotemeyer M, Shima D, Hörtensteiner S, Geisler M, Weder B, Rea PA, Rentsch D, Schroeder JI, Lee Y, Martinoia E (2010) Arsenic tolerance in Arabidopsis is mediated by two ABCC-type phytochelatin transporters. Proc Natl Acad Sci U S A 107:21187–21192

Song ZP, Shao HF, Huang HG, Shen Y, Wang LZ, Wu FY, Han D, Song JY, Jia HF (2017) Environ Exp Bot 137:158–165

Spuches AM, Kruszyna HG, Rich AM, Wilcox DE (2005) Thermodynamics of the as(III)–thiol interaction: arsenite and monomethylarsenite complexes with glutathione, dihydrolipoic acid, and other thiol ligands. Inorg Chem 44:2964–2972

Srivastava S, D'Souza SF (2009) Increasing sulfur supply enhances tolerance to arsenic and its accumulation in Hydrilla verticillata (L.f.) Royle. Environ Sci Technol 43(16):6308–6313

Srivastava S, D'Souza SF (2010) Effect of variable sulfur supply on arsenic tolerance and antioxidant responses in Hydrilla verticillata (L.f.) Royle. Ecotoxicol Environ Saf 73(6):1314–1322

Steffens JC (1990) The heavy metal-binding peptides of plants. Ann Rev Plant Physiol Plant Mol Biol 41:553–575

Strickman RJ, Mitchell CPJ (2017) Accumulation and translocation of methylmercury and inorganic mercury in Oryza sativa: an enriched isotope tracer study. Sci Total Environ Ann 574:1415–1423

Swain EB (2007) Socioeconomic consequences of mercury use and pollution. Ambio 36(1):45–61

Takahashi H, Kopriva S, Giordano M, Saito K, Hell R (2011) Sulfur assimilation in photosynthetic organisms: molecular functions and regulations of transporters and assimilatory enzymes. Annu Rev Plant Biol 62(1):157–184

Thomas X, Troncy J (2009) Arsenic: a beneficial therapeutic poison - a historical overview. Adler Museums Bull 35:3–13

Tian M, Hui M, Thannhauser TW, Pan S, Li L (2017) Selenium-induced toxicity is counteracted by sulfur in broccoli (Brassica oleracea L. var. italica). Front Plant Sci 8:1425

Venkatachalam P, Srivastava AK, Raghothama KG, Sahi SV (2009) Genes induced in response to mercury-ion-exposure in heavy metal hyperaccumulator Sesbania drummondii. Environ Sci Technol 43:843–850

Wallschlaeger D, Feldmann J (2010) Formation, occurrence, significance and analysis of organoselenium and organotellurium compounds in the environment. In: Sigel H, Sigel A (eds) Organometallics in environment and toxicology. Wiley, Hoboken, pp 319–364

Watanabe T, Kouho R, Katayose T, Kitajima N, Sakamoto N, Yamaguchi N, Shinano T, Yurimoto H, Osaki M (2014) Arsenic alters uptake and distribution of sulphur in Pteris vittata. Plant Cell Environ 37(1):45–53

Wood BA, Feldmann J (2012) Quantification of phytochelatins and their metal(loid) complexes: critical assessment of current analytical methodology. Anal Bioanal Chem 402:3299–3309

Wood BA, Miyashita S, Kaise T, Raab A, Meharg AA, Feldmann J (2011) Arsenic is not stored as arsenite–phytochelatin complexes in the seaweeds Fucus spiralis and Hizikia fusiforme. Environ Chem 8:30–43

Xu X, Zhao J, Li Y, Fan Y, Zhu N, Gao Y, Li B, Liu H, Li YF (2016) Demethylation of methylmercury in growing rice plants: an evidence of self-detoxification. Environ Pollut 210:113–120

Xu X, Meng B, Zhang C, Feng X, Gu C, Guo J, Bishop K, Xu Z, Zhang S (2017a) Qiu G (2017): the local impact of a coal-fired power plant on inorganic mercury and methyl-mercury distribution in rice. Environ Pollut 223:11–18

Xu X, McGrath SP, Zhao F-J (2017b) Rapid reduction of arsenate in the medium mediated by plant roots. New Phytol 176:5574–5579

Yamazaki Y, Iwasaki K, Mikami M, Yagihashi A (2010) Distribution of eleven flavor precursors, S-alk(en)yl-L-cysteine derivatives, in seven Allium vegetables. Food Sci Technol Res 17(1):55–62

Yang X, Lu Y, Zhong J, Qian Y, Zhao Z, Liu X (2017) The positive effect of sulfur on selenium detoxification under selenite condition in wheat. Commun Soil Sci Plant Anal 48(13):1564–1573

Ye W-l, Wood BA, Stroud JL, Andralojc J, Raab A, McGrath SP, Feldmann J, Zhao F-J (2010) Arsenic speciation in phloem and xylem exudates of castor bean. Plant Physiol 154:1505–1513

Zhang H, Xu W, Guo J, Zhenyan H, Ma M (2005) Coordinated responses of phytochelatins and metallothioneins to heavy metals in garlic seedlings. Plant Sci 169:1059–1065

Zhang LH, Hu B, Li W, Ce RH, Deng K, Li H, Yu FY, Ling HQ, Li YJ, Chu CC (2014) OsPT2, a phosphate transporter, is involved in the active uptake of selenite in rice. New Phytol 201:1183–1191

Zhao F-J, Ma JF, Meharg A, McGrath SP (2008) Arsenic uptake and metabolism in plants. New Phytol 181:777–794

Zhao F-J, McCrath SP, Meharg AA (2010a) Arsenic as a food chain contaminant: mechanisms of plant uptake and metabolism ad mitigation strategies. Ann Rev Plant Biol 61:535–559

Zhao F-J, Ago Y, Mitani N, Li R-Y, Su Y-H, Yamaji N, McGrath SP (2010b) The role of the rice aquaporin Lsi1 in arsenite efflux from roots. New Phytol 186:392–399

Zhu YG, Pilon-Smits EAH, Zhao FJ, Williams PN, Meharg AA (2008) Selenium in higher plants: understanding mechanisms for biofortification and phytoremediation. Trends Plant Sci 14:436–442

Chapter 5
Metallomics in Fish

Camila Pereira Braga, Jiri Adamec, and Pedro de Magalhães Padilha

Abstract Metallomics allows the integration of traditionally analytical studies with inorganic and biochemical studies. The study of metallomics in living organisms allows us to obtain information about how the metal ion is distributed and coordinated with proteins, the essentiality and/or toxicity, and the individual concentrations of metal species, thus contributing to elucidation of the physiological and functional aspects of these biomolecules. In this context, several lines of research have appeared in the literature with different terms and approaches. For example, *metallomic*, which deals with the characterization of the total metal/metalloid species present in an organism; *metalloprotein*, which deals with the characterization of the total elements present in a specific site of an organism (cellular behavior, protein, metalloprotein); and *metallomic*, which deals with a more in-depth study of metallome. In this area, information is sought on the interactions and functional connections of metal/metalloid species with genes, proteins, metabolites and other biomolecules of the organism and, therefore, the elucidation of the biological role exerted by the metal ions bound to the biomolecules. In this chapter, we will describe techniques used in animal studies.

Keywords Animal · Fish · Metallomic · Metalloprotein

C. P. Braga · J. Adamec
Department of Biochemistry, University of Nebraska-Lincoln, Lincoln, NE, USA

P. de Magalhães Padilha (✉)
Department of Chemistry and Biochemistry, Institute of Bioscience, São Paulo State University (UNESP), Botucatu, São Paulo, Brazil
e-mail: padilha@ibb.unesp.br

M. A. Z. Arruda (ed.), *Metallomics*, Advances in Experimental Medicine and Biology 1055, https://doi.org/10.1007/978-3-319-90143-5_5

Abbreviations

2D-PAGE	Two-dimensional polyacrylamide gel electrophoresis
AAS	Atomic absorption spectrometry
ESI	Electrospray ionization
ESI-MS/MS	Electrospray ionization tandem mass spectrometry
FAAS	Flame atomic absorption spectrometry
GFAAS	Graphite furnace atomic absorption spectrometry
ICP-MS	Inductively coupled plasma source mass spectrometry
IEF	Isoelectric focusing
LC	Multidimensional liquid chromatography
MALDI	Matrix assisted laser desorption ionization
MALDI-TOF /MS	Time-of-flight mass spectrometry coupled to laser-assisted matrix
Mm	Molecular mass
MS	Mass spectrometry
PAGE	One-dimensional polyacrylamide gel electrophoresis
pI	Isoeletric point
SDS-PAGE	Sodium dodecyl sulfate polyacrylamide gel electrophoresis
SEC	size-exclusion chromatography

5.1 Introduction

Metal ions are important in biological processes by stabilizing proteins and nucleic acids and participating in enzymatic reactions (Szpunar 2004; Barnham and Bush 2014). Metal deficiencies result in anemia (iron [Fe] and copper [Cu]), bone deterioration (calcium [Ca]), growth retardation (zinc [Zn]), skin changes (Zn), and brain and heart diseases (Cu); some metals are considered carcinogens (arsenic [As], cadmium [Cd], chromium and nickel); and the presence and or excessive presence of metals can be associated with nephrotoxicity (Cd and uranium) and neurotoxicity (aluminum, mercury [Hg] and manganese [Mn]) (Lippard 1994; Szpunar 2004). Therefore, elucidation of the biological essentiality and toxicity of metals has become important in varied fields of research (Haraguchi 2004).

The aim of metallomics is to verify the distribution of metal and/or metalloid and elucidate the physiological and functional aspects of proteins, genes and metabolites containing metals of biological interest (Haraguchi 2004). Therefore, in metallomics, metallomes are defined as metalloproteins, metalloenzymes and metal-containing biomolecules representing the main target of metallomics studies (Haraguchi 2004). Metallomics information can be classified as qualitative (identification of individual metal species), quantitative (determination of concentrations of these species), and comparative (monitoring changes of the metallomes of a given organism under the influence of external stimulus) (Haraguchi 2004).

Metalloproteins are characterized by high affinity of the metal–protein interaction. They are also are known as metalloenzymes when they exert catalytic activity using the metal incorporated into their active site as part of the catalytic process, such as alcohol dehydrogenase, catalase, carbonic anhydrase, DNA polymerase, ferritin, transferrin and RNA polymerase (Haraguchi 2004; Dydio et al. 2016). In an attempt to protect organisms from metal toxicity, some metalloenzymes—such as superoxide dismutase, glutathione peroxidase, glutathione reductase, glutaredoxins, thioredoxins and peroxiredoxins—are induced so they can be used as biomarkers of metal toxicity, thus contributing to elucidation of the physiological and functional aspects (López-Barea and Gómez-Ariza 2006). Metalloproteomic studies consist of a selectivity component (separation technique to isolate the target species from matrix), a sensitivity component (sensitive detector to quantify the elements), and a structural component (specific detector for molecules) (Gao et al. 2003; da Silva et al. 2010).

The most used techniques in the fractionation/purification stage of the biomolecules are one- (PAGE) or two-dimensional polyacrylamide gel electrophoresis (2D-PAGE) and multi-dimensional liquid chromatography (LC). In the step of quantitative qualitative determination and speciation analysis, techniques employed include x-ray fluorescence with synchrotron radiation (SR-XRF), flame atomic absorption or graphite furnace atomic absorption spectrometry (FAAS/GFAAS), and inductively coupled plasma source mass spectrometry (ICP-MS). To characterize the biomolecules (metalloproteins and/or metal-binding proteins), the most commonly used techniques are time-of-flight mass spectrometry coupled to laser-assisted matrix desorption ionization (MALDI TOF MS) and electrospray ionization tandem mass spectrometry (ESI-MS/MS).

5.2 Metallomic Studies in Fish

Fish live in different habitats and can adapt to changes in the environment; they are also an important source of proteins, minerals and vitamins (Beyer et al. 1996). Through ingestion, biomagnification and bioconcentration, fish can accumulate chemicals (Van der Oost et al. 2003), and thus they provide information about environmental contamination by known toxic or new chemicals (Beyer et al. 1996; Van der Oost et al. 2003). In fish, metal-bound proteins can influence total metal levels, thus reflecting toxic exposure (Hauser-Davis et al. 2012). Thus, metalloproteins in fish have been used as biomarkers for environmental contamination showing ecological importance. In addition, another group of small proteins that show ecological importance, known as metallothioneins, are important in the detoxification of essential (Zn and Cu) and toxic metals (As, Cd and Hg) (Sun and Chai 2010). Cysteine residues present in metalloproteins and metallothioneins have a high affinity to bind with heavy metals, such as Pb^{2+}, Hg^{2+}, Zn^{2+}, Cd^{2+}, $Fe^{2+/3+}$, As^{3+}, Cu^{+} and Ag^{+} (Huang et al. 2016).

Mercury (Hg) is one of the main toxic metals involved in environmental contamination. It is a naturally occurring metal found in the earth's crust and, in small quantities, in air, soil and water, where it assumes various chemical forms, such as metallic or elemental Hg (Hg°); inorganic metal as mercuric salts ($HgCl_2$, HgS) and mercurous chloride (Hg_2Cl_2); and organic Hg bound to carbon radicals (methylmercury [MeHg] and ethylmercury [EtHg]) (Clarkson and Magos 2006). Due to Hg's toxic effects on the environment and health (loss of skin sensitivity, loss of muscle coordination, deafness and death), countries around the world have been acting to minimize the risks of Hg contamination.

5.2.1 The Problem of Hg in the Amazon Region

Past gold-mining activity in the Amazon region is associated with Hg contamination. Therefore, the development of hydroelectric complexes in the Madeira River basin has aroused scientific discussions about the possible remobilization and bioavailability of Hg species in reservoirs in the Amazon region. However, with the construction and implementation of hydroelectric complexes in the Madeira River basin, the Hg dynamics in this environment can be altered due to environmental changes (Bastos et al. 2006). Hg concentrations have remained within natural levels in Amazonian rivers; however, it is known that the transformations and processes of remobilization of previously unavailable mercurial species can change their chemical form, thus making them available to aquatic biota (Pfeiffer et al. 1993; Bastos et al. 2006).

Bioaccumulation along the trophic chain is responsible for the high Hg content in Amazonian fish. In the algae and aquatic plants of Amazonian rivers, Hg concentrations are greater than those in the rivers themselves; therefore, fish that feed exclusively on these algae and plants have greater concentrations of Hg (Akagi et al. 1995; Aula et al. 1995). This happens along the trophic chain through to top predatory fish, which have concentrations 1 million times greater than those found in river waters. Among the species of Amazonian fish, some are of great commercial and ecological interest. In addition, fish have a great importance for human consumption because fish is the main source of protein and represents the daily food of the riverine population.

In this context, the identification of biomarkers of Hg toxicity and Hg-species toxicity applicable to environmental and human health surveillance in the areas of direct and indirect influence of hydroelectric complexes is important in order to contribute to estimation of the magnitude of total or partial Hg exposure.

5.2.1.1 Fish Studies: Biomarkers of Hg Toxicity in the Amazon Region

Studies are being conducted to study possible biomarkers of Hg toxicity in fish using 2D-PAGE, analytical technique for the quantification of elements (atomic absorption spectrometry [AAS]), and MS.

In 2D-PAGE under denaturing conditions, only stable links of proteins having metal ions are uncompromised due to protein conformation. Proteins exhibit functional groups with sulfur atoms (thiol groups) in their peptide fragment—which have characteristics of soft bases and bind preferentially to metals and metalloids with characteristics of soft acids, such as Hg in the form R-Hg^+—even when they are submitted to protein-denaturing conditions, under which such metal–protein structure is maintained. The connections in the case of metal-binding proteins should also be stable because the denaturing conditions did not break these connections. In this case, Hg can be bound to nitrogen (N), which is an intermediate base and can form stable bonds with an element having the characteristics of either a hard acid or a soft acid (in this case, Hg).

5.2.1.2 Protein Fractionation Using 2D-PAGE

2D-PAGE is an extremely powerful technique that allows simultaneous separation and resolution of several proteins in two dimensions according with the isoelectric point (pI) and the molecular mass (Mm) (O'Farrell 1975). 2D-PAGE consists of five main steps: sample preparation, first dimension, second dimension, protein detection, scanning and image processing (Gorg et al. 2000).

For each spot on the gel to represent one or more proteins, the proteins in the sample must be denatured, disaggregated, reduced and solubilized. The first dimension is isoelectric focusing, in which the proteins are separated according to their pI in strips containing gels with a pH gradient. In the second dimension, which involves sodium dodecyl sulfate–PAGE, the proteins are separated based on their molecular mass. Each spot on the gel can be considered as an orthogonal coordinate of a protein that migrated specifically as a function of its pI (x-axis) and its Mm (y-axis). The appearance or disappearance of spots may provide information about stage-specific proteins, whereas spot intensity provides quantitative information regarding differential protein expression.

After the second-dimension step is performed, the gels are stained and scanned using a specific scanner, and the intensity of each spot is converted into an image using software to analyze the gels images obtaining the number of spots present in the gel as well as pI and Mm information from each spot.

5.2.1.3 AAS

AAS is a technique in which free gaseous atoms absorb electromagnetic radiation at a specific wavelength, thus producing a corresponding measurable signal that is proportional to the concentration of the free atoms present in the optical pathway (Garcia and Baez 2012). AAS consists of a source of radiation, a sample-introduction system, an atomization system, a monochromator, a detection system, and a reading; all of these are connected to a computerized system for the control and processing of data (Jenniss et al. 1999).

The two types of atomizers most used in AAS are FAAS and GFAAS. FAAS is used for elemental analyzes at mg L^{-1} levels (using some milliliters of the sample), whereas GFAAS is used at low concentrations, e.g. μg L^{-1} (using microliters of the sample) (Garcia and Baez 2012). Analysis of biological samples can pose some problems: The analyte may be present in different oxidation states; the analyte may be combined with different anions or bound to proteins or other organic binders; and the organic portion of the sample may cause large problems. To minimize interferences in the analysis of biological samples, the procedure commonly performed is wet digestion/mineralization of the organic portion of the sample using strong acids.

5.2.1.4 MS

Mass spectrometers consist of an ion source, a mass analyzer, an ion detector, and a data-acquisition component. It is an analytical instrument capable of converting neutral molecules into ions in gaseous form and separating them according to their mass/charge ratio using electromagnetic fields. The most commonly used ionization techniques in protein analysis are MALDI and ESI (Aebersold and Mann 2003).

5.2.1.5 Studies

A preliminary study developed by Moraes et al. (2012), using 2D-PAGE and SR-XRF to study fish from the Madeira River Basin in the Amazon region of Brazil, found a protein spot with Hg in muscle samples of dourada (*Brachyplatystoma rousseauxii* [predator species]) and muscle samples of pacu (*Mylossoma* and *Myleus spp.* [omnivorous with a tendency to herbivory]) (Moraes et al. 2012).

The same group continuing the metallomic study in fish from Madeira River Basin in the Amazon region of Brazil developed a method to determine Hg concentration using GFAAS involving acid mineralization, a chemical modifier (copper nitrate) and a permanent modifier (sodium tungstate) (Moraes et al. 2013). Using those methods, the investigators found Hg muscle concentrations (mg kg^{-1}) in dourada, pacu and jaraqui of 0.331 ± 0.007, 0.081 ± 0.002, and 0.132 ± 0.004, respectively (Moraes et al. 2013).

In studies published in 2015, the results of protein fractionation were shown with posterior analysis of the presence of Hg in muscle samples of dourada (*Br. rousseauxii*) and in muscle and liver samples of tucunaré (*Cichla* spp.) from the Madeira River, Brazil (Braga et al. 2015; Vieira et al. 2015). The methods used were 2D-PAGE to determine protein fractionation, GFAAS to determine Hg concentration and ESI–MS/MS to determine protein characterization (Braga et al. 2015; Vieira et al. 2015). The study of dourada identified parvalbumin (isoforms: alpha and beta), protein NLRC5, 39S ribosomal protein L36 mitochondrial, N-alpha-acetyl transferase 20, Mth938 domain-containing protein, ubiquitin-40S ribosomal protein S27a as potential candidates for biomarkers of Hg toxicity (Braga et al. 2015), whereas in tucunaré (*Cichla* spp.), seven proteins with characteristics of Hg biomarkers were found, including parvalbumin and its isoforms, ubiquitin-40S and ribosomal-S27, zinc finger, BTB domain containing protein 24, and protein of double-specificity phosphatase 22-B (Vieira et al. 2015).

Vieira et al. (2017), using the same methodology, analyzed liver tissue samples of dourada (*Br. rousseauxii*) and pacu (*Mylossoma duriventre*) and identified possible biomarkers of Hg toxicity. In this study, proteins were found, including isoforms of parvalbumin, ubiquitin-40S ribosomal protein S27a, brain-specific angiogenesis inhibitor 1-associated protein, 2-like protein 2, and betaine-homocysteine S-methyltransferase 1 exhibited be Hg-associated proteins.

5.2.1.6 Other Studies Involving Hg Toxicity

Korbas et al. (2008), using zebrafish (*Danio rerio*), a common model employed to study toxicology, investigated organic Hg exposure at the body and molecular level using larval stage. SR-XRF mapping was used to generate the image with metal localization, and the results showed that the accumulation of MeHg and EtHg directly affected ocular tissue. In addition, compared with other tissues analyzed, the investigators found four times higher Hg in the layer of the eye lens (1 $\mu g/cm^2$) (Korbas et al. 2008). This was associated with organic Hg being able to generate visual defects with partial or complete loss of vision.

In a preliminary study performed by Kutscher et al. (2012)— using a method combining size-exclusion chromatography, LC coupled to ICP-MS, and ESI-MS/MS—investigated MeHg-binding proteins in certified tuna fish muscle (BCR-464) and their association with Hg toxicity (Kutscher et al. 2012). This study found a high molecular-weight protein as MeHg binding and skeletal muscle myosin heavy chain; using LC-ICP-MS, it was possible to detect Hg-containing peptides (Kutscher et al. 2012).

5.2.2 Other Studies in Fish

A preliminary study using liver tissue of Nile tilapia (*Oreochromis niloticus*) employed 2D-PAGE electrophoresis; SR-XRF to map proteins spots with Ca, Fe and Zn; and FAAS to quantify Ca, Fe and Zn (Lima et al. 2010). The results showed a range in protein spots of Ca, Fe and Zn of 1.08 to 5.80, 2.02 to 8.03, and 1.60 to 8.55 mg g^{-1}, respectively (Lima et al. 2010). Studying the same species and using similar methodology (in this studying using GFAAS) to analyze Mn and Zn in the protein spots, the investigators found a range from 0.8 to 2.6 mg g^{-1} and 1.0 to 6.3 mg g^{-1}, respectively (Santos et al. 2011).

In study published in 2015 involving muscle of Nile tilapia (*Or. niloticus*), 2D-PAGE was used for protein fractionation, FAAS and GFAAS to determine Ca, Cu, Zn and Mn concentration, and ESI–MS/MS for protein characterization (Cavecci et al. 2015). The results showed non-specific binding—such as the Zn found in parvalbumin-2—which can be a non-specific binding protein that forms a metal-binding protein rather than a metal cofactor. The Ca and Zn detection in hemoglobin subunit alpha was also associated with non-specific binding regarding the terminal N atoms present in the hemoglobin chains (Cavecci et al. 2015).

5.3 Metallomic Studies in Rat

To understand the mechanisms of pathogenesis and the complications of some diseases, such as diabetes mellitus type 1 (DM1), animal models have been extensively used. Effective methods for inducing experimental diabetes, such as the administration of β-cytotoxic chemical agents like aloxane and streptozotocin (STZ), are effective ways to promote DM1 (Rees and Alcolado 2005). It is suggested that STZ acts by stimulating the production of reactive oxygen species (ROS) along with increased lipid peroxidation and decreased activity of antioxidant enzymes. In addition, one of the major toxic effects of STZ involves alteration of the DNA structure, which is fragmented by ROS, thus compromising insulin biosynthesis and secretion (Xiang et al. 2010).

Studies were developed using plasma and liver from DM1 animals, 2D-PAGE, GFAAS and FAAS (for quantitative determination of Cu, Mg, Se and Zn in the spots that showed differences of expression), and protein characterization by ESI/MS-MS. In total, 35 proteins were found to be differently expressed in the plasma, indicating (1) alpha-1-macroglobulin and haptoglobulin as potential biomarkers of controlled type 1 diabetes (i.e. insulin treated); and (2) 2′-deoxynucleoside 5′-phosphate N-hydrolase 1, transmembrane protein 11, serum amyloid P-component, vitamin D-binding protein and biliverdin as possible candidates for biomarkers of uncontrolled type 1 diabetes (Braga et al. 2017). In addition, these studies showed the presence of Cu, Mg, Se and Zn in spots relating some metal-binding interactions that can be involved in the disease progression and complications in cases of DM1 (Braga et al. 2017).

5.4 Metallomic Studies in Bovines

Different factors—such as color, water-retention capacity, palatability and nutritional value—vary between bovine breeds and species, and these influence the quality of the meat. Meat quality has a direct relation with its degree of tenderness, which is of extreme importance because it leads to an increase in consumption and its commercialization. Studies have been developed to understand what leads to variation in meat tenderness.

In this context, using muscle tissue from Nellore breed (*Bos indicus*) and methodologies, such as 2D-PAGE, SR-XRF to map protein spots with Ca, and ESI–MS/MS to characterize protein, the investigators found two proteins, which were detected as Ca, pyruvate kinase and albumin with differences in animals having tough meat, animals having tender meat, and Piedmontese breed animals (*B. taurus*) used as comparative model (Baldassini et al. 2015).

References

Aebersold R, Mann M (2003) Mass spectrometry-based proteomics. Nature 422:198–207

Akagi H, Malm O, Kinjo Y et al (1995) Methylmercury pollution in the Amazon, Brazil. Sci Total Environ 175:85–95

Aula I, Braunschweiler H, Malin I (1995) The watershed flux of mercury examined with indicators in the Tucuruí reservoir in Pará, Brazil. Sci Total Environ 175:97–107

Baldassini WA, Braga CP, Chardulo LAL et al (2015) Bioanalytical methods for the metalloproteomics study of bovine longissimus thoracis muscle tissue with different grades of meat tenderness in the Nellore breed (Bos indicus). Food Chem 169:65–72

Barnham KJ, Bush AI (2014) Biological metals and metal-targeting compounds in major neurodegenerative diseases. Chem Soc Rev 43:6727–6749

Bastos WR, Gomes JPO, Oliveira RC et al (2006) Mercury in the environment and riverside population in the Madeira River basin, Amazon, Brazil. Sci Total Environ 368:344–351

Beyer J, Sandvik M, Hylland K et al (1996) Contaminant accumulation and biomarker responses in flounder (Platichthys flesus L.) and Atlantic cod (Gadus morhua L.) exposed by caging to polluted sediments in Sorfjorden, Norway. Aquat Toxicol 36:75–98

Braga CP, Bittarello AC, Padilha CCF et al (2015) Mercury fractionation in dourada (Brachyplatystoma rousseauxii) of the Madeira River in Brazil using metalloproteomic strategies. Talanta 132:239–244

Braga CP; Vieira JC; Leite AL et al (2017) Metalloproteomic and differential expression in plasma in a rat model of type 1 diabetes. Int J Biol Macromol 104:414–422.

Braga CP, Vieira JCS, Grove RA et al (2017) A proteomic approach to identify metalloproteins and metal-binding proteins in liver from diabetic rats. Int J Biol Macromol 96:817–832

Cavecci B, De Lima PM, De Queiroz JV et al (2015) Metalloproteomic profile determination of muscle samples from Nile tilapia (Oreochromis niloticus) using AAS and ESI-MS/MS after 2D-PAGE separation. J Braz Chem Soc 26:239–246

Clarkson TW, Magos L (2006) The toxicology of mercury and its chemical compounds. Crit Rev Toxicol 36:609–662

da Silva MAO, Garcia JS, GHMF S et al (2010) Evaluation of sample preparation protocols for proteomic analysis of sunflower leaves. Talanta 80:1545–1551

Dydio P, Key HM, Nazarenko A, et al (2016) An artificial metalloenzyme with the kinetics of native enzymes. Science (80-) 354:102–106

Gao Y, Chen C, Zhang P et al (2003) Detection of metalloproteins in human liver cytosol by synchrotron radiation X-ray fluorescence after sodium dodecyl sulphate polyacrylamide gel electrophoresis. Anal Chim Acta 485:131–137

Garcia R, Baez AP (2012) Atomic absorption spectrometry (AAS). In: At. Absorpt. Spectrosc, pp 1–12

Gorg A, Obermaier C, Boguth G et al (2000) The current state of two-dimensional electrophoresis with immobilized pH gradients. Electrophoresis 21:1037–1053

Haraguchi H (2004) Metallomics as integrated biometal science. J Anal At Spectrom 19:5

Hauser-Davis RA, De Campos RC, Ziolli RL (2012) Fish metalloproteins as biomarkers of environmental contamination. Rev Environ Contam Toxicol 218:101–123

Huang S, Liu X, Wang D et al (2016) Structural basis for the selective Pb(II) recognition of Metalloregulatory protein PbrR691. Inorg Chem 55:12516–12519

Jenniss SW, Katz SA, Lynch RW (1999) Applications of atomic spectrometry to regulatory compliance monitoring. ACH-Models Chem 136:55–68
Korbas M, Blechinger SR, Krone PH et al (2008) Localizing organomercury uptake and accumulation in zebrafish larvae at the tissue and cellular level. Proc Natl Acad Sci U S A 105:12108–12112
Kutscher DJ, Sanz-Medel A, Bettmer J (2012) Metallomics investigations on potential binding partners of methylmercury in tuna fish muscle tissue using complementary mass spectrometric techniques. Metallomics 4:807–813
Lima PM, Neves RDCF, Dos Santos FA et al (2010) Analytical approach to the metallomic of Nile tilapia (Oreochromis niloticus) liver tissue by SRXRF and FAAS after 2D-PAGE separation: preliminary results. Talanta 82:1052–1056
Lippard SJ (1994) Metals in medicine. In: Bioinorganic chemistry. University Science Books, Mill Valley, pp 505–584
López-Barea J, Gómez-Ariza JL (2006) Environmental proteomics and metallomics. Proteomics 6(Suppl 1):S51–S62
Moraes PM, Santos FA, Padilha CCF et al (2012) A preliminary and qualitative metallomics study of mercury in the muscle of fish from Amazonas, Brazil. Biol Trace Elem Res 150:195–199
Moraes PM, Santos FA, Cavecci B et al (2013) GFAAS determination of mercury in muscle samples of fish from Amazon, Brazil. Food Chem 141:2614–2617
O'Farrell PH (1975) High resolution two-dimensional electrophoresis of proteins. J Biol Chem 250:4007–4021
Pfeiffer WC, Lacerda LD, Salomons W, Malm O (1993) Environmental fate of mercury from gold mining in the Brazilian Amazon. Environ Rev 1:26–37
Rees DA, Alcolado JC (2005) Animal models of diabetes mellitus. Diabet Med 22:359–370
Santos FA, Lima PM, Neves RCF et al (2011) Metallomic study on plasma samples from Nile tilapia using SR-XRF and GFAAS after separation by 2D PAGE: initial results. Microchim Acta 173:43–49
Sun H, Chai Z-F (2010) Metallomics: an integrated science for metals in biology and medicine. Annu Reports Sect "A" Inorganic Chem 106:20–38
Szpunar J (2004) Metallomics: a new frontier in analytical chemistry. Anal Bioanal Chem 378:54–56
Van der Oost R, Beyer J, Vermeulen NPE (2003) Fish bioaccumulation and biomarkers in environmental risk assessment: a review. Environ Toxicol Pharmacol 13:57–149
Vieira JCS, Cavecci B, Queiroz JV et al (2015) Determination of the mercury fraction linked to protein of muscle and liver tissue of Tucunaré (Cichla spp.) from the Amazon region of Brazil. Arch Environ Contam Toxicol 69:422–430
Vieira JCS, Braga CP, de Oliveira G et al (2017) Identification of protein biomarkers of mercury toxicity in fish. Environ Chem Lett 15:717–724
Xiang F, Lu X, Strutt B et al (2010) NOX2 deficiency protects against streptozotocin-induced β-cell destruction and development of diabetes in mice. Diabetes 59:2603–2611

Chapter 6
The Use of Stable Isotopic Tracers in Metallomics Studies

Maria Montes-Bayón and Jörg Bettmer

Abstract Mass spectrometry represents an essential technique in Metallomics studies. It permits the identification of metal-containing molecules as part of the metallome as well as their quantification at low concentration levels. The technique becomes even more powerful in combination with the use of isotopically enriched species. Provided that they are stable, these isotopically labelled species can be easily distinguished from their natural counterparts by mass spectrometric techniques. This capability permits that these species are used for accurate and precise quantitative experiments and/or metabolic studies applying inductively coupled plasma as ionization source. In this chapter, we present the different concepts of using stable isotope tracers and isotope dilution analysis as quantification strategy. Besides some fundamental aspects, various examples from Metallomics studies, for instance, on the preparation of isotopically enriched metalloproteins and determination by isotope dilution analysis or the exploration of the biological pathways of Se species, are shown in order to demonstrate the usefulness of isotopes.

Keywords Isotopes · ICP-MS · Isotope dilution analysis · Tracer studies · Metalloproteins · Metallometabolites

Abbreviations

AQUA	Absolute quantification
CRM	Certified reference material
DNA	Deoxyribonucleic acid
DOM	Dissolved organic matter
ESI	Electrospray ionization
FAZ	Fractional absorption of zinc

M. Montes-Bayón (✉) · J. Bettmer
University of Oviedo, Department of Physical and Analytical Chemistry, Oviedo, Spain
e-mail: montesmaria@uniovi.es; bettmerjorg@uniovi.es

M. A. Z. Arruda (ed.), *Metallomics*, Advances in Experimental Medicine and Biology 1055, https://doi.org/10.1007/978-3-319-90143-5_6

GC-MS	Gas chromatography-mass spectrometry
HPLC	High-performance liquid chromatography
ICP	Inductively coupled plasma
ICP-MS	Inductively coupled plasma-mass spectrometry
IDA	Isotope dilution analysis
IPD	Isotope pattern deconvolution
LA-ICP-MS	Laser ablation-inductively coupled plasma-mass spectrometry
LC-MS	Liquid chromatography-mass spectrometry
LC-MS/MS	Liquid chromatography-tandem mass spectrometry
MALDI	Matrix-assisted laser desorption and ionization
MC-ICP-MS	Multicollector-inductively coupled plasma-mass spectrometry
MELD	Model for end-stage liver disease
MeSeCys	Se-methyl selenocysteine
MS	Mass spectrometry
NTI-MS	Negative thermal ionization-mass spectrometry
PCR	Polymerase chain reaction
RSD	Relative standard deviation
SeCys	Selenocysteine
SeMet	Selenomethionine
SOD	Superoxide dismutase
TIMS	Thermal ionization mass spectrometry

6.1 Introduction

6.1.1 The Possibilities of Using Stable Isotope Tracers in Metallomics Studies

6.1.1.1 From Nonmetals to Metal/Metalloid Tracers

The use of isotopic labels has been successfully applied for a number of years for the identification and quantification of important biomarkers in the study of biological systems. Mainly, the concept is based on the observation that chemically identical but isotopically unique chemicals will exhibit identical behavior in chromatographic and mass spectrometric analyses. Thus, molecules containing elements with different (stable) isotopic signatures that can be obtained through a number of chemical labelling and chemical label-free methods are used for this aim. Commonly, molecules containing labels of ^{13}C, ^{15}N, ^{18}O, etc. have been widely obtained and used for a number of years in metabolic experiments performed in bacteria and mammals already in the 1950s. For instance, stable isotope-labelled fatty acids and amino acids were applied to animals and in rare cases also to man. Absorption and metabolism were followed, for example, by appearance of the label in individual compartments and in metabolic end products such as water or small gaseous compounds, such as CO_2, N_2, and H_2.

The availability of mass spectrometry (MS) after the introduction of ESI and MALDI around the year 2000 inspired the design of in vivo labelling experiments with cells or whole organisms, with the aim to analyze molecular alterations in different chemical compartments. As stated in a recent review article (Lehmann 2017), for the first time, with LC-MS/MS an analytical system was available with a molecular specificity and data collecting capacity that had a chance to cope with the complexity of a cell or an organism. Within this frame, stable isotope labelling was found to be essential for quantitative LC-MS/MS analyses. This situation in combination with analytical progress in other fields (PCR, DNA sequencing, microarray technology, immunofluorescence techniques) finally laid the ground for the field of comprehensive protein analysis, named *proteomics*.

Since biomarkers take many forms as cell responses to stimuli and can be manifested during transcription, translation, and/or metabolic processing, researchers have relied upon mixed-isotope labelling coupled with MS to perform relative quantification of biomarkers between two or more biological samples. This has been applied in the field of "quantitative proteomics" either at the peptide (using the AQUA peptides) or the protein level using stable isotope-labelled intact proteins. With the advent of mass spectrometry-based proteomics, new concepts for in vivo labelling were developed tailored to detect proteome expression differences following a stimulus or a genetic manipulation.

But not only tracing or metabolic experiments have made use of stable isotopes. The use of C, H, O, or N isotopically labelled compounds as internal standards to improve the quantification of small organic molecules has been increasing since the year 2000. This has been specifically the case in clinical chemistry where quantifications with both high precision and accuracy (trueness) are required. Therefore isotope dilution in combination with GC-MS or LC-MS has been extensively applied in this field, covering quantification of biogenic amines, steroid hormones, amino acids, drugs, and drug metabolites, among others. In particular, newborn screening for inborn metabolic errors is firmly established in numerous developed countries, and it is usually conducted by isotope dilution analysis with isotopically enriched amino acids. Forensic analyses and doping control are also growing areas of isotope dilution applications.

From the experience acquired over years in the field of "organic" mass spectrometry, the extension of the work to isotopically labelled hetero-elements (defining as such, those elements besides C, H, O, and N) occurred naturally. Such development has been associated to the use of inductively coupled plasma (ICP) as ionization source for mass spectrometry. ICP-MS does not provide structural information since the molecules are disrupted into their atomic components once they reach the ionization source. Nevertheless, this, a priori, disadvantage represents the important strength of ICP-MS since it reduces the sample complexity and it has led to an innovative research approach named Metallomics. Similar to other "omics" discipline, e.g., proteomics or genomics, Metallomics is defined as the global investigation and characterization of the entirety of metal and metalloid species such as metalloproteins or other metal-containing biomolecules (*the metallome*) within a defined biological entity. Metallomics is a transdisciplinary research area with an impact on geochemistry, clinical biology and pharmacology, plant and animal physiology, and nutrition.

Thus, the isotopically labelled elemental species have been used in the field of Metallomics with two different purposes (as previously in the fields of proteomics and metabolomics): (i) as quantification tool (by applying isotope dilution analysis, IDA, as the main quantification strategy) and (ii) in the form of metabolic tracers to address the fate of labelled elements or elemental species in biological systems. As in the case of organic mass spectrometry, the use of isotopically enriched elemental species permits to have the ideal "internal standard" for quantification. The major asset of isotope dilution is its high metrological quality over the traditional methods of standard additions, internal and external calibration. Isotope dilution can be considered as an evolution of internal standard methods that are very advantageous to correct for matrix effects and (partial) analyte losses (Meija and Mester 2008). In addition, the feature of ICP-MS to provide "virtually" species-independent ionization will permit the application of other quantification strategies such as the species-unspecific isotope dilution which is an outstanding feature of this technique. Such quantification possibilities open new and extraordinary venues in the field of Metallomics studies (Heumann 2004; Bettmer 2010; García Alonso and Rodríguez González 2013). These different quantification aspects will be covered in the next section. Regarding the use of isotope labelling for stable isotope tracing experiments, several studies can be found in combination with ICP-MS as detector for metabolic tracing experiments in cell, fish, or plant models. For instance, the suitability of a food fortification procedure has been conducted by comparing zinc absorption in adults from rice porridges prepared from a hydroponically biofortified rice labelled intrinsically with ^{70}Zn with the same variety but fortified at point of use with $^{70}ZnSO_4$ to the same total zinc content. Or the differential incorporation of several isotopically labelled inorganic Se species commonly used as foodstuff enrichment sources in root and leaf extracts of ryegrass (*Lolium perenne L.*) which had been co-exposed to two labelled Se species. Most recent publications have revealed the use of isotopically enriched hetero-elements as tracers has permitted to extend the traditional speciation experiments even beyond, for instance, in the case of nanoparticles' fate in environmental and biological systems. Intrinsic labelling of metal-containing engineered nanoparticles with stable isotopes seems to be a useful tool for the highly sensitive and selective detection of these species in the environment and organisms, thus enabling tracing of their transformation, uptake, distribution, and clearance. Figure 6.1 shows, schematically, the different levels of labelling experiments that can be used for metabolic experiments.

All these aspects of the applications of isotopically labelled species in Metallomics studies will be illustrated in the next sections in order to point out the principles of the methodologies as well as a summary of the most remarkable existing literature and the unexplored possibilities for this type of applications in the growing field of Metallomics.

6.1.1.2 The Use for Quantification: Isotope Dilution Analysis (IDA)

The quantification of any compound bound to metals represents a highly challenging task in Metallomics studies. The great variety of binding forms, in which the metal can be found in biological systems, requires sophisticated strategies to

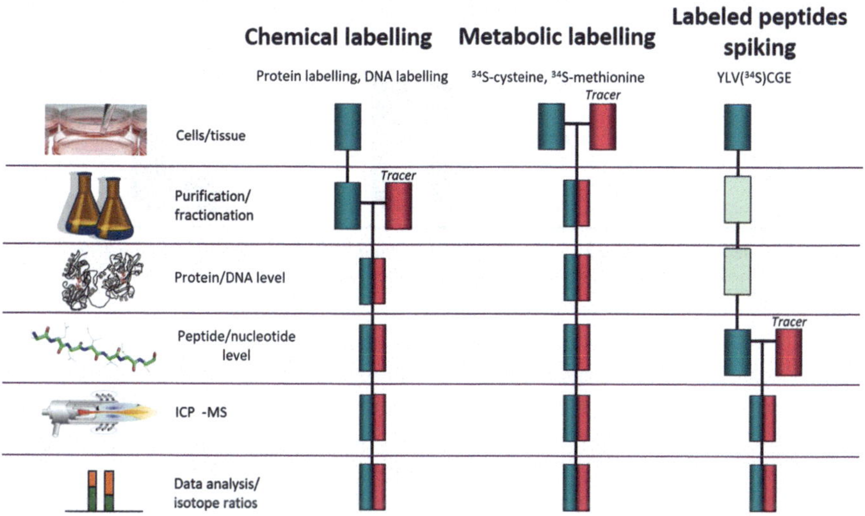

Fig. 6.1 Different possibilities of labelling to conduct Metallomics experiments: at the purification level, at the metabolic level, or at the quantification level. (Modified from Bettmer et al. 2009)

characterize them completely and finally to quantify them. Usually, not all compounds forming the metallome are quantified, but a number of relevant or selected ones. Among traditional quantification strategies, isotope dilution analysis (IDA) is considered as a method providing highly accurate results. Potential sources of error are usually well understood and can be controlled. Therefore, IDA is often named a "definitive method" (Heumann 1992).

The application of IDA of metals requires certain basic prerequisites. First of all, the element of interest needs to have at minimum two stable isotopes.[1] This excludes per se important metals from being acquirable with IDA – Na, Al, Sc, Mn, Co, As, Y, Nb, Rh, I, Cs, and Au – and the four lanthanides Pr, Tb, Ho, and Tm. All other metals can be principally accessed. Secondly, the metal of interest needs to be available in an isotopic composition different from the natural one. The most capable composition consists of the depletion of the naturally most abundant isotope and the enrichment of one isotope with naturally low abundance, so-called spike or spike solution. Virtually all metals highly enriched in one isotope are commercially available. The third requirement is that the mass spectrometric method needs to be free of interferences. With respect to ICP-MS – as probably most often used quantification technique for metals in Metallomics studies – spectral interferences on two isotopes of the element of interest have to be eliminated completely as they can influence the measurement of the isotope ratio (R) of the sought metal. This requires either collision/reaction cell instruments or those that provide an enhanced mass resolving power.

The basic principle of IDA is represented in Fig. 6.2. Briefly, a sample containing the sought metal in its natural isotopic composition with unknown concentration

[1] In a few cases, a long-lived radioactive isotope can be also used for IDA, e.g., ^{129}I or ^{60}Co.

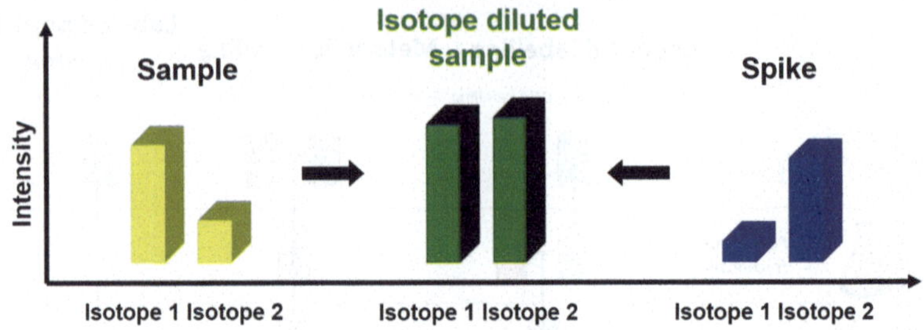

Fig. 6.2 Basic concept of isotope dilution analysis

(c_{sample}) is mixed with a spike solution of known isotopic composition (different from the natural, e.g., enriched in the naturally lowest abundant one) and concentration (c_{spike}).

After a complete mixing, the resulting isotope ratio of the mixture (R_m) is established between the one of the metal being present in the sample and the one of the added spike, reflecting the unknown concentration of the metal in the sample by the following equation (Eq. 6.1). Detailed description of the mathematical derivation of the isotope dilution equation can be found in the literature (Heumann 1992; Rodríguez-González et al. 2005).

$$c_{sample} = c_{spike} \frac{m_{spike}}{m_{sample}} \frac{M_{sample}}{M_{spike}} \left(\frac{A^2_{spike} - R_m A^1_{spike}}{R_m A^1_{sample} - A^2_{sample}} \right) \tag{6.1}$$

With

m_{sample}, m_{spike}: the mass of the sample and the mass of the added spike.

M_{sample}, M_{spike}: the atomic masses of the element in the sample and the spike, respectively.

$A^{1,2}_{sample,\ spike}$: the isotope abundances of isotope 1 and 2 in the sample and spike, respectively.

Equation 6.1 reveals that the measurement of R_m is sufficient for the determination of the unknown concentration of the analyte. It is recommended to use exactly the same mass spectrometer for the determination of the isotope ratios and the isotopic abundances in order to consider potential mass bias and detector dead time effects. Once these parameters are controlled, the isotope ratio R_m is equal to the ratio of signal intensities (I) of the isotopes 2 and 1 that are measured during the experiment. The isotope dilution analysis bears the advantages of highly accurate and precise results provided that a complete mixture between the analyte in the sample and the added spike has occurred. This can be explained by the fact that important sources of errors influencing the quality of analytical results are virtually eliminated due to the measurement of isotope ratios. After a complete isotope mixing, any analyte losses during sample preparation and measurement do not affect the quality of the results as the analyte present in the sample and the added spike are affected in the same manner. Instrumental changes in sensitivity do not influence the determination of isotope ratios. Therefore, the use of an isotopically enriched spike is considered as the perfect internal standard for the determination of a sought element. The main application field of IDA is the determination of total element concentrations, but it has also turned out to be very useful for specific compounds, e.g., elemental species. Therefore, it is not surprising that IDA has entered the field of Metallomics studies due to its ease of use and expected analytical figures of merit that permit high-quality investigations to follow a sought metal in biological systems. Two main analytical strategies have been applied to support Metallomics studies that will be explained in the following two sections: "Species-Specific IDA" and "Species-Unspecific IDA."

6.1.1.3 Species-Specific IDA

The basic idea of species-specific IDA is the conceptual application of IDA for the accurate and precise determination of a specific compound or compounds. The first examples making use of elemental mass spectrometry dated back to the late 1980s. Inorganic selenium species were separated by ion chromatography and detected offline by thermal ionization mass spectrometry (TIMS) (Heumann and Großer 1989). A few years later, the first application with online coupling of HPLC and ICP-MS appeared on the example of iodine species (Heumann et al. 1994).

In order to apply species-specific IDA, several prerequisites need to be fulfilled. Like for IDA of the elements, in general, the metal requires minimum two stable isotopes. As the sought analyte is a metal-containing compound, this compound needs to be available in an isotopic composition different to the natural one. Some low-molecular metal compounds enriched in one isotope are commercially available (e.g., methyl-Hg, butyl-Sn compounds, etc.), but usually it requires the lab-made synthesis of the sought compound. In some cases, this might be challenging because the synthesis is usually carried out at the low mg range (Demuth and Heumann 2001). In the case of metalloproteins and other metal-containing complexes, the exact identity has to be guaranteed (Deitrich et al. 2007).

Such a prepared spike, in this case species-specific, needs to be characterized in terms of isotopic composition, purity, and concentration. This spike can be then added in a known amount to a sample with the sought species with unknown concentration. After complete mixing between the spike and the species in the sample, any analyte losses, species degradation, or transformation does not influence the accuracy of the analytical result. Those processes might affect the species in both isotopic compositions to the same extent. The next step is to separate the sought species from other compounds containing the same metal as they would disturb the measurement of R_m. Most common techniques are gas and liquid chromatography and electrophoretic and field-flow fractionation techniques and can be easily coupled online to ICP-MS. In the case of slab gel electrophoresis, sample introduction into the ionization source can be realized by laser ablation. In general, transient signals are produced, e.g., as an eluting peak from a chromatographic column.

By continuous monitoring of two isotopes following peak integration, the isotope R_m can be determined and the corresponding concentration results from Eq. 6.1. Therefore, species-specific IDA denotes a specific case of total elemental determination using IDA. The main limitation is availability of isotopically enriched species. Low-molecular metal-containing species, provided that they are stable, are relatively easy to be synthesized. In the case of high-molecular species, e.g., metalloproteins, the synthesis and purification requires more sophisticated strategies. Although a huge number of metal-containing compounds are principally accessible by species-specific IDA, an alternative strategy has been provided for those applications where isotopically enriched compounds are not available.

6.1.1.4 Species-Unspecific IDA

Species-unspecific IDA, also known as post-column IDA, was developed on the basis of HPLC coupled to ICP-MS (Rottmann and Heumann 1994a, b). First examples were dedicated to the analysis of metals associated to dissolved organic material (DOM). Due to its attractiveness and universal applicability, the concept of species-unspecific IDA has entered various applications (Heumann 2004; Rodríguez-González et al. 2005; García Alonso and Rodríguez González 2013).

The basic idea consists of a separation of the species of interest, e.g., by means of liquid or gas chromatography, or electrophoresis. A solution (in the case of gas chromatography a volatile species in inert gas) is continuously added containing the element in an isotopically altered composition to the eluent of the separation unit (post-column). In contrast to species-specific IDA, this spike can have any chemical form available of the sought element and does not have to be identical to the species of interest. For instance, the determination of iron-containing proteins can be achieved by the use of an isotopically enriched Fe(III) solution by monitoring two relevant Fe isotopes (Del Castillo Busto et al. 2006). However, it has to be guaranteed that the chemical form of the spike and the sought compound, e.g., a metalloprotein, show identical or very similar transport and ionization efficiencies. Figure 6.3a represents a typical chromatogram on sulfur-containing proteins using $^{34}SO_4^{2-}$ for post-column isotope dilution (Zinn et al. 2008).[2] As the spike (in a known concentration and isotopic composition) is continuously added to the eluent of the chromatographic system, the trace for ^{34}S shows higher signal intensity than the naturally higher abundant ^{32}S (natural abundances of ^{32}S and ^{34}S are 94.93% and 4.29%, respectively). The sulfur-containing compounds eluting from the column, of course, consist of a natural isotopic composition; a fact that is reflected in the observed signal intensities (Fig. 6.3a). Out of the recorded intensity chromatogram, one can easily transfer it to an isotope ratio chromatogram[3] (Fig. 6.3b). During no specific sulfur-containing compounds elutes from the column, the

[2] Although sulfur is shown in this example, the principle is the same for any other metal suitable for species-unspecific IDA.

[3] In this specific case, the isotope ratio of ^{34}S and ^{32}S was used.

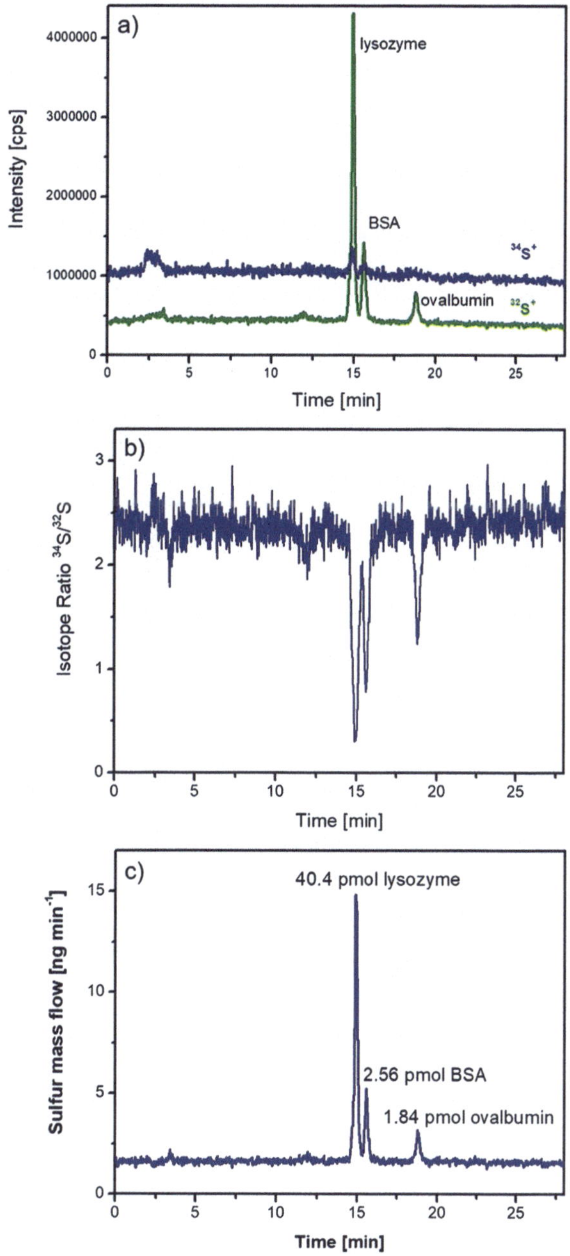

Fig. 6.3 (**a**) Intensity chromatogram of three proteins monitoring ^{32}S and ^{34}S, (**b**) $^{34}S/^{32}S$ isotope ratio chromatogram, and (**c**) resulting mass flow chromatogram of S. (Adapted from Zinn et al. 2008)

observed time-dependent isotope ratio (R_t) results from the one of the spike (R_{sp}) and the blank contribution of the mobile phase. Once a sulfur-containing compound elutes from the column, its natural isotopic composition gets mixed with that one of the spike solution (isotope mixing). This results in a change of the observed isotope ratio into the direction of the natural one (in this case $R_{nat}(^{34}S/^{32}S) = 0.0452$) depending on the present mass of the sought element in the eluting compounds (Fig. 6.3b).

In order to extract quantitative information for each compound detected in a chromatogram, the isotope ratio chromatogram needs to be converted into a so-called mass flow chromatogram (Fig. 6.3c). For this, Eq. 6.2 is applied in each point of the chromatogram.

$$\mathrm{Mf}_{\mathrm{sample}} = \mathrm{Mf}_{\mathrm{spike}} \frac{M_{\mathrm{sample}}}{M_{\mathrm{spike}}} \left(\frac{A^2_{\mathrm{spike}} - R_t A^1_{\mathrm{spike}}}{R_t A^1_{\mathrm{sample}} - A^2_{\mathrm{sample}}} \right) - \mathrm{Mf}_{\mathrm{blank}} \quad (6.2)$$

where

$\mathrm{Mf}_{\mathrm{sample}}$, $\mathrm{Mf}_{\mathrm{spike}}$, and $\mathrm{Mf}_{\mathrm{blank}}$ are mass flow of the sample, the spike, and the blank, respectively, and R_t, time-dependent isotope ratio (isotope 1/isotope 2).

With a known mass flow of the spike (it can be usually calculated from the concentration of the spike and its flow rate) and of the blank, the remaining unknown is the time-dependent isotope ratio (R_t). It is therefore continuously determined during the chromatographic run as shown in Fig. 6.3b. The mass of the sought element present in a specific compound detected is now calculated by integration of the peak, e.g., lysozyme in the chromatogram. The product of mass flow and time (equivalent to the area of the integrated peak) will result in the mass of the element in this compound that eluted from the chromatographic or electrophoretic column. In order to correlate the detected mass to the initially present concentration in the sample, the recovery rate of each compound has to be known or to be determined. The reason is that the isotope mixing, an indispensable requirement for IDA, occurs post-column. Therefore, the determination of the column recovery is absolutely needed in order to guarantee accuracy and precision of the obtained results using species-unspecific IDA (Meija and Mester 2008; Bettmer 2010).

As a spike of any chemical composition can be used in its isotopically enriched form, species-unspecific IDA can be denoted as a generic approach for the determination of metallo-compounds. Therefore, its widely spread applicability in Metallomics studies can be found on metal-containing metabolites, metalloproteins, and metalloenzymes.

6.1.1.5 Examples on Metallomics Studies Making Use of IDA

Traditionally, species-related IDA methods were proposed for low-molecular compounds of elements like Se, Hg, Pb, and Sn (Demuth and Heumann 2001; Ruiz Encinar et al. 2004; Monperrus et al. 2005; Poperechna and Heumann 2005).

These compounds can be clearly considered as forming part of the metallome in a certain biological system.

Se-containing amino acids and other Se-containing metabolites have been in the focus of extensive research. Selenium is an essential element and considered to be cancer-protective agent. In order to determine its various chemical forms (Gammelgaard et al. 2008; Fairweather-Tait et al. 2010), IDA methods, mainly with detection by ICP-MS, have gained great interest (Pedrero and Madrid 2009). Species-specific IDA was applied to the determination of selenomethionine in selenium-enriched yeast (Hinojosa Reyes et al. 2004). ^{77}Se-enriched methionine was synthesized in *S. cerevisiae* in a multiple-step procedure and used for further quantification studies of selenomethionine in yeast samples. A similar approach was used for the determination of selenomethionine in human serum (Ruiz Encinar et al. 2004). The application of a modified capillary HPLC-ICP-MS interface allowed lowering the absolute detection limits to the 75 fg level. By nano HPLC-ICP-MS, accurate quantification of selenium-containing peptides was achieved by species-unspecific IDA (Giusti et al. 2005).

An important task in Metallomics studies is the quantification of metal-containing biopolymers like metalloproteins and metalloenzymes. During the analysis, it is especially critical and important to maintain the analyte intact, e.g., any metal loss will affect the accuracy of ICP-MS detection. Sample preparation and separation methods need to be adjusted to the integrity of the sought metalloproteins. Besides non-denaturing electrophoretic techniques, the most commonly used separation techniques are size-exclusion, ion-exchange, and ion-pairing chromatography (Montes-Bayon et al. 2006). As mentioned in item 6.1.1.4, species-unspecific IDA does only require a spike solution with the isotopically enriched element but in any chemical form available and maintainable. Therefore, its application to structurally challenging biomolecules like proteins is relatively straightforward. Many studies have been shown on the determination of selenoproteins (Hinojosa Reyes et al. 2003; Jitaru et al. 2008; Kirby et al. 2008) and on the determination of metallothioneins (Schaumlöffel et al. 2002). In the latter work, the combination of capillary electrophoresis with species-unspecific ICP-MS turned out to be an important quantitative tool for the studies of various metallothionein isoforms. The simultaneous introduction of spike solutions of Cu, Zn, Cd, and S allowed the authors to quantify the different protein forms and, moreover, to determine the stoichiometric composition through the sulfur-to-metal ratio (Schaumlöffel et al. 2002).

Generally more accurate is the application of species-specific IDA. In the case of metalloproteins, this requires their synthesis in an isotopically altered form. In a few examples, the removal of the naturally present metal ions followed by the incubation with a spike solution can be sufficient. Examples are superoxide dismutase and transferrin as summarized in Table 6.1. Other metalloproteins require the use of biochemical procedures, usually involving the cloning of a certain gene for the (over)expression of the sought protein in cell cultures or *Escherichia coli*. The metalloprotein is then expressed in the presence of a spike solution of required metal (Table 6.1).

Table 6.1 Examples on metalloproteins that have been synthesized in an isotopically enriched form

Metalloprotein	Enriched isotopes	Method of synthesis	Reference
Rusticyanin	^{65}Cu	Protein was expressed cell cultures, isolated, purified, and incubated with an excess of Cu^{2+} solutions	Harrington et al. (2005)
Ferritin	^{57}Fe	Cloning and overexpressing the *Phaseolus vulgaris* ferritin gene pfe in *Escherichia coli* in the presence of $^{57}FeCl_2$	Hoppler et al. (2008)
Superoxide dismutase	^{65}Cu, ^{68}Zn	Cu and Zn as metal cofactors of the enzyme were removed under different conditions and replaced by $^{65}Cu^{2+}$ and $^{68}Zn^{2+}$	Deitrich et al. (2007)
Human transferrin	^{57}Fe	Apo-transferrin was saturated with an excess of $^{57}Fe^{3+}$	Del Castillo Busto et al. (2006)
Plastocyanin	^{63}Cu	Cloning, heterologous overexpression, purification, and reconstitution of a variant of the type 1 copper protein plastocyanin from *Synechocystis* in *Escherichia coli* in the presence of $^{63}Cu^{2+}$	Hann et al. (2006)
Hemoglobin	^{57}Fe	Preparation of ^{57}Fe-enriched protoheme and apo-hemoglobin followed by reconstitution of hemoglobin	Brauckmann et al. (2016)
SEPP1 and GPX3	^{76}Se	Incubation of *Escherichia coli* after vector cloning in the presence of ^{76}Se-methionine in *E. coli*	Konopka et al. (2016)
Carbonic anhydrase	^{67}Zn, ^{68}Zn	Incorporation into carbonic anhydrase: effects on isotope enrichment	Pessôa and Arruda (2017)

Analytically well documented is the analysis of human transferrin glycoforms in a certified reference material (CRM 470, human serum) and various serum samples (Del Castillo Busto et al. 2006). Transferrin is the main iron-transporting protein in serum, and their glycoforms deficient in carbohydrate (CDT, carbohydrate-deficient transferrin) are used as clinical biomarkers. Both IDA strategies, species-specific and -unspecific, were compared for the achievable accuracy and precision using anion-exchange chromatography. In this case, an excellent agreement was found that also revealed a quantitative recovery of the analytes from the column.

Another well-studied metalloprotein (metalloenzyme) in this context is superoxide dismutase 1 (SOD1) containing one Cu and one Zn ion in each subunit. In 2007, the first report appeared on the preparation and characterization of bovine SOD enriched in ^{65}Cu and ^{68}Zn (Deitrich et al. 2007). This spike was later used for the species-specific IDA using non-denaturing gel electrophoresis coupled to laser ablation-ICP-MS (Deitrich et al. 2010). The enriched SOD remained stable during the entire method and tested it successfully in complex matrices like bovine liver. However, detection limits were still improvable. Better detection limits could be obtained by anion-exchange chromatography coupled to ICP-MS as shown on the analysis of SOD in red blood cells (Nuevo Ordóñez et al. 2011).

Based on these and other related studies, the application of IDA methods for the analysis of metalloproteins has captured great interest by European metrological institutes. The aim is to provide measurement procedures to achieve comparable and traceable results (Swart 2013; Swart and Jakubowski 2016). An example is the extension of SOD determinations by the use of a species-specific double and triple IDA (Gleitzmann et al. 2016). The calculated mass fractions of SOD1 in human red blood cells showed excellent agreement and showed again the potential of IDA methods in the accurate and precise determination of metalloproteins and metalloenzymes.

6.1.1.6 Stable Tracers for Metallomics Experiments

Enriched stable isotopes are nowadays indispensable tools for experiments regarding the fate of minerals and trace elements in biological systems. Their use has increased rapidly since the first experiments conducted in 1960s. Today stable isotope tracers are used in various scientific areas, including Metallomics, to investigate pathways of minerals, trace elements, and heteroatom-containing biomolecules in diverse biological systems, ranging from estimating human absorption of essential elements such as iron to studies regarding the bioaccumulation of mercury in aquatic food webs or tracking the fate of nanoparticles in the environment.

Most of the isotope tracer experiments conducted nowadays are based on the previous radioisotope tracer experiments and, somehow, similar as recently revised in an interesting review article (Stürup et al. 2008). However, the use of enriched stable isotopes tracers has the important advantage of dealing with non-toxic species that can be applied in experiments involving living organisms without any toxicological concerns. On the other hand, the main drawbacks of enriched stable isotopes is that, unlike radioisotopes, they are often affected by the presence of a significant amount of the selected isotope as part of the natural abundance of the element present in the biological system under investigation. Hence, a relatively large amount of the enriched stable isotope tracer has to be administered to induce a detectable shift in the tracer isotope signature in the biological sample of interest. In addition, the sensitivity that can be achieved with regular radioactivity detectors to perceive low levels of emitting material is usually very high in comparison to what can be achieved with the regular ICP-MS systems. Considering that the tracers undergo a significant dilution on their way to the compartment or organism of interest, the experiments have to be carefully designed using an adequate concentration of isotopic tracers that permits to detect changes on the isotopic signatures of the elements of interest with conventional ICP-MS instruments.

In general terms, the modalities of label isotope experiments that can be applied for Metallomics studies can be divided into two different categories based on the experimental type of information that is intended to be extracted out of the performed experiment:

- *Qualitative isotope distribution information*: in this case, the objective of this experiment is to follow the distribution of an isotope tracer in a given system and determine the compartments or organisms into which the tracer is incorporated by measuring the variation on the measured isotope ratio with respect to the control sample.
- *Quantitative isotope information*: these experiments involve, additionally, the calculation of a mass balance of the tracer isotopes added to the given system (this is the determination of the absolute concentration of the incorporated tracer in every specific compartment).

The choice of one or the other depends on the biological process under investigation and on the information required. Indeed, qualitative isotope experiments are by their nature simpler than quantitative experiments, but they permit to monitor the fate of an element or more often a specific biomolecule containing the heteroelement of interest in a biological system. In addition, qualitative isotope distribution experiments are ideal for investigating species transformations during metabolism or the transfer of isotopically labelled compounds through a biological system. Regarding the quantitative experiments, the main consideration is that since they are mass-balanced experiments, the isotope tracer flux in and out of the system under evaluation should be estimated, and such scenario requires the use of "closed" systems for which the fluxes in and out can be controlled. This is relatively complex to achieve and, in this case, animals and humans are ideal models to conduct quantitative isotope balance experiments, as excretion can be estimated by collecting urine and feces. Potted plants are also well-established models since the uptake of nutrients and atmospheric deposition can be controlled in growth chambers and even volatile species production can be obtained with the adequate setup. The extension on the use of these well-established models out of the laboratory requires the preparation and maintenance of artificial biological systems. Such scenario turns into a very complex process involving not only addition and measurement of the isotope tracer but also the control and determination of all of the other parameters defining the biological system under investigation.

In general terms, some other considerations that have to be taken into account when designing a Metallomics experiment using isotopic tracers (qualitative and quantitative) are detailed below:

(a) The cost: the use of enriched isotopes with a low natural abundance is typically more expensive than those with a high natural abundance due to the difficulty of preparation, which inevitably increases the cost of the experiments.
(b) Uncertainty considerations in qualitative studies are less critical than in quantitative isotope experiments, and factors like mass bias, detector dead-time, etc. can often be ignored, as long as the mass spectrometric method applied is interference-free and the isotope ratios can be measured with a precision of better than 1% RSD. However, in the case of quantitative experiments, careful uncertainty measurements have to be taken into account which implies complicated mathematics to be applied in the calculation of the concentrations.

(c) The methodologies for quantitative experiments make use of advanced matrix algebra to resolve the contributions of several different enriched stable isotope tracers to the isotope pattern including isotope pattern deconvolution (IPD) using multiple linear regression analysis. A recent contribution of a transgenerational marking study where the transfer of a Sr isotope double spike (^{84}Sr and ^{86}Sr) from female spawners of the common carp (*Cyprinus carpio L.*) to the center of the otoliths of their offspring was studied by LA-MC-ICPMS refers to the requirement of a full metrological protocol for data processing during isotope pattern deconvolution of the enriched Sr isotope double spike (Zitek et al. 2014). In this example the authors include, as specific aspect, the control of the blanks (considering the presence of Kr, Sr, and Rb), the variation of the natural Sr isotopic composition in the sample, the mass bias effect, the interferences (Rb), and the total combined uncertainty.

Regarding the nature of the tracer to be used in the experiments, two different scenarios can be considered: (i) the use of pure isotopically labelled elements and (ii) the use of pure isotopically labelled elemental species. Both possibilities, with the corresponding advantages and limitations, will be briefly discussed in the following section.

6.1.1.7 Isotopically Labelled Elements

Isotopically labelled elemental species (normally minerals and trace elements) have been used in biological systems for the evaluation of their biosynthetic incorporation into the studied model organism. One of the inherent problems with enriched stable isotope labels is the lack of "isotopically pure" tracers; there will always be traces of other isotopes. In fact, most commercially available enriched stable isotopes are 95–99% pure (in the best case), and although the content is certified, the uncertainty over the certification of minor isotopes is often very large (up to 100% in some cases). Thus, a previous step for characterization of the isotopic abundances and concentration of the tracer to be used is mandatory before starting any experiment. Such process is often expensive and time-consuming. This is a problem, in particular, for quantitative isotope experiments, as it cannot then be assumed that only one (pure) isotope is added to the biological system. Thus, this factor must be taken into account, so more isotopes have to be measured and a mass balance for each isotope of the element is needed. In the same way, the uncertainty over the certification of the enriched isotopes needs to be included in the overall expanded uncertainty of the experimental results.

Biosynthetic incorporation of isotopic elemental tracers in biological systems has been conducted in a number of applications. Perhaps the studies of elemental incorporation in plants and, in lesser extent, nutritional studies have been the most successful ones. Elements like Se, Zn, and Fe have been used in the labelling of plants with enriched isotopes taken up through the roots using one or more than one of the isotopically labelled elements. Some examples of these particular applications will be given in the next sections.

6.1.1.8 Isotopically Labelled Elemental Species

In this case, all the considerations that have been documented for the isotopically labelled elements still apply although additional aspects have to be taken into account. The most troublesome fact of using isotopically labelled species is the need for having these species commercially available, with high purity (not only regarding isotopic composition but also regarding the possible presence of other species). In addition, species stability referring to the absence of modification of the structure due to other processes than the process under study needs to be taken into account. Very few isotopically labelled species are nowadays commercially available in the market to conduct this type of experiments; among them some organometallic species (no so much in use for Metallomics experiments) like tributyltin or methyl mercury can be found. In addition, organometalloid species like Se-labelled Se-methionine can be also purchased.

Apart from those, labelled elements in their different oxidation states like Cr(III)/Cr(IV), Fe(II)/(III), or Se(IV)/Se(VI) can be also obtained in a relatively easy way. If the labelled species is not commercially available, the experiment has to be initiated by its synthesis with the corresponding increment of cost and feasibility of the study. Recent experiments involve also the use of isotopically labelled metallic NPs for addressing the fate of this emerging species in biological or environmental media. In this case, many other important considerations on the characterization of starting isotopically enriched material have to be taken into account like size and shape, etc. This will be further explained in the following sections.

6.2 Applications on the Use of Stable Tracers for Metallomics

In the next sections, some relevant examples that have been selected from the literature related to the application of the use of isotopically labelled elements and elemental species will be revised.

6.2.1 Iron and Iron Species in Plant and Human Tissues

In agricultural practice, the deficiency of Fe in crops (also called Fe chlorosis) is characterized by a marked decrease in leaf chlorophyll that, in turn, results in losses in crop quality and yield; the most common cause of Fe deficiency is low soil Fe bioavailability, due to the occurrence of this metal in oxy-hydroxide insoluble forms. The efficiency of Fe(III) chelates such as Fe(III)-ethylenediamine di(o-hydroxyphenylacetic) acid (o,oEDDHA) as an Fe fertilizer is due to the remarkable stability over a wide range of pH values and the low reactivity in soils. However, this compound presents two different stereoisomers that might affect the way for its uptake by plants. Thus, a new approach to study plant Fe uptake and translocation using this compound is the use of multiple-stable isotope tracer methodologies, which also provide the possibility to carry out long-term experiments and provide relatively "closed" systems for quantitative calculations. The aim of the work was to address the uptake, movement, and distribution of Fe and ligands from racemic and meso Fe(III)-o,oEDDHA chelates, when applied simultaneously to the roots of Fe-deficient plants (Orera et al. 2010). For this aim, sugar beet plants grown in nutrient solution were treated with a mixture of racemic and meso-o,oEDDHA isomers, each one labelled with a different Fe stable isotope (^{54}Fe or ^{57}Fe). Then, the contents of the Fe stable isotopes in different plant parts were determined by ICP-MS, whereas those of the ligands (racemic and meso-o,oEDDHA) were determined by HPLC-ESI/TOFMS. The authors could established that the meso isomer appeared to be the major contributor to the exceptional efficiency of Fe(III)-o,oEDDHA to deliver Fe to plants in nutrient solution, with rates of xylem transport and total uptake about twofold higher than those found for the racemic isomer. Both isomers of the chelating agent were incorporated and distributed by plants at similar rates, in amounts one order of magnitude lower than those of Fe.

Similarly, iron uptake and accumulation in the brain of rodents was evaluated using a similar working principle. Iron deposits in the brain are a common hallmark of Alzheimer's disease and Parkinson's disease. This spurred the hypothesis that iron may play a functional role in the pathogenesis of neurodegenerative disorders through free radical damage. Short-term studies using radiotracers suggested that brain iron uptake is small as compared to other tissues in adult rodents. This led to the assumption that brain iron uptake must also be marginal in humans after brain development is complete. Thus, in these experiments, the authors applied a novel approach to determine directly the fraction of iron that was transferred over time from diet to the brain and other organs in adult rats by adding a known amount of a stable iron isotope (^{57}Fe) with drinking water to adult rats over 4 months (Chen et al. 2014). In this case the isotope ratio measurements were taken using negative thermal ionization mass spectrometry (NTI-MS) instead of ICP-MS. In addition, an enriched ^{58}Fe label was used for quantification in order to determine the amount of natural iron in the sample following isotope dilution principles. Isotopic enrichment of a tissue sample was calculated as the percentage change of the $^{57}Fe/^{56}Fe$ isotope ratio in the sample compared to natural iron. Transfer rates from diet to tissues were calculated as total moles of tracer recovered in the sample/organ relative to total moles of tracer consumed with the drinking water. From the amount of dietary iron consumed and the transfer rate, the amount of non-tracer iron taken up from the diet and transferred into the different body tissues was also calculated.

Using this approach, a lot of physiological data could be extracted in a single experiment. For instance, isotopic enrichments could be calculated and were found to range from 27.0% for blood down to 5.32% for muscle. This can be explained by considering that the blood has the highest turnover rate of all body iron pools due to constant renewal of red blood cells and hemoglobin synthesis.

In contrast to the blood, iron turnover of muscle tissue is low. The authors explain this effect due to the fact that iron in muscle is mainly present as myoglobin for oxygen storage and is only released/replaced when muscle cells are catabolized. In addition, an efficiency of iron transfer from the feed to the liver (0.132 ± 0.019%) was about ten times lower than for blood (1.48 ± 0.11%). Iron uptake of the heart and kidney was lower by three orders of magnitude as compared to the blood and even lower for the brain. However, differences between tissues were much smaller when the fraction of iron taken up from the diet was expressed relative to total tissue iron content.

The most significant finding of the conducted experiments was that iron uptake by the adult brain in a mammalian model was small as compared to other tissues but considerable when measured relative to brain iron content in normal adult rats. About 9% of brain iron could be traced back to the diet over a period of 4 months, a time span that can be compared to 7–10 human years. This became possible by using a stable iron isotope for tracing instead of short-living radiotracers and by measuring directly the fraction of iron in the brain that originates from diet.

Finally, iron labelling shows also a great potential in pharmacokinetic experiment as observed in a recent publication aimed to determine Fe-species absorption by administration of hemin to rat models. The iron in heme is in its oxidized state which is a highly unstable form. In its reduced state, named hemin, Fe is in the form of iron(III) protoporphyrin. Thus, the use of hemin to evaluate iron species absorption instead of heme is very meaningful considering the instability of the heme molecule. This molecule was labelled on ^{58}Fe and extrinsically administrated to rats intravenously to obtain the concentration-time profiles of this compound in rat plasma. This study represented a proof of concept for the methodology to be applied in further pharmaceutical applications (Zhao et al. 2014).

6.2.2 *Zinc in Biological Studies*

Over approximately the past 20 years, there has been a steadily growing body of experience and expertise in the application of zinc stable isotope techniques to investigate whole body human zinc homeostasis and physiology (Krebs and Hambidge 2001). Of greater fundamental importance, zinc has three stable isotopes for which the natural abundance is sufficiently low to allow their utilization as "tracers." These are ^{67}Zn (natural abundance 4.1%), ^{68}Zn (18.8%), and ^{70}Zn (0.6%). The availability of these three stable isotopes of zinc in low natural abundance concentration makes it possible to administer all three tracers essentially simultaneously via different routes. For instance, dual stable isotope tracer techniques to measure fractional absorption of zinc (FAZ) require only a single measurement in plasma or urine of both orally and intravenously administered zinc tracers. In this experiment, only a fraction of the oral zinc dose is enriched with the tracer, and a different zinc stable isotope is injected intravenously. Urine monitoring possibly with multiple venipunctures and/or long fecal collection periods are needed depending on the endpoints measured (Tran et al. 2015). Analysis of two stable isotopes in urine permits calculation of FAZ; the following equation was used:

$$\text{FAZ}\left(\text{urine; \%}\right) = \text{enrichment}\left(\text{oral / intravenous}\right) \times \text{dose}\left(\text{intravenous / oral}\right).$$

The following assumptions need to be taken to use the previous equation in order to use this strategy for estimation of FAZ:

(i) That both isotopes (oral and intravenous) enter the plasma at the same time.
(ii) That the isotope enrichment in plasma and urine are the same after 40 h (enough time has elapsed since isotope administration so that the intravenous and oral tracer enrichments are decaying proportionally).
(iii) That urine enrichments of both isotopes are similar for plasma.
(iv) That the extrinsic label isotope reflects that of the intrinsic zinc in the diet.

Using the multiple spiking techniques, it has been possible to establish that while the fractional absorption of zinc from a meal depends on the quantity of exogenous zinc and on specific dietary factors such as phytic acid, the fractional absorption does not appear to be dependent on the size of the rapidly exchanging pool of the host. In contrast, the quantity of endogenous zinc excreted via the intestine is positively correlated with both the amount of absorbed zinc and the zinc "status" of the host, and thus this process has an equally critical role in maintaining zinc homeostasis. Thus the challenge remains for research at the whole body level to carefully characterize zinc distribution and exchange under diverse circumstances, while research at the cellular level must elucidate the regulatory processes and the factors to which they respond.

Closely related to these studies is the evaluation and use of Zn-fortified crops as means of increasing Zn uptake in humans. Measuring zinc absorption from biofortified crops is an important first step before demonstrating their efficacy in improving zinc status. Some studies have reported zinc absorption from staple foods, including wheat, rice, beans, millet, and maize, which were genetically biofortified with zinc by plant breeding. Some of these studies used extrinsic labelling of the biofortified and control meals, assuming that the food's endogenous zinc and the extrinsic zinc label mix in a common pool in the gastrointestinal tract. However, there are several reasons why isotope exchange with native zinc, either with or without biofortification, may not be complete. The exchange may be influenced by physiologic factors, such as residence time of the meal in the stomach or small intestine, the physical state of the meal (liquid or solid), the physical state of the isotopic label, and the tightness with which the native zinc is bound to other food constituents. Thus, some recent experiments compare zinc absorption in adults from rice porridges prepared from a hydroponically biofortified rice labelled intrinsically with ^{70}Zn with zinc absorption from a control rice of the same variety and phytic acid content, fortified at point of use with $^{70}ZnSO_4$ to the same total zinc content (Brnić et al. 2016). The observed results showed that zinc absorption from rice biofortified with zinc hydroponically turned out to be similar to zinc absorption from rice fortified with zinc immediately before consumption. Such finding indicated that zinc entering plants as a result of soil fertilization could be well absorbed by humans and that biofortification by this route is likely to be an efficacious intervention strategy.

Very recent experiments have also used Zn isotopic tracers to address the fate of ZnO nanoparticles in environmental samples. Some authors suggested few years ago that stable isotopes tracing may prove valuable for monitoring nanoparticles in the field and addressing emerging research questions but very few studies have been conducted so far. As previously explained, besides the costs, there are several technical challenges in creating stable isotope-labelled nanoparticles and tracing their fate in biota including: (1) availability of suitable isotopically enriched precursor for the synthesis of the material; (2) quantitative measurement of the newly accumulated metal, distinguishing that from background concentrations, particularly while working at low exposure concentrations with essential metals like Zn; (3) detection of the tracer in the animal tissue following short-time exposures at environmentally realistic concentrations. Despite these limitations, a recent publication reports on the synthesis of isotopically unique ^{67}ZnO nanoparticles and the advantage of using these to determine whether ^{67}Zn-enriched nanoparticles could be detected in animal tissues after short dietary exposures (Dybowska et al. 2011). For this aim, freshwater snails (*L. stagnalis*) were fed with diatoms mixed with different amounts of the synthesized ^{67}ZnO nanoparticles (made by forced hydrolysis) and then fed them unlabelled food to clear their gut. The ^{67}Zn retained after gut clearance was used to define assimilation. They also fed snails diatoms labelled with ^{67}Zn to show how a tracer can improve experimental sensitivity, thereby overcoming the limitations imposed by the high natural concentrations of total Zn in animals and the environment. The obtained results revealed that a Zn concentration as low as 1 mg g^{-1} can be detected in freshwater snails exposed to about 15 mg g^{-1} of Zn when an enriched stable isotope of Zn is used as a tracer (^{67}Zn). In contrast, exposures 333 times higher were required to detect Zn uptake if no tracer was used. Thus, the enriched stable isotope technique was found to be sufficiently sensitive to determine the uptake rate of Zn in *L. stagnalis* at an exposure equivalent to the lowest concentrations of Zn that might be expected in environmental media. The authors concluded that without the tracer, uptake rates could only be determined at concentrations equivalent to some of the most contaminated Zn conditions (or much longer exposure times), reducing the feasibility of studying a range of environmentally realistic exposure conditions.

6.2.3 Selenium Experiments

Selenium has been an element widely studied also on the frame of isotopically enriched labelled studies. Selenium is both essential and toxic for mammals; the range between the two roles is narrow and not only dose dependent but also related to the chemical species present in foodstuff. First experiments were conducted already in the year 2006 by the group of Suzuki in which the metabolic pathway for the formation of Se-methyl selenocysteine (MeSeCys) was precisely traced by referring to those for selenomethionine (SeMet) and selenite by applying a tracer method involving multiple Se stable isotopes (Suzuki et al. 2006). Monomeric selenoamino acids are major forms in some selenium-accumulators, such as SeMet accumulators (grains such as wheat and mushroom) and MeSeCys-accumulators (garlic and onion). These selenium species get converted into the common intermediate selenide for the synthesis of either selenoproteins for utilization or selenosugar for excretion. Therefore, this work aimed to determine the precise mechanism underlying the transformation of MeSeCys into selenide in comparison with that of SeMet and selenite. For this aim, male Wistar rats were depleted of endogenous natural abundance selenium with a single ^{80}Se-enriched isotope, and then ^{76}Se-MeSeCys, ^{77}Se-SeMet, and ^{82}Se-selenite were orally administered simultaneously. Organs and body fluids were obtained after a number of days and subjected to speciation analysis.

MeSeCys and SeMet were detected in organs in their intact forms, suggesting that selenoamino acids are absorbed and delivered to organs in their intact forms, as in the case of general amino acids, and then transformed for utilization and excretion. However, MeSeCys and SeMet were not detected in the urine, suggesting that selenoamino acids are not excreted into urine in their intact forms. The SeMet-specific delivery and presence in the pancreas were also observed for MeSeCys at a comparable level to in the case of SeMet. As for the transformation of MeSeCys and SeMet into selenide, MeSeCys is considered to be transformed into methylselenol through the β-lyase reaction and then demethylated into selenide for the syntheses of selenoproteins for utilization and selenosugar for excretion. On the other hand, SeMet is suggested not to be transformed into methylselenol, but into selenide through the trans-selenation reaction followed by the β-lyase reaction. The authors suggest that SeMet and MeSeCys transformed into selenide through the β-lyase reaction but differently; SeMet is transformed into SeCys through the trans-selenation reaction and then into selenide through the β-lyase reaction for SeCys, while MeSeCys is transformed into methylselenol through the β-lyase reaction for MeSeCys and then through the demethylation reaction into selenide.

Similarly, a recent work on a non-hyperaccumulator plant, ryegrass (*Lolium perenne L.*), was chosen as model system for being an easy and fast to grow plant. Moreover, ryegrass is one of the most important forage varieties in terms of contribution to pasture composition, and an increase of Se accumulation was previously evidenced in greenhouse experiments, as well as nonnegative effect on plant yield, when Se was supplied as selenite or selenate forms at low concentration. In this work (Di Tullo et al. 2015), the authors used two stable isotopes, ^{77}Se and ^{82}Se, to label, respectively, selenite and selenate sources added to hydroponic growth media (double spike) in order to monitor respective bio-incorporation of these two species in roots and leaves of ryegrass. HPLC-ICP-MS was used for Se species determination after water extraction and enzymatic hydrolysis of freeze-dried samples. Through the use of isotopically enriched tracers, it was shown that inorganic selenium could be taken up and bio-transformed, i.e., converted into organic compounds and/or incorporated into Se-containing proteins. In general, total selenium measurement showed that supplementation with Se(VI) resulted in similar total Se level in whole plant compared to that enriched with Se(IV). However, allocation in plant compartments and metabolism of Se depends on the nature of initial inorganic species.

6.3 Isotopic Signatures as Intrinsic Tracers in Metallomics Studies

Isotope fractionation is the general term referring to altered isotope abundances when no isotopically labelled species has been added into the sample. It has been reported that isotope fractionation can occur in the physical and chemical transformation of many elements, including Fe, Zn, Cu, and Cd within biological processes. Therefore, variations in isotope ratios of the heavy elements could provide new insights into biochemical processes, and they have been recently investigated as potential tools for diagnosis. For instance, iron stable isotopes now begin to serve as tracers of nutrient uptake paths, processes, and efficiencies in humans. In particular, Fe isotope fractionation effects may serve as a tool to identify long-term differences in dietary Fe absorption between individuals. Iron isotope ratios have been commonly measured to study the absorption and utilization of dietary Fe in humans by applying isotopically enriched Fe as tracers. Sophisticated mass spectrometric techniques, mostly based on the use of multicollector (MC-ICP-MS), allow the measurement of small changes in Fe isotope ratios produced by natural processes. Variations in the $^{56}Fe/^{54}Fe$ isotope ratio of 1 to 2‰ can be induced in natural samples both by microbial activity and abiotic processes. Therefore, the accuracy and precision of such measurements have to be carefully addressed in this kind of experiments.

Also, all humans contain a specific iron stable isotope signature in their blood. Human blood and muscle tissue are enriched in the light iron isotope, ^{54}Fe, by 1 or 2‰ over the heavy iron isotope, ^{56}Fe, when compared to initial surveys of the human diet. Similarly, the accurate measurement of Fe isotopic fractionation has permitted to establish that men and women differed in the Fe isotope composition of their blood. This can be ascribed to different reasons, including the fact that iron losses from the body are biased in favor of the discharge of heavier Fe isotopes and that light Fe isotopes are preferentially absorbed in the intestine (Walczyk and von Blanckenburg 2002).

Similar studies have been also conducted in plants. The isotope fractionation associated with the growth of higher plants is substantial and also generates systematic plant-specific patterns. Some of the work published in this regard hypothesizes that these patterns mirror the two different strategies that plants have developed to incorporate iron from the soil: reduction of Fe(III) in soils by plants results in the uptake of iron, which is depleted in ^{56}Fe relative to ^{54}Fe when compared to the available Fe in soils; complexation with siderophores by plants results in the uptake of iron that is heavier than that in soils. Furthermore, younger parts of strategy I plants get increasingly depleted in heavy isotopes as the plant grows, while strategy II plants incorporate nearly the same isotope composition throughout. These results point to an entirely different translocation mechanisms between strategy I and II plants. Thus, the conclusion is that plant metabolism represents an important cause of isotopic variation in the biogeochemical cycling of Fe (Guelke and von Blanckenburg 2007).

Besides iron, isotopic fractionation of copper (Cu) and zinc (Zn) during plant uptake has been also reported. The results show that the fractionation patterns of these two micronutrients are decoupled during the transport from nutrient solution to the root. In roots, an enrichment of the heavier isotopes for Zn has been found, but an enrichment of isotopically light Cu was simultaneously found, suggesting a reduction of Cu(II) possibly at the surfaces of the root cell plasma membranes. This observation applies for both graminaceous and non-graminaceous species and confirms that reduction is a predominant and ubiquitous mechanism for the acquisition of Cu into plants similar to the mechanism for the acquisition of iron (Fe) by the strategy I plant species (Jouvin et al. 2012).

Recent studies in mammals reveal also variations in Cu, Fe, and Zn natural isotope compositions in organs, body fluids, diets, and feces of mice and sheep. Large and systematic isotope variability was observed, notably in the δ^{66}Zn in the liver and δ^{65}Cu in kidneys, but significant differences exist between mice, sheep, and humans, especially in the δ^{66}Zn value of blood. The results are interpreted with reference to current knowledge about the fact that the light isotopes preferentially fractionate into "softer" bonds involving sulfur such as cysteine and glutathione, whereas heavy isotopes fractionate into "harder" bonds involving nitrogen (histidine) and even more oxygen, notably hydroxides, phosphates, and carbonates. Bonds involving the reduced forms Cu^{+} and Fe^{2+} are enriched in the light isotopes relative to bonds involving the oxidized Cu^{2+} and Fe^{3+} forms. The isotopically heavy Cu in the kidneys may reflect isotope fractionation during redox processes and may be relevant to ascorbate degradation into oxalate (Balter et al. 2013).

Very recent publications try to correlate isotopic fractionation results with certain disease conditions, in particular for the diagnosis and prognosis of liver cirrhosis. Significant differences in the isotopic composition of Cu were observed between the reference population and the patients. A wide spread in δ^{65}Cu was observed within the cirrhosis population and could be linked to the severity of the disease. Patients with end-stage liver disease showed a significantly lighter serum Cu isotopic composition. Many clinical parameters used for the diagnosis and monitoring of liver diseases, i.e., the levels of aspartate aminotransferase, De Ritis ratio, prothrombin and international normalized ratio, albumin, bilirubin, Na, and C-reactive protein, correlated well with the δ^{65}Cu values, as did the ceruloplasmin level and the ceruloplasmin/Cu concentration ratio. The isotopic composition of serum Cu appeared to reveal the synthetic and hepatocellular function of the liver synergistically with inflammation and fluid retention in the cohort studied. A relevant relationship was also observed between δ^{65}Cu and scores of mortality risk, such as the model for end-stage liver disease (MELD). Thus, the isotopic composition of serum Cu is suggested as potential new approach for the prognosis of liver disease, and although further investigation is required, for evaluation of the mortality risk in end-stage liver disease and prioritization of liver transplants (Costas-Rodríguez et al. 2015). In this vein, a decrease of $\delta^{65}Cu$ in the serum of cancer patients has been also assigned to the extensive oxidative chelation of copper by cytosolic lactate. The potential of Cu isotope variability as a new diagnostic tool for breast and colorectal cancer seems strong (Télouk et al. 2015). Therefore, the measurement of isotope fractionation of essential elements in body fluids or tissues seems to hide an extraordinary potential for Metallomics experiments in human metabolism and diagnosis.

References

Balter V, Lamboux A, Zazzo A, Télouk P, Leverrier Y, Marvel J, Moloney AP, Monahan FJ, Schmidt O, Albarède F (2013) Contrasting Cu, Fe, and Zn isotopic patterns in organs and body fluids of mice and sheep, with emphasis on cellular fractionation. Metallomics 5:1470–1482

Bettmer J (2010) Application of isotope dilution ICP-MS techniques to quantitative proteomics. Anal Bioanal Chem 397:3495–3502

Bettmer J, Montes Bayón M, Ruiz Encinar J, Fernández Sánchez ML, Fernández de la Campa MR, Sanz Medel A (2009) The emerging role of ICP-MS in proteomic analysis. J Proteome 72:989–1005

Brauckmann C, Frank C, Schulze D, Kaiser P, Stosch R, Swart C (2016) Preparation and characterisation of an ^{57}Fe enriched haemoglobin spike material for species specific isotope dilution mass spectrometry. J Anal At Spectrom 31:1846–1857

Brnić M, Wegmüller R, Melse-Boonstra A, Stomph T, Zeder C, Tay FM, Hurrell RF (2016) Zinc absorption by adults is similar from intrinsically labeled zinc-biofortified rice and from rice fortified with labeled zinc sulfate. J Nutr 146:76–80

Chen JH, Singh N, Taya H, Walczyk T (2014) Imbalance of iron influx and efflux causes brain iron accumulation over time in the healthy adult rat. Metallomics 6:1417–1426

Costas-Rodríguez M, Anoshkina Y, Lauwens S, Van Vlierberghe H, Delanghe J, Vanhaecke F (2015) Isotopic analysis of Cu in blood serum by multi-collector ICP-mass spectrometry: a new approach for the diagnosis and prognosis of liver cirrhosis? Metallomics 7:491–498

Deitrich C, Raab A, Pioselli B, Thomas-Oates JE, Feldmann J (2007) Chemical preparation of an isotopically enriched superoxide dismutase and its characterization as a standard for species-specific isotope dilution analysis. Anal Chem 79:8381–8390

Deitrich CL, Braukmann S, Raab A, Munro C, Pioselli B, Krupp EM, Thomas-Oates JE, Feldmann J (2010) Absolute quantification of superoxide dismutase (SOD) using species-specific isotope dilution analysis. Anal Bioanal Chem 397:3515–3524

Del Castillo Busto ME, Montes-Bayón M, Sanz-Medel A (2006) Accurate determination of human serum transferrin isoforms: exploring metal-specific isotope dilution analysis as a quantitative proteomic tool. Anal Chem 82:8218–8226

Demuth N, Heumann KG (2001) Validation of methylmercury determinations in aquatic systems by alkyl derivatization methods for GC analysis using ICP-IDMS. Anal Chem 73:4020–4027

Di Tullo P, Versini A, Bueno M, Le Hécho I, Thiry Y, Biron P, Castrec-Rouelle M, Pannier F (2015) Stable isotope tracing: a powerful tool for selenium speciation and metabolic studies in non-hyperaccumulator plants (ryegrass *Lolium perenne L.*). Anal Bioanal Chem 407:9029–9042

Dybowska AD, Croteau M-N, Misra SK, Berhanu D, Luoma SN, Christian P, O'Brien P, Valsami-Jones E (2011) Synthesis of isotopically modified ZnO nanoparticles and their potential as nanotoxicity tracers. Environ Pollut 159:266–273

Fairweather-Tait SJ, Collings R, Hurst R (2010) Selenium bioavailability: current knowledge and future research requirements. Am J Clin Nutr 91:1484S–1491S

Gammelgaard B, Gabel-Jensen C, Stürup S, Hansen HR (2008) Complementary use of molecular and element-specific mass spectrometry for identification of selenium compounds related to human selenium metabolism. Anal Bioanal Chem 390:1691–1706

García Alonso JI, Rodríguez González P (2013) Isotope dilution mass spectrometry. RSC publishing, Cambridge, pp 12–108

Giusti P, Schaumlöffel D, Ruiz Encinar J, Szpunar J (2005) Interfacing reversed-phase nanoHPLC with ICP-MS and on-line isotope dilution analysis for the accurate quantification of selenium-containing peptides in protein tryptic digests. J Anal At Spectrom 20:1101–1107

Gleitzmann J, Raab A, Schulze D, Wätzig H, Feldmann J, Swart C (2016) Accurate and precise quantification of Cu,Zn-SOD in human red blood cells using species-specific double and triple IDMS. J Anal At Spectrom 31:1922–1928

Guelke M, von Blanckenburg F (2007) Fractionation of stable iron isotopes in higher plants. Environ Sci Technol 41:1896–1901

Hann S, Obinger C, Stingeder G, Paumann M, Furtmüller PG, Koellensperger G (2006) Studying metal integration in native and recombinant copper proteins by hyphenated ICP-DRC-MS and ESI-TOF-MS capabilities and limitations of the complementary techniques. J Anal At Spectrom 21:1224–1231

Harrington CF, Vidler DS, Watts MJ, Hall JF (2005) Potential for using isotopically altered metalloproteins in species-specific isotope dilution analysis of proteins by HPLC coupled to inductively coupled plasma mass spectrometry. Anal Chem 77:4034–4041

Heumann KG (1992) Isotope dilution mass spectrometry (IDMS) of the elements. Mass Spectrom Rev 11:41–67

Heumann KG (2004) Isotope-dilution ICP-MS for trace element determination and speciation: from a reference method to a routine method? Anal Bioanal Chem 378:318–329

Heumann KG, Großer R (1989) Negative thermal ionization mass spectrometry of selenium. Part 2. Selenite and selenate species determination in ground waters with isotope dilution technique. Fresenius Z. Anal Chem 332:880–883

Heumann KG, Rottmann L, Vogl J (1994) Elemental speciation with liquid chromatography–inductively coupled plasma isotope dilution mass spectrometry. J Anal At Spectrom 9:1351–1355

Hinojosa Reyes L, Marchante-Gayón JM, García Alonso JI, Sanz-Medel A (2003) Quantitative speciation of selenium in human serum by affinity chromatography coupled to post-column isotope dilution analysis ICP-MS. J Anal At Spectrom 18:1210–1216

Hinojosa Reyes L, Moreno Sanz F, Herrero Espílez P, Marchante-Gayón JM, García Alonso JI, Sanz-Medel A (2004) Biosynthesis of isotopically enriched selenomethionine: application to its accurate determination in selenium-enriched yeast by isotope dilution analysis-HPLC-ICP-MS. J Anal At Spectrom 19:1230–1235

Hoppler M, Meile L, Walczyk T (2008) Biosynthesis, isolation and characterization of ^{57}Fe-enriched *Phaseolus vulgaris* ferritin after heterologous expression in *Escherichia coli*. Anal Bioanal Chem 390:53–59

Jitaru P, Cozzi P, Gambaro A, Cescon P, Barbante C (2008) Simultaneous speciation analysis of glutathione peroxidase, selenoprotein P and selenoalbumin in human serum by tandem anion exchange-affinity HPLC and on-line isotope dilution ICP-quadrupole MS. Anal Bioanal Chem 391:661–669

Jouvin D, Weiss DJ, Mason TFM, Bravin MN, Louvat P, Zhao F, Ferec F, Hinsinger P, Benedetti MF (2012) Stable isotopes of Cu and Zn in higher plants: evidence for Cu reduction at the root surface and two conceptual models for isotopic fractionation processes. Environ Sci Technol 46:2652–2660

Kirby JK, Lyons GH, Karkkainen MP (2008) Selenium speciation and bioavailability in biofortified products using species-unspecific isotope dilution and reverse phase ion pairing-inductively coupled plasma-mass spectrometry. J Agri Food Chem 56:1772–1779

Konopka A, Winter D, Konopka W, del Castillo Busto ME, Nunez S, Goenaga-Infante H, Fisicaro P, Lehmann WD (2016) [Sec-to-cys]selenoprotein - a novel type of recombinant, full-length selenoprotein standard for quantitative proteomics. J Anal At Spectrom 31:1929–1938
Krebs NE, Hambidge KM (2001) Zinc metabolism and homeostasis: the application of tracer techniques to human zinc physiology. Biometals 14:397–412
Lehmann WD (2017) A timeline of stable isotopes and mass spectrometry in the life sciences. Mass Spectrom Rev 36:58–85
Meija J, Mester Z (2008) Paradigms in isotope dilution mass spectrometry for elemental speciation analysis. Anal Chim Acta 607:115–125
Monperrus M, Tessier E, Veschambre S, Amouroux D, Donard O (2005) Simultaneous speciation of mercury and butyltin compounds in natural waters and snow by propylation and species-specific isotope dilution mass spectrometry analysis. Anal Bioanal Chem 381:854–862
Montes-Bayon M, Pröfrock D, Sanz-Medel A, Prange A (2006) Direct comparison of capillary electrophoresis and capillary liquid chromatography hyphenated to collision-cell inductively coupled plasma mass spectrometry for the investigation of Cd-, Cu- and Zn-containing metalloproteins. J Chromatogr A 1114:138–144
Nuevo Ordóñez Y, Deitrich CL, Montes-Bayón M, Blanco-González E, Feldmann J, Sanz-Medel A (2011) Species specific isotope dilution versus internal standardization strategies for the determination of Cu, Zn-superoxide dismutase in red blood cells. J Anal At Spectrom 26:150–155
Orera I, Rodríguez-Castrillo JA, Moldovan M, García-Alonso JI, Abadia A, Abadia J, Álvarez-Fernández A (2010) Using a dual-stable isotope tracer method to study the uptake, xylem transport and distribution of Fe and its chelating agent from stereoisomers of an Fe(III)-chelate used as fertilizer in Fe-deficient strategy I plants. Metallomics 2:646–657
Pedrero Z, Madrid Y (2009) Novel approaches for selenium speciation in foodstuffs and biological specimens: a review. Anal Chim Acta 634:135–152
Pessôa GS, Arruda MAZ (2017) Incorporation of ^{67}Zn and ^{68}Zn into carbonic anhydrase: effects on isotope enrichment and enzymatic aspects. J Anal At Spectrom 32:1473–1480
Poperechna N, Heumann KG (2005) Species-specific GC/ICP-IDMS for trimethyllead determinations in biological and environmental samples. Anal Chem 77:511–516
Rodríguez-González P, Marchante-Gayón JM, García Alonso JI, Sanz-Medel A (2005) Isotope dilution analysis for elemental speciation: a tutorial review. Spectrochim Acta B: At Spectrom 60:151–207
Rottmann L, Heumann KG (1994a) Development of an on-line isotope dilution technique with HPLC/ICP-MS for the accurate determination of elemental species. Fresen J Anal Chem 350:221–227
Rottmann L, Heumann KG (1994b) Determination of heavy metal interactions with dissolved organic materials in natural aquatic systems by coupling a high-performance liquid chromatography system with an inductively coupled plasma mass spectrometer. Anal Chem 66:3709–3715
Ruiz Encinar J, Schaumlöffel D, Ogra Y, Lobinski R (2004) Determination of selenomethionine and selenocysteine in human serum using speciated isotope dilution-capillary HPLC–inductively coupled plasma collision cell mass spectrometry. Anal Chem 76:6635–6642
Schaumlöffel D, Prange A, Marx G, Heumann KG, Brätter P (2002) Characterization and quantification of metallothionein isoforms by capillary electrophoresis–inductively coupled plasma-isotope-dilution mass spectrometry. Anal Bioanal Chem 372:155–163
Stürup S, Hansen HR, Gammelgaard B (2008) Application of enriched stable isotopes as tracers in biological systems: a critical review. Anal Bioanal Chem 390:541–554
Suzuki KT, Doi C, Suzuki N (2006) Metabolism of ^{76}Se-methylselenocysteine compared with that of ^{77}Se-selenomethionine and ^{82}Se-selenite. Toxicol Appl Pharmacol 217:185–195
Swart C (2013) Metrology for metalloproteins-where are we now, where are we heading? Anal Bioanal Chem 405:5697–5723
Swart C, Jakubowski N (2016) Update on the status of metrology for metalloproteins. J Anal At Spectrom 31:1756–1765

Télouk P, Puisieux A, Fujii T, Balter V, Bondanese VP, Morel A-P, Clapisson G, Lamboux A, Albarede F (2015) Copper isotope effect in serum of cancer patients. A pilot study. Metallomics 7:299–308

Tran CD, Gopalsamy GL, Mortimer EK, Young GP (2015) The potential for zinc stable isotope techniques and modelling to determine optimal zinc supplementation. Nutrients 7:4271–4295

Walczyk T, von Blanckenburg F (2002) Natural iron isotope variations in human blood. Science 295:2065–2066

Zhao D, Zhang Y, Wang Y, Xu C, Dong C, Li C, Ren S, Zhang W, Lu Y, Dai Y, Chen X (2014) Pharmacokinetics study of hemin in rats by applying ^{58}Fe-extrinsically labeling techniques in combination with ICP-MS method. J Pharm Biomed Anal 88:331–336

Zinn N, Krüger R, Leonard P, Bettmer J (2008) μLC coupled to ICP-SFMS with post-column isotope dilution analysis of sulfur for absolute protein quantification. Anal Bioanal Chem 391:537–543

Zitek A, Irrgeher J, Cervicek M, Horsky M, Kletzl M, Weismann T, Prohaska T (2014) Individual-specific transgenerational marking of common carp, *Cyprinus carpio L.*, using ^{86}Sr/^{84}Sr double spikes. Mar Freshwater Res 65:978–986

Chapter 7
Bioimaging Metallomics

Valderi Luiz Dressler, Edson Irineu Müller, and Dirce Pozebon

Abstract This chapter focuses on bioimaging in metallomics, which involves metal and metalloids distribution in animal tissues. It starts with laser ablation-inductively coupled plasma-mass spectrometry followed by secondary ion mass spectrometry, synchrotron-based X-ray fluorescence, and electron microscopy, including transmission electron microscopy, scanning electron microscopy, and energy-dispersive X-ray spectroscopy. The basic principles of these techniques and their application for qualitative and quantitative imaging of elements are presented. Sample preparation for bioimaging is briefly discussed. The usefulness of element bioimaging is demonstrated for cells and several animal tissues, including the brain, liver, hair, eye, teeth, and bone. As such, this chapter addresses the state of the art in bioimaging metallomics.

Keywords Metallomics · Bioimaging · LA-ICP-MS · SIMS · SRXRF · SEM · TEM · EDX

Abbreviations

2D	Two dimensional
2D-PAGE	Two-dimensional polyacrylamide gel electrophoresis
3D	Three dimensional
AD	Alzheimer disease
AmD	Amyloid deposits
ATP	Adenosine triphosphate
BNCT	Boron neutron capture therapy
BSA	Bovine serum albumin

V. L. Dressler (✉) · E. I. Müller
Universidade Federal de Santa Maria, Santa Maria, Rio Grande do Sul, Brazil

D. Pozebon
Universidade Federal do Rio Grande do Sul, Porto Alegre, Rio Grande do Sul, Brazil
e-mail: dircepoz@iq.ufrgs.br

M. A. Z. Arruda (ed.), *Metallomics*, Advances in Experimental Medicine and Biology 1055, https://doi.org/10.1007/978-3-319-90143-5_7

CCD	Charge-coupled device
CRM	Certified reference material
CT	Computed tomography
Da	Dalton
DNA	Deoxyribonucleic acid
DOTA	Tetraazacyclododecane-1,4,7,10-tetraacetic acid
DTPABMA	2-[bis[2-[carboxymethyl-[2-(methylamino)-2-oxoethyl]amino] ethyl]amino]acetic acid
EDX	Energy-dispersive X-ray spectroscopy
EEL	Energy loss
EELS	Electron energy loss spectroscopy
EF	Energy-filtered
EFTEM	Energy-filtered transmission electron microscopy
EM	Electron microscopy
ESI-MS	Electrospray-mass spectrometry
FEG	Field emission gun
FWHM	Full width at half maximum
GE	Gel electrophoresis
HPF	High-pressure freezing
ICP	Inductively coupled plasma
ICP-MS	Inductively coupled plasma mass spectrometry
ID	Isotope dilution
IEF-GE	Isoelectric focusing gel electrophoresis
IPG	Immobilized pH gradient strips
IR	Infrared
IS	Internal standard
LA	Laser ablation
LA-ICP-MS	Laser ablation-inductively coupled plasma mass spectrometry
LA-MS	Laser ablation mass spectrometry
LMIG	Liquid-metal ion gun
LOD	Limit of detection
MC-ICP-MS	Multicollector-inductively coupled plasma mass spectrometry
MeCAT	Metal-coded affinity tag
METTEM	Metal-tagging transmission electron microscopy
MLEM	Maximum likelihood expectation maximization
MRI	Magnetic resonance image
MSI	Mass spectrometry imaging
ND-IEF-GE	Non-denaturing isoelectric focusing gel electrophoresis
NINA	Nanoimaging and nanoanalysis
NPs	Nanoparticles
ODU	Optical density unit
PAGE	Polyacrylamide gel electrophoresis
QDs	Quantum dots
QMS	Quadrupole mass spectrometer

RAPTA	Ru-arene bound to 1,3,5-triaza-7-phosphotricyclo-[3.3.1.1] decane
REEs	Rare earth elements
ROI	Regions of interest
RSF	Relative sensitivity factor
SDS-PAGE	Sodium dodecyl sulphate-polyacrylamide gel electrophoresis
SEC-ICP-MS	Size exclusion chromatography-inductively coupled plasma mass spectrometry
SEM	Scanning electron microscopy
SF-ICP-MS	Sector field-inductively coupled plasma mass spectrometry
SIMS	Secondary ionization mass spectrometry
SN-ICP-MS	Solution nebulization inductively coupled plasma mass spectrometry
SNR	Signal-to-noise ratio
SOD	Superoxide dismutase
SRXRF	Synchrotron X-ray fluorescence
STEM	Scanning transmission electron microscopy
STEM-EELS	Scanning transmission electron microscopy-electron energy loss spectroscopy
TEM	Transmission electron microscopy
ToF	Time of flight
ToF-ICP-MS	Time of flight-inductively coupled plasma mass spectrometry
ToF-SIMS	Time of flight-secondary ionization mass spectrometry
UV	Ultraviolet
UHV	Ultra high vacuum
VSOP	Very small iron oxide particles
WD	Wilson's disease
WDS	Wavelength-dispersive spectrometry
XAFS	X-ray absorption fine structure
XANES	X-ray absorption near edge structure spectroscopy
XAS	X-ray absorption spectroscopy
XRF	X-ray fluorescence

7.1 Introduction

Essential and nonessential elements are present in a broad range of concentrations in the human body. Issues related to function, speciation, and interactions of the elements have been investigated, and many discoveries have occurred thanks to the development of sensitive and selective analytical techniques for the determination of elements and molecules in different specimens. In this context, laser ablation-inductively coupled plasma-mass spectrometry (LA-ICP-MS), secondary ion mass spectrometry (SIMS), transmission electron microscopy (TEM), scanning electron

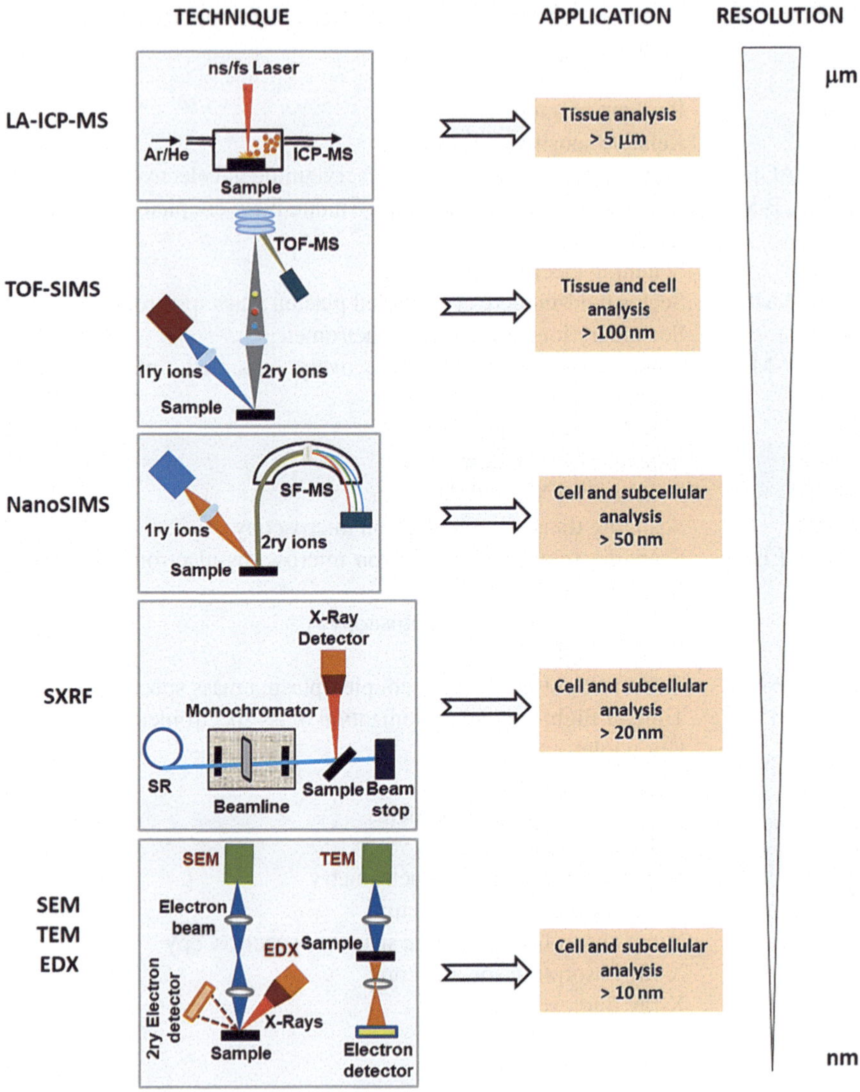

Fig. 7.1 Principles and characteristics of techniques for bioimaging metallomics

microscopy (SEM), energy-dispersive X-ray spectroscopy (EDX) and synchrotron-based X-ray fluorescence (SXRF) have played an important role. These techniques allow for in situ determination of elements in biological tissues and cells (bioimaging) with high spatial resolution. These techniques are complementary in metallomics studies and are excellent tools to investigate biochemical functions, biosorption, and bioaccumulation processes of metals at the cellular and subcellular levels. The main characteristics and applications of these techniques in bioimaging metallomics are presented in this chapter. The principles of these techniques are schematized in Fig. 7.1.

7.2 Laser Ablation-Inductively Coupled Plasma-Mass Spectrometry

Inductively coupled plasma–mass spectrometry (ICP-MS) is well known as a multi-elemental and quantitative analysis technique with large linear response, isotopic analysis capability, and high sensitivity, with limit of detection (LOD) typically in the ng L^{-1} level. The ICP-ion source is versatile to associate with different sample introduction systems, including laser ablation (LA). Laser ablation-inductively coupled plasma-mass spectrometry (LA-ICP-MS) provides direct analysis of solids where the ablated material is carried to the ICP by a gas stream, usually Ar or Ar mixed with He or N_2. The carrier gas passes through the LA cell that is connected to the injector tube in the plasma torch by a transfer line, typically a Tygon or polytetrafluoroethylene (100 cm length × 0.6 cm internal diameter) tube. Once the sample vapour is into the plasma, the analyte is ionized and then detected.

In LA-ICP-MS, the plasma works on dry conditions that reduce interference by polyatomic ions in comparison to wet plasma. However, this interference still exists for a number of biologically relevant elements (isotopes), caused by major elements in the sample, plasma gas (Ar), and gas from the atmosphere, namely, CO_2, N_2, and O_2 (Lear et al. 2012). These interferences can be reduced or eliminated by employing quadrupole instruments (QMS) equipped with dynamic or collision reaction cells (Lear et al. 2012; Austin et al. 2009; Reifschneider et al. 2013a), sector field ICP-MS (SF-ICP-MS) and multicollector ICP-MS (MC-ICP-MS) instruments (Zoriy et al. 2008; Becker et al. 2007).

Laser ablation enables in situ analysis with spatial resolution normally at 10 to 200 μm for nanosecond (ns) laser. Higher spatial resolution (in nm) is achieved using a femtosecond (fs) laser (Fernandez et al. 2007), with low sample uptake per laser shot (fg to pg), making LA-ICP-MS a quasi-nondestructive technique. Bioimaging (element distribution in a biological sample) by means of LA-ICP-MS was introduced by Wang in 1994 and revisited by Kindness in 2003 (Kindness et al. 2003). Subsequently, bioimaging using LA-ICP-MS has grown substantially with the pioneering effort of Becker (see Becker et al. (2014) and others cited therein).

LA-ICP-MS has been widely applied for qualitative and quantitative imaging. Although bioimages obtained by LA-ICP-MS do not have the fine spatial resolution of scanning electron microscopy (SEM), transmission electron microscopy (TEM), secondary ion mass spectrometry (SIMS), or X-ray microanalysis (μ-XRF), LA-ICP-MS is highly sensitive and can be used for isotopic analysis. The technique enables multiple elemental images without the use of radioactive labelling and can be combined with immunohistochemistry, isoelectric focusing electrophoresis, and polyacrylamide gel electrophoresis (PAGE) to measure element and protein distributions in biological tissues (Pozebon et al. 2014).

7.2.1 Instrumentation

A laser ablation system consists of a high-power laser source, an ablation chamber, and a CCD camera.

Lasers with wavelengths from infrared (IR) to ultraviolet (UV) have been investigated for sample ablation and hyphenation with ICP-MS (Gunther and Hattendorf 2005). A trend to change from IR (1064 nm) and visible (694 nm and 532 nm) to UV wavelengths (266 nm, 213 nm, and 193 nm) has occurred; shorter wavelength lasers have demonstrated better ablation characteristics for analyte detection by ICP-MS.

Elemental fractionation (sum of all non-stoichiometric effects during the ablation process, aerosol transport, and ionization in the ICP source) is one of the main difficulties of obtaining quantitative results in LA-ICP-MS. Laser wavelength, laser pulse duration (fs to ns), laser power, laser fluence, and spot size are parameters that influence fractionation. In general, elemental fractionation is reduced with the reduction of laser wavelength and pulse duration due to reduced thermal alteration of the sample material. Claverie et al. (2009) investigated the aerosol produced by fs laser ablation of polyacrylamide gels for Se detection in selenoproteins. Ablation of narrow lanes at high repetition rates showed that up to 77% of particles were below 1 μm and the transport efficiency of particles coming from the deepest ablation was lower. Laser fluence was not found to influence particle size distribution but allowed for an increase in the amount of particles ablated.

The LA cell must accommodate the sample in order to avoid the entrance of air and losses of ablated aerosol. Laser ablation chambers with different geometry have been proposed, with internal volumes in the range of 3 to 100 cm^3. The volume and geometry of the ablation chamber and the transfer line influence sample transport efficiency and signal profile because they affect the dispersion of the sample vapour and, consequently, the resolution of the signal and magnitude of the signal-to-background ratio. The influence of the sample characteristics must also be considered, and manufacturers of LA systems provide cells to accommodate samples of different sizes, although standard ablation chambers can accommodate samples of 2 to 3 cm in diameter.

Particle dispersion and long washout time of the ablation chamber can worsen image resolution. To minimize the aerosol dispersion and the limitation of the scanning rate of mass analysers, a small-inner-volume LA cell was developed (Wang et al. 2013). The ablation chamber was based on a constant laminar flow and well-controlled delivery of the laser-ablated aerosol into the transport system, leading to minimized tailing of the aerosol washout and improving separation of the signals. By employing such an LA cell and SF-ICP-MS instrument, high-resolution imaging of a cell biomarker (tagged with Ho) in breast cancer tissue with a subcellular resolution (1 μm) was achieved (Wang et al. 2013). Van Malderen et al. (2016a) published a review covering developments in the design of rapid response ablation chambers for LA-ICP-MS and their impact on bioimaging applications. They focused on LA cells capable of rapid pulse response for a single shot <10 ms (1% full width at half maximum [FWHM]), which is needed for rapid bioimaging analysis and subcellular mapping.

Refrigerated ablation chambers have been developed for biological tissue analysis (Konz et al. 2014a; Zoriy et al. 2005; Feldmann et al. 2002), where the sample (frozen sections, usually) can be analysed directly, allowing for preservation of its integrity and better reproducibility, which is very important for obtaining bioimages without drift or blurring. Hamilton et al. (2016) observed that the Na, P, and Mn distribution images in vegetal and animal tissues ablated at -4 °C were better than those at room temperature. The better resolution was attributed to lower water load of the cooled chamber and/or less energy dissipation/heating effects into the laser ablation spot vicinity, without side effects on the ablation process/crater formation. Bioimages of ^{63}Cu, ^{56}Fe, and ^{64}Zn in native eye tissue at -5 °C appeared to be less blurred, with higher sample integrity (Konz et al. 2014b).

Full ionization of ablated particles and efficient transport of ions to the mass analyser limit overall sensitivity (Wang et al. 2013) even for fast-washout LA cells. For higher multiplexed applications at subcellular resolution, simultaneous isotope detection, analysis speed, and sensitivity are required. Time-of-flight (ToF) mass analysers provide fast and quasi-simultaneous real-time full-elemental-mass range isotope detection, being quite suitable to handle short transient signals, despite the worse sensitivity of ToF that is one order of magnitude lower in comparison to quadrupole mass analysers. This approach is appropriate in multiprotein imaging at a subcellular scale, similar to mass cytometry (Giesen et al. 2014).

Mass cytometry was proposed by Bandura et al. (2009). They developed a cytometry-ToF-ICP-MS instrument (CyToF system) that can detect several isotopes simultaneously at a high mass-spectrum acquisition frequency. It measures the abundance of lanthanide elements tagged to antibodies, and amplification is achieved by a polymeric metal-chelating reagent or metal nanoparticles. Imaging mass cytometry – an approach that combined CyToF, immunocytochemistry, and immunohistochemistry with a high-resolution LA system and a low-dispersion LA chamber – to image 32 proteins and their modifications simultaneously at a resolution of 1 μm was described by Giesen et al. (2014). Human formalin-fixed, paraffin-embedded breast cancer and human mammary epithelial samples were analysed, revealing substantial tumour microenvironment heterogeneity. For analysis, the cell sample or tissue section was labelled with antibodies that were selected to target proteins and protein modifications relevant to breast cancer. Before staining, antibodies were tagged with a unique lanthanide element. The ablated material was then transported by a mixed Ar and He stream to the CyToF system. The signals of individual laser shots at 20 Hz were fully separated, and the LOD was approximately six ion counts that correspond to ~500 molecules.

7.2.2 *Calibration*

Since the introduction of LA-ICP-MS for solid sample analysis by Gray (1985) in 1985, one of the main challenges of the technique's application has been full quantitative analysis. Therefore, strategies for element quantification have involved internal standardization, external calibration with matrix-matched standards, standard addition, isotope dilution (ID), and methods such as film coating and printing.

An internal standard (IS) can compensate for fluctuations in laser power and fluence, differences in sample thickness, sample mass ablated and transported, as well as instabilities and drifts taking place in the ICP source. Isotopes of C and S naturally present in biological samples have been mostly used as IS in biological matrices analysis using LA-ICP-MS. Although the ionization potentials of C and S are relatively high, S isotopes are subject to spectral interferences, and both elements may not be uniformly distributed within the biological tissue, reasonable results were obtained when C or S was used as IS (Sela et al. 2011).

Sulphur is a keratin matrix element homogeneously distributed in hair, and the isotopes ^{34}S and ^{32}S have been effectively used as IS in the analysis of hair (Sela et al. 2007; Pozebon et al. 2008). ^{13}C has been used as IS in leaves analysis (Wu et al. 2013a) to compensate the water content effect and the possible heterogeneity of the materials. A linear relationship among water and ^{13}C signal was observed and the same was for ^{33}S and ^{34}S, which reinforced that they could be used as IS to compensate for the different water content in the samples. ^{13}C was also used as IS to cancel out variations in ablation, transport, and ionization efficiency in element quantification in animal tissues (Feldmann et al. 2002; Hare et al. 2009; Egger et al. 2014; Lum et al. 2016). However, use of ^{13}C as IS is not a consensus and evidence that carbon might not be an efficient internal standard for quantitative analysis by ICP-MS has been found (Marshall et al. 1991; Todoli and Mermet 1998; Frick and Günther 2012; Frick et al. 2015; Limbeck et al. 2015). Carbon may be released as gaseous species (CO_2 and CO) or solid particles from the ablated sample. As such, the transport of C to the ICP and respective ionization would differ remarkably from the analyte. Thus, C cannot completely compensate for the effects abovementioned, affecting the accuracy and precision of the results. Other drawbacks are that carbon has a high ionization potential in comparison to most elements, and, therefore, it behaves differently into the plasma. It is worth citing that the background signal of carbon is very high due to the atmospheric plasma and the element is not homogeneously distributed in some samples. The isotopes ^{43}Ca or ^{44}Ca, naturally present in the samples, were used as IS in analysis of teeth layers (Kang et al. 2004; Hoffmann et al. 2000; Farell et al. 2013) and bone (Bellis et al. 2006; Castro et al. 2010; Ranaldi and Gagnon 2009), despite that Ca varies significantly within the tooth (Arora et al. 2011). However, it is difficult to have another element naturally present in biological samples that fulfils the requirement of an ideal IS.

The isotopes ^{232}Th and ^{238}U were used as IS for Cu, Fe, Zn, P, and S imaging in the human brain (Farell et al. 2013). Matrix-matched standards spiked with the analytes, and IS elements were prepared and used for calibration. Although ^{232}Th and ^{238}U have different ionization potential and mass than the analytes, the authors stated they were feasible for human brain imaging. Austin et al. (2011) compared the performance of raw analytical data and data normalized to ^{13}C, ^{52}Cr, ^{53}Cr, ^{89}Y, and ^{101}Ru for imaging experiments. They observed that ^{13}C is a suitable IS, which exhibited a linear response with respect to the mass ablated, and apparent independence from the adjacent ^{14}N mass peak when the ^{13}C signal in the sample was at least 6% of the raw signal. They also observed that ^{13}C can compensate for variations in sample drying, which occurs during long hours (5–10 h) of imaging experiments.

Internal standardization based on Ir-spiked ink on the sample surface was investigated by Moraleja et al. (2016) to evaluate the distribution and accumulation of the cytostatic drugs cisplatin, carboplatin, and oxaliplatin in rat kidney. A conventional inkjet printer was used to print 4 μm sagittal sections of kidney. A reproducible and homogeneous deposition of the ink along the tissue was observed, but the ink was partially absorbed on top of the tissue. This approach provided a pseudo-internal standardization because the ablation of the sample and IS took place subsequently and not simultaneously, which would worsen the image resolution despite the homogeneous distribution of Ir along the sample surface. However, due to the complete ablation of the tissue, transport of the aerosols that originated from the IS and transport of the sample tissue to the ICP took place practically at the same time. Satisfactory bioimages were obtained, enabling a reliable comparison of the kidney from rats treated with different Pt-based drugs, even for tissues analysed on different days. On-tissue internal standardization with Rh was investigated for Fe imaging in sheep brain. To this end, standards were prepared by immersion of homogenized sheep brain mounted on a glass slide in Fe solution (containing Rh as IS) for 30 min (O'Reilly et al. 2014). The precision was improved approximately twofold when the ^{56}Fe intensity was normalized to that of ^{103}Rh, measured by an SF-ICP-MS instrument operated at medium resolution ($m/\Delta m$ = 4000). The data were cross-validated by μ-XRF and the average Fe concentration found in selected regions of the tissue corroborated those LA-ICP-MS. Konz et al. (2013) described Au sputtering over the sample surface to produce a film and used ^{197}Au as IS instead of ^{13}C for imaging of Mg, Fe, and Cu in histological eye tissue. Considering the type of sample analysed, the authors stated that Au sputtering is a more reliable, precise, and robust methodology than internal standardization with ^{13}C.

In summary, internal standardization usually improves precision and accuracy in LA-ICP-MS, but it is challenging to choose an adequate IS.

7.2.3 Matrix-Matched Standards

Calibration for biological sample analysis using LA-ICP-MS has mostly relied on in-house prepared matrix-matched standards, which usually consist of tissue homogenates spiked with defined amounts of analytes in solution followed by homogenization and mounting on a glass slide. Standards and samples should be as similar as possible to compensate for fractionation effects. Becker et al. (2005a, b) pioneered quantitative element imaging in brain tissue by LA-ICP-MS using matrix-matched standards for calibration. Standards prepared with certified reference materials (CRMs) can be used for matrix-matched calibration in LA-ICP-MS, but matrices available are quite limited. This imposes severe restrictions, with respect not only to sample composition but also to elements with certified concentrations.

When standards are prepared from CRMs, the different water content among CRMs and fresh or frozen tissues (water content around 70–95%) is a concern. Depending on the case, the water content in the CRM must be adjusted to that in the sample (Kindness et al. 2003; Feldmann et al. 2002; Jackson et al. 2006).

Matrix-matched standards in different forms have been prepared and applied for element imaging using LA-ICP-MS. Spiked whole blood or blood serum with known amounts of analyte was used by Pugh et al. (2011) for quantitative imaging of Sr, Gd, and Pt in thin animal tissue sections. The spiked blood or blood serum was placed in a customized frozen block of carboxymethyl cellulose, which was subsequently frozen and cryocut to sections with 20–60 μm thickness. Santos et al. (2009) prepared standards of powdered mussel tissue, dried snail, and rat brain tissues for Cu, Zn, Cd, Hg, and Pb imaging in marine snail. The tissues were spiked with the analytes and pressed into pellets. Calibration using the snail tissue pellets yielded better results owing to matrix similarity. Semi-quantitative results were obtained using one-point calibration with mussel tissue. Certified bone meal was used to prepare standards in the form of pellets for calibration and subsequent Ba/Ca imaging in teeth tissues (Hare et al. 2011). Hair strands enriched with the analyte (Sela et al. 2007; Pozebon et al. 2008) or powdered hair standards placed on carbon tabs were employed (Sela et al. 2007) for calibration and quantitative element determination in hair. Standards of rat brain tissue spiked with elements of interest and encapsulated in a sol-gel matrix produced by tetraethyl orthosilicate were prepared by Sela et al. (2011) for element imaging in animal tissue. Hare et al. (2009) and Lear et al. (2012) prepared standards of homogenized chicken breast. The tissue was spiked with elements of interest, homogenized, pressed into moulds, frozen, and cryocut to produce multipoint calibration curves for element quantification in mouse brain. Reifschneider et al. (2015) proposed egg yolk as matrix matching for Tm imaging in mouse tissues. Egg yolk spiked with Tm was heated up to 90 °C for 10 min to form a solid, which was sectioned by cryocutting, fixed on a glass slide, and then ablated. By calibration with such standards, quantitative and spatially resolved data were obtained for tumour cells and macrophages labelled with thulium-1,4,7,10-tetraazacyclodoecane-α,α,α,α-tetramethyl-1,4,7,10-tetraaceticacid (Tm(DOTMA)), complementing information given by magnetic resonance image (MRI).

A dried droplet supported on appropriate materials is another calibration approach with increased interest, representing an easy-to-handle alternative to typically matrix-matched tissue standards. This strategy combines the simplicity, easy handling, and wide analyte concentration range, provided that "chromatographic" effects are evaluated and circumvented. This strategy of calibration was applied for Mn, Ni, Cu, and Zn imaging in human malignant mesothelioma biopsy where Au deposited onto a filter paper was used as IS (Bonta et al. 2016) and in Pt imaging in human malignant pleural mesothelioma (Bonta et al. 2014). Dried-droplet calibration was also conducted for quantitative imaging of elements in mouse kidney, as well as traumatic brain injury model tissue sections, by laser ablation mass spectrometry (LA-MS). Mass spectrometry imaging (MSI) employing LA provides quantitative data by comparing signals of Na in tissue samples to those obtained in calibration standards of the target element applied to adjacent control tissue sections. Both LA-ICP-MS and LA-MS were employed to image a rat brain injury, revealing accumulations of Na and Ca in the impact area and its peripheral regions (Shariatgorji et al. 2016).

Microarray plates (110 μm in diameter, spaced 185 μm apart, with more than 1000 wells cm^{-2}) were used as microcontainers for holding spiked gelatin as matrix-matched standards in single-cell analysis. This approach allowed for evaluation of the mean concentration of Cu in the cell population across different exposure levels and for examination of the cellular distribution of Cu within the populations. Subcellular LA-ICP-MS imaging was demonstrated to corroborate synchrotron radiation confocal X-ray fluorescence (SR-XRF) microimaging of single cells investigated under in vivo conditions (Van Malderen et al. 2016a, b).

Cold-curing resin blocks spiked with platinum acetylacetonate prior to curing were used for calibration. After washing with water/methanol/ethanol and incubation in isobutanol, samples were infiltrated with Technovit, a cold-polymerizing resin poly-2-hydroxyethyl methacrylate, and then polymerized and cured. The polymer block with the tissue was sectioned, and the obtained tissue slices were placed on object slides and ablated. Platinum distribution at different time intervals after cisplatin treatment of mice was determined quantitatively in different tissues including the cochlea, testis, and kidney (Reifschneider et al. 2013b).

7.2.4 *Isotope Dilution and Alternative Methods of Calibration*

Total consumption nebulizers or those that produce desolvated aerosol are usually employed for calibration with solutions in LA-ICP-MS. With such nebulizers, the ablation chamber and transport line do not get wet, which would hinder the analysis. Desolvation systems produce aerosol more comparable to the LA-aerosol, reducing interferences mainly by oxides and hydroxides.

Online ID and LA-ICP-MS were applied for quantitative Fe imaging in sheep brain tissues; a solution of enriched ^{57}Fe was added to the dry aerosol before entering the ICP, via a total consumption nebulizer (flow rate of 8 μL min^{-1}). An SF-ICP-MS instrument operated at medium mass resolution was employed, resolving polyatomic interference on ^{56}Fe and ^{57}Fe. Recoveries of 80 to 109% of the expected Fe concentration in the sheep brain were obtained for sample-to-standard ratios of 1:1 to 1:5 (Douglas et al. 2016).

Calibration with solutions has been investigated for quantitative imaging of elements in rat brain. To this end, the aerosol from the ablation chamber and that from a nebulizer adapted to a desolvation system was introduced into the ICP (Pozebon et al. 2010). Both aerosols were mixed in the injector tube inside a special ICP torch. Matrix matching was performed by solution nebulization of a series of aqueous standards with simultaneous laser ablation of brain homogenate, followed by parallel nebulization of 2% (v v^{-1}) HNO_3 and laser ablation of the brain sample. The ratio of the slope of calibration curve obtained by laboratory standards prepared with mouse brain homogenate spiked with elements of interest at defined concentrations and that based on aqueous standards was used to correct the difference of sensitivity among solution nebulization and laser ablation.

A micronebulizer inserted directly into a cooled laser ablation chamber was applied for solution-based calibration in human brain tissue analysis for U and Th imaging (Becker et al. 2005a). During laser ablation of thin (20 μm) sections of the brain, standard solutions with increasing analyte concentration were nebulized and the aerosol introduced in the ablation chamber. For the correction of different element sensitivities in nebulization ICP-MS and LA-SF-ICP-MS, a correction factor was obtained by the ratio concentration of element determined by solution-based calibration in LA-SF-ICP-MS/true concentration of element in the sample.

7.2.5 Calibration for Nanoparticles Imaging

The main difficulties of nanoparticles (NPs) imaging in biological materials using LA-ICP-MS are the NPs' heterogeneity with respect to size, low concentration, and lack of standards. Matrix-matched NP standards to establish calibration curves for each specific type of NPs are needed for real samples analysis (Jimenez et al. 2016). Nanoparticle quantification in a single cell by LA-ICP-MS depends on cell cycles; NP sedimentation and aggregation; size, shape, and charge of NPs; laser-NPs interaction; fractionation effects on the aerosol transport; and particle ionization inside the ICP. Therefore, quantitative imaging of NPs by LA-ICP-MS is still challenging not only in single cells but also in biological tissue.

Zebrafish larvae exposed to SeCd/ZnS quantum dots (QDs), which are small semiconductor nanoparticles, and to ionic Cd were analysed by LA-ICP-MS. For LA, frozen zebrafish larvae were placed over a polycarbonate plate previously covered with a thin gold layer to serve as IS. In larvae exposed to ionic Cd, the metal was located mainly in the eyes area, and in the larvae incubated with QDs, it appeared to be more homogeneously distributed along the larvae surface. These results suggested that QDs were not assimilated in the organism, and the accumulated Cd could correspond to the fraction of ionic Cd present in the solution, but in this case, it would be localized in the eyes region. Another possibility would be that QDs could be more adsorbed over the zebrafish larvae peel than absorbed (Zarco-Fernandez et al. 2016).

The incorporation of Au-, Ag-, and Al_2O_3-NPs in *Danio rerio* (*D. rerio*) embryos and *Daphnia magna* (*D. magna*) was evaluated by Böhme et al. (2015). Calibration was carried out using matrix-matched standards with NPs of similar size embedded in agarose gel. For NPs higher than 100 nm, a size-dependent sensitivity in LA-ICP-MS was observed, which was linked to segregation in the ablation process, changes in NP size distribution, and gravitational settling in the transport line or incomplete ion formation in the ICP. In some cases, NP melting can lead to their agglomeration and give rise to errors. Thus, the NPs quantified/mapped would be different than those really taken up by the organisms.

Very small iron oxide particles (VSOP) doped with Eu were injected intravenously into mice and mapped in liver and atherosclerotic plaques using LA-ICP-MS (Scharlach et al. 2016). Calibration with Eu-VSOP enabled quantitative Fe and particles imaging in tissue sections. It was considered a new tool to provide specific and quantitative analysis of particle distribution at the tissue level, allowing correlation with endogenous Fe content in aortic roots sections. It was suggested that Prussian blue staining for histological evaluation of super paramagnetic iron oxide particles accumulation can lead to overestimation of particle uptake in tissues with high endogenous Fe. Thus, the proposed LA-ICP-MS methodology is advantageous for high imaging resolution.

7.2.6 Single-Cell Analysis

Cell tracking studies using LA-ICP-MS are increasing owing to the high sensitivity of ICP-MS, high specificity, and several labelling possibilities, even with a limited lateral resolution of commercially available setups. Mapping and quantitation of trace elements in single cells require high sensitivity, selectivity, and spatial resolution. Single-cell shots at high repetition rates are commonly used to ablate the whole cell and improve bioimage quality. Although the limited resolution of the LA-ICP-MS technique, with the adjustment of instrumental parameters, it is possible to map element distribution at the cellular and subcellular levels. In this context, LA-ICP-MS was utilized in bioimaging of intracellular NPs distribution in cellular substructures of individual cells. Silver and gold NPs in individual fibroblast cells were quantified based on the number of NPs at the single-cell level. Subcellular mapping, enabling differentiation of NPs in the cytosol from those in the cell nucleus region, was achieved by improving the spatial resolution through optimization of the ablation/scanning parameters. With this study, single 3T3 cells, incubated with Ag or Au NPs, showed a strong dependency of particle uptake on concentration and incubation conditions. Quantification of Ag and AuNPs in single cells was possible based on a matrix-matched calibration using nitrocellulose membranes doped with NPs suspension. The results provide insight into NP/cell interactions and have implications for the development of analytical methods in tissue diagnostics and therapeutics (Drescher et al. 2012).

Calibration in single-cell analysis using LA-ICP-MS is challenging, since it is necessary to produce small standards, similar to cells, in order to avoid differences in the volume ablated and to provide similar mass load and responses.

Human $CD4^+$ T cells were labelled with the Gd-based MRI contrast agents Gd-DTPABMA (Dotarem) and Gd-DOTA (Omniscan), which are both in routine clinical use. Cells (50,000) were plated onto slides where single-cell identification was demonstrated by ablating 25 μm diameter areas. This ensured sampling of the complete cell while minimizing the chance of overlap onto neighbouring cells. Only single cells were targeted; cells present in clusters or closer than 25 μm apart were disregarded. Passive loading of up to 108 Gd atoms per cell has occurred. In mixed preparations of labelled and unlabelled cells, LA-ICP-MS was capable of enumerating labelled cells at close to the predicted ratio (Managh et al. 2013).

Human regulatory macrophages (Mregs) labelled with Au after incubation with 50 nm Au-NPs were injected in mice. Au-labelled human Mregs could be mostly detected in the lung, liver, and spleen, while low Au signal was also detected in the kidney, heart, brain, and intestine. An 8 μm laser spot size was sufficient to allow detection of Au-labelled Mregs, whereas the spot size was small enough to allow spatial resolution of adjacent cells; a 55 μm diameter laser beam size ensured complete ablation of an individual cell, while reducing the chance of inadvertently ablating neighbouring cells (Managh et al. 2014).

7.2.7 Metal-Binding Protein Analysis

The LA-ICP-MS technique has been applied in metalloproteomics studies where metals bound to proteins are separated using gel electrophoresis (GE). Spots with a protein assignment in the gel are analysed using LA-ICP-MS, while the metalloprotein could be identified by mass spectrometry. Nevertheless, the lack of suitable calibration standards and the metal loss during separation were still topics of concern in quantitative analysis. Separation techniques used in this field and their impact on the dissociation of metal-protein complexes, losses induced by both the separation system, and disequilibrium effects were discussed by Hagège et al. (2015).

Metal-coded affinity tags based on click chemistry (MeCAT-Click) were used to analyse the proteome of *Escherichia coli* (*E. coli*) in response to heat stress (He et al. 2016). The proteins extracted from the cells were differentially labelled, with Lu and Tm, according to their reaction to iodoacetamide and Ln-DOTA-azide. Cysteine-containing proteins from two different biological states of *E. coli* were specifically tagged and separated by a two-dimensional gel electrophoresis (2D-GE) and analysed by the LA–ICP–MS. The signal intensities of the Tm and Lu were compared in order to quantify different sample states. Based on this information, the electrospray-mass spectrometry (ESI-MS) technique was applied in order to identify the proteins with a difference in abundance. The sensitivity of the LA–ICP–MS allowed finding one upregulated protein that was nearly invisible with Ag staining. The advantage of the MeCAT-Click approach is the possibility to locate low-abundant proteins with a differential expression.

Copper- and Zn-containing proteins in the plankton were separated by the native polyacrylamide gel electrophoresis (PAGE) prior to spots sampling by the LA-ICP-MS (Jiménez et al. 2013). Although the protein separation was efficient, the metal detection was unsuccessful as the metal-proteins binding were lost during migration in the gel along the electrophoretic process or even during the staining process. By the blue native PAGE, two bands correspondent to proteins 60 kDa and 45 kDa were identified in addition to another one that was poorly resolved between 15 and 35 kDa. Other elements probably associated with proteins in the plankton (Cd, Co, Mn, Ni, and Pb) were investigated, but they were not detected as their concentrations were too low. Cuproprotein behaviours in the waterborne oyster after Cu exposure were evaluated (Xu et al. 2014a). Non-denaturing 2D-PAGE, ICP-MS, LA-ICP-MS, and ESI-MS were employed in a synergistic way. After exposure, the Cu distribution in the gills and the digestive gland was first measured by the solution nebulization-ICP-MS after acid decomposition. Then, the cuproproteins in the cytosols were analysed by the size exclusion chromatography-ICP-MS (SEC-ICP-MS) and the non-denaturing-PAGE-LA-ICP-MS, with a prior identification by the ESI-MS/MS. The regulation of cuproproteins was finally evaluated by the SDS-PAGE. An increase of the Cu level was observed for extracellular superoxide dismutase (Cu/Zn-SOD) and L-ascorbate oxidase. The high Cu uptake disturbed the metabolic pathway of the Zn, Ni, Co, Fe, and Mn towards a lower content, which was generally in the gills and the digestive gland. On the other hand, enrichment of the Cu that was found in the digestive gland and the cytosol was further identified as Cu/Zn-SOD, which was responsible for the Cu homeostasis and detoxification.

Non-denaturing isoelectric focusing gel electrophoresis (ND-IEF-GE) associated with the LA-ICP-MS was tested to identify the U-protein complexes in the biological samples. The possibility of determining the intact U-protein complexes has been demonstrated by means of an in vitro synthesized U-bovine serum albumin (U-BSA) and U-transferring complexes. The optimized methodology was applied for the in vivo U-protein complexes quantitation in the digestive gland cytosols of crayfish exposed to waterborne-depleted U. The cytosol samples and calibrants were separated by the ND-IEF-GE, and the resulting immobilized pH gradient strips (IPG) were scanned by the LA-ICP-MS. The calibration solutions were mixed with a bovine serum albumin (BSA) for a final volume and incubated before being submitted to the ND-IEF-GE; for each prepared IPG strip, three parallel scans were ablated. The linearity of the U-calibration curve was satisfactory and used for quantification of the in vivo U-protein complexes. In spite of the poor precision of the data due to the heterogeneity of the U-BSA complex located from the centre to the edge of the IPG strip (3 mm wide), the method could be applied for quantitation of U complexed with proteins. The U content in protein complexes found by the LA-ICP-MS and the solution nebulization-ICP-MS (SN-ICP-MS) for raw digested samples were in agreement. According to an analysis using the SEC-ICP-MS, U in the cytosols of the crayfish digestive gland mainly existed as U-protein complexes. Evidence was found of the in vivo interaction between the U and the Fe proteins (e.g. ferritin) in the presence of excessive UO_2^{2+} in the aquatic environment. Uranylated proteins detected in the IEF electropherograms were subjected to a tryptic digestion before the analysis of the ESI-MS/MS for identification of U-protein targets where 42 potential protein targets of U were detected (Xu et al. 2014b).

Selenoproteins from the cytoplasm human cell extracts, which were separated on the electrophoretic strips by iso-electrofocusing, were detected using the LA-ICP-MS. The proteins were identified in the non-ablated parts of the gel by ESI-MS/MS. The proteins LOD was down to 1 ng mL^{-1} (as Se), corresponding to about 0.2 fg of Se. The same method was applied to the selenoprotein speciation in human cell lines from the kidney, liver, skin, and prostate. A comprehensive screening of the entire biological selenoproteomes expressed in cell lines without the use of radioactive ^{75}Se was also possible. The principal proteins found included selenoprotein 15, glutathione peroxidase 1, glutathione peroxidase 4, thioredoxin reductase 1, and thioredoxin reductase 2 (Bianga et al. 2014).

7.2.8 Sample Preparation and Image Generation

The LA-ICP-MS technique was employed to produce images of element distributions in animal tissues. In general, the information gained with imaging of major components as well as trace elements in animal tissue samples could help to understand biological processes and clarify questions in environmental and biomedical fields. More information about studies concerning bioimages could be found in Becker et al. (2014) and Pozebon et al. (2014, 2017).

Both the thickness of the tissue and the laser parameters have to be optimized for each specific application. For the analysis of dried tissues, the laser ablation process could be performed at room temperature. However, the application of cooled laser ablation chambers has proved to be advantageous, not only for the analysis of cryopreserved materials but also for dried biological samples (Zoriy et al. 2005; Becker et al. 2005a).

Either the paraffin-embedded or the native frozen (stored at temperatures about −70 °C) sections were employed in general to fix the animal tissue on glass microscope slide for further LA-ICP-MS analysis. The tissues samples were sliced to about a 20–200 μm thickness and deposited on the slide. Imaging of the native cryosections (20–30 μm thick) was introduced by Becker et al. (2014). Nevertheless, the thinner sections were better for calibration, and so the trend is ~ 6 μm or lower thickness. Fixed and cryoprotected tissues of the murine brain were investigated by Hare et al. (2014) with respect to losses of elements and sample contamination, and they concluded that metal leaching is highly metal specific and that preparatory steps for cryosectioning of the brain tissue should be approached according to this specificity.

Two-dimensional images could be created by repeatedly scanning a specimen. Paul et al. (2015) developed a method for 3D images of the mouse brain by serial sectioning and subsequent volume reconstruction based on an image registration. However, such a method is not readily applicable to all sample types. Atlases depicting quantitative Fe, Cu, and Zn distribution (3D images) in the mouse brain (cerebrum and brainstem) were constructed from the analysis by the LA-ICP-MS of 46 thin sections of the sample. In spite of a lengthy analysis (158 h), the study improved the visualization of the metals distribution in the brain and could help to elucidate the role in neurobiology (Hare et al. 2012).

For element imaging, laser ablation is destructive to the specimen. Quantification using this technique usually requires laboratory matrix-matched standards. Matrix-matched standards and samples for imaging by the LA-ICP-MS analysis must be analysed under identical experimental conditions. For quantification, average ion responses of the individual raster is plotted against calibration curves, whereas the analyte is usually normalized to that of an IS as discussed before. To obtain images of element distribution, the tissue section is ablated line by line, covering the entire section area. Appropriate laser scan speeds may be calculated by considering the relationship between the laser scan speed, the laser spot diameter, and the total scan cycle of the mass spectrometer (Bonta et al. 2015).

The time-dependent ion intensity raw data obtained from the mass spectrometer is usually transformed into a spreadsheet format using Excel or other software for further data analysis. In general, the software for imaging takes into account parameters such as line length, number of lines, data offset at the start and at the end, the diameter of the laser beam, and the distance between the lines. The pixel size of the resulting images is dependent on the line distance, the number of spots, and the spot diameter. Pixel dimensions for biological tissue imaging has been reviewed by Becker et al. (2014). Images of the elements distribution were obtained using commercial software (Zoriy et al. 2006) or were otherwise written in-house (Becker et al. 2010; Pornwilard et al. 2013; Osterholt et al. 2011; Paul et al. 2012; Pessôa et al. 2016; Uerlings et al. 2016; Hugo et al. 2016).

Improved spatial resolution could be achieved through the use of a narrow laser spot diameter and a low laser scan rate but at the expense of time and sensitivity. The sample section thickness is also a critical experimental parameter when the 3D image of the tissue section is intended. The measurement time for imaging of the biological tissues could take a few minutes to several hours, depending on the size of the tissue area being analysed and laser parameters.

Images of the trace elements distribution in the tissue sections by the LA-ICP-MS has been typically performed using spatial resolutions (spot size) of 30 to 50 μm and above. The limiting factor for increasing the resolution is the resulting sensitivity of the ICP-MS technique, where signal-to-noise ratio is poor due to inherent background spectral interferences and a reduced sample amount introduced into the ICP with a decreasing laser beam diameter. Tissue sections of about 2–3 cm in diameter could be analysed by using standard ablation chambers. In the case of hair, it could be directly analysed. The length will depend on the size of the ablation chamber, but different segments could be analysed separately and combined with the data. The length profiles of the detected elements were usually plotted starting from the hair root to the tip, where distribution patterns of the elements were observed along the hair length. In this way, it was possible to monitor the cisplatin intake of the human body in anticancer therapy, where the variation of the Pt along the hair length with reference to each cisplatin dose was clearly observed (Pozebon et al. 2008). In a study about grizzly bears contamination (Noël et al. 2014), the Hg concentration along the entire length of individual grizzly bear hair demonstrated a relationship between the Hg accumulation rate in the hair and the Hg intake through the fish diet, as well as element elimination from the body and metabolism. It was found that 70% of the coastal grizzly bears sampled had Hg levels exceeding the neurochemical effect level proposed for polar bears.

7.3 Secondary Ion Mass Spectrometry

Secondary ion mass spectrometry (SIMS) plays an important role in imaging for metallomics application due to the high resolution and capability of mapping metals and metalloids in biological tissues, compounds, cells, and subcellular structures. The principle of the SIMS is the use of a primary ion source to sputter neutral particles, elements, and molecular ions from a sample surface. The negative or positive ions formed are then extracted and analysed by a mass spectrometer, whereas removal of the surface atoms or molecules through the energy loss of the primary ions occurs in the form of a collision cascade (Wedlock and Berners-Price 2011a).

The first experiments using SIMS for elemental imaging were conducted by Castaing and Slodzian (1962) in the 1960s, but the application of the technique in life sciences wasn't discussed until the 1980s. Since then, SIMS instrumentation has progressed, making elemental imaging with this technique more applicable to life sciences. Commercial SIMS spectrometers are equipped with quadrupole mass analysers, time-of-flight mass analysers (ToF-SIMS), or double-focusing sector field mass analysers (NanoSIMS) (Pacholski and Winograd 1999). The ToF-SIMS and the NanoSIMS instruments have been mostly employed, while the NanoSIMS enabled a spatial resolution as low as 50 nm, and the ToF-SIMS detected very low amounts of compounds, ranging from a few to 10,000 Da with a spatial resolution of about 100 nm.

A beam of energetic (0.5–20 keV) primary ions is used to sputter the sample surface under a high vacuum (about 10^{-10} Torr). The primary ion beam used to generate secondary ions from the sample depends on the sample matrix and on the element or molecule of interest. The primary ion beam could be composed of positive ions of a certain element (Ga^+, Cs^+, Bi^+, Au^+, O_2^+, Xe^+, or Ar^+) or a negative (O^-, O_2^-). An oxygen ion beam was often employed for most metals because it favours the formation of positive ions (Williams 1985; Boxer et al. 2009; Katz and Newman 1987). The primary ions were generated with a duoplasmatron of Ar, O_2, Xe, etc. or from a liquid-metal ion gun (LMIG) that consisted, namely, of Au, Cs, Ga, or Bi. The duoplasmatron could operate with almost any gas, but O_2 is commonly used because it enhances the yield of Al, Si, and rare earth elements (REEs) secondary ions. The duoplasmatron is appropriate to extract the O^-, O_2^-, or O_2^+ depending on the selected electrical polarity. The LMIG primary source enhanced the yield of secondary ions in comparison with the duoplasmatron sources. In addition, the LMIG could provide primary ion beams spot sizes down to 50 nm in diameter as well as high current densities (1 A cm^{-2}). These characteristics made it possible for high spatial resolution and depth profiling by using a single ion source. High ion current densities allowed in-depth profiling, where high sample sputter rates were achieved and bulk analysis was possible. In addition, in this condition, the sensitivity is improved. In contrast, by using the LMIG beam with a low current density (<0.1 nA), it is possible to sputter only the first or second atom or molecule layer from the sample surface. This condition enabled a superficial mapping of the elements or molecules, mainly of biological materials, but at a cost of sensitivity reduction (Pacholski and Winograd 1999; Boxer et al. 2009; Katz and Newman 1987).

Each metal used as the LMIG source leads for different characteristics of the sample analysis, which depended on the primary ion mass and respective energy. For example, a Cs-LMIG produced a high yield of C, O, and S secondary ions with a low lateral resolution (>1 μm), while a resolution of about 30 nm was achieved with a Ga-LMIG. In general, the total secondary ion yield increased with the primary ion energy increase, whereas the production of heavier secondary ions increased with the primary ion mass increase at a constant energy. However, there was no universal primary ion source, and the SIMS instruments could be equipped with one or more sources, enabling a surface, thin film, and bulk analysis as well as depth profiling.

The primary ion beam could reach the sample at different angles. Sputtering increased with the angle of incidence increase, but secondary ion yield decreased. Depth resolution increased with the angle of incidence increase because the sputtering variation was lower and the length of the ionized species decay decreased. Due to these facts, the angle of incidence was typically 60°.

Charges accumulate on the surface of insulating materials (samples) when they are submitted to the primary ion beam, which is caused by the uncompensated loss of the secondary electrons during sputtering. To circumvent this effect, insulating material was metalized or flooded with electrons because the primary and secondary ions and electron fluxes produced a net electric charge on the sample surface. This charge deflected the primary ion beam (especially when the low-energy primary ions were used) away from the region of interest. As a consequence, the emission of the electrons and the ions from the surface were hindered. In addition, the surface potential variations affected the energy distribution of the emitted secondary ions and their transmission to the mass spectrometer. However, such compensation is not necessary if the sample material is conductive because the charge is compensated by the electric current flowing through the sample to the ground (Waugh et al. 1984; Boxer et al. 2009; Katz and Newman 1987).

The secondary ions generated were conducted to the mass spectrometer where they were analysed according their mass to charge ratios (m/z). A Faraday cup, an ion-sensitive image amplifier, an electron multiplier, or a CCD device can be used as ion detection systems to record the magnitude of the secondary ion signal at the exit of the mass analyser. Depending on the instrument, components, such as ion lenses, deflectors, diaphragms, and hexapoles, were installed in the primary and secondary ion paths in order to shape and focus the primary ion beam onto the sample or the secondary ions to the mass analyser. Signal intensities were dependent on the product of the sputter yield and the ionization probability, which could vary over several orders of the magnitude. With NanoSIMS instruments, the ions are counted by several parallel detectors that enable simultaneous detection, which is a requisite for high-precision isotopic ratio measurements. Thus, the NanoSIMS links high-resolution microscopy with highly sensitive isotopic/elemental analysis, providing spatially resolved information not only with respect to molecular composition of the materials but also isotopic and elemental (Pacholski and Winograd 1999).

In order to analyse a given region of the sample surface, the primary ion beam was scanned over the surface of the sample. The ion beam that strikes the sample may render intrinsic damage of the sample surface. However, for molecular systems, this damage resulted in a loss of molecular-specific signals when the number of incident ions exceeded approximately 1% of the number of surface molecules. At higher amounts of primary ions, chemical fragments began to accumulate on the surface and the information provided by the mass spectra became limited to elemental species and small molecular ions. In summary, the surface or structure damage may lead to incorrect results, and, therefore, it was necessary to adjust the current density of the primary ion beam and the sampling time. As a result of such, the primary ion source was operated in a static mode (static SIMS) or a dynamic mode (dynamic SIMS), where the ion beam reached the sample continuously or pulsed, respectively. Each mode exerted a different extent of surface damage and, thus, chosen for different purposes. In static mode, a low-energy and low-density ion beam (lower than 10^{12} ions cm^{-2} to minimize the interaction of the primary ions with the molecules on the top monolayer) was employed, whereas only a sample layer of 1–2 nm was removed. In contrast, a high ion density and energy beam were used in a dynamic SIMS. Consequently, the dynamic SIMS was more destructive, and interactions of the primary ions occurred with the species in the deeper sample layers (10–20 μm). By means of the dynamic SIMS, a volume of 100–1000 μm^3 of the sample material could be removed. This amount was much higher than that removed by means of static SIMS. These differences made the static SIMS and the dynamic SIMS more appropriate for qualitative and quantitative element imaging, respectively (Amstalden van Hove et al. 2010; Pacholski and Winograd 1999; Boxer et al. 2009; Katz and Newman 1987).

In order to obtain chemical information from the original sample surface, the static mode ToF-SIMS was widely employed for the atomic monolayer and surface molecular analyses. When the SIMS was used in a dynamic mode, the sample surface could be eroded in a controlled form along with depth-profiling achieved, and layers of up to a few μm thick could be depth profiled. The dynamic SIMS could penetrate deeper into the specimen and produce smaller fragments (Lobinski et al. 2006) As a result of such, it was more appropriate for elemental analysis with the LOD of about 10 ng g^{-1}.

By combining one mass analyser (ToF, sector field, or quadrupole) with a specific primary ion source, the surface analysis with a high spatial resolution was achieved using the SIMS. For the imaging of metals and metalloids, a highly focused ion beam was rastered over the surface of interest. At each pixel point, it is possible to collect ions of a single m/z, a few ions with specified m/z, or a whole mass spectrum ranging from a few to 1000 m/z. After rastering a specific area of the sample, images of the distributed elements or compounds could be constructed given the resolution determined by the ion beam diameter, which typically ranged from 50 to 100 nm.

7.3.1 Sample Preparation

In the high-resolution view, the SIMS is a valuable tool in biology and medicine with respect to characterizing elements and molecules in biological tissues and cellular and subcellular scales – but the sample must be solid and stable in vacuum conditions. However, due to the high vacuum requirements for the SIMS operation, complexity of the biological materials, and the fragile nature of biological cells, the sample could not be directly analysed, and special care had to be taken to maintain the structural and chemical integrity of the cells and tissues. As a consequence, the sample preparation depended on the type of biological system and the SIMS instrument employed for analysis. Incorrect sample preparation could result in artefacts, structural damage, and element diffusion and loss, rendering the elemental mapping imprecise or even useless. This was more critical in cell analysis, especially for elements with a high diffusion rate such as Na and K. Therefore, different sample preparation procedures for SIMS analysis have been proposed (Wedlock and Berners-Price 2011a), most of them also used for scanning electron microscopy (SEM) or transmission electron microscopy (TEM). Semi-thin (300–500 nm thickness) or thin (60–80 nm thickness) sections of fixed biological materials were commonly used in the SIMS analysis. Nevertheless, the SIMS could also be applied to native tissue mounted onto conventional glass slides (Qin et al. 2011; Grovenor et al. 2006; Cunha et al. 2016).

Samples could be fixed using chemicals or high-pressure freezing (HPF) with subsequent dehydration and infiltration with a resin or through a cryogenic procedure. It was assumed that freezing helped to preserve the chemical integrity of the biological sample, as the fixation and dehydration occurred at a low temperature that could potentially preserve the hydration shell. The use of HPF was challenging for the cell analysis, as cells do not always freeze well and cryoprotective additives (such as sugars, polymers, and proteins) were required, which could alter the SIMS ion yields. For analysis of animal tissues like brain, resin-embedded samples to thin sections or semi-thin sections, loaded onto TEM grids, or silicon wafers, or glass slips, and finally gold-coated to avoid a charge build-up on the sample surface (Grovenor et al. 2006; Quintana et al. 2006, 2007).

A chemical fixation preserved the internal cellular structures by physically changing the cell using reagents like glutaraldehyde, paraformalin/formaldehyde, or trehalose. After this treatment, the cellular compartments containing chemical information were preserved, but the integrity of the cellular membrane was often compromised, and the distribution of the diffusible ions was not retained. Thus, the chemical fixation of the cell could not be appropriate for highly diffusible elements like Na, K, and Ca, since redistribution of these elements was observed in the cell components. However, the chemical fixation was suitable when the aim was to determine the elements bounded to molecules, the DNA-interacting drugs, and other covalent tracers that were not highly diffusible ions (Grovenor et al. 2006).

Cryogenic fixation involves frozen-hydration, freeze-drying, and freeze-fracturing. Frozen-hydration comprised off lash-freezing the sample in a liquid nitrogen-cooled propane prior to freezing in liquid nitrogen. This process avoided physical damage of the sample from water crystallization and also preserved the chemical integrity of the cells and tissues. Freeze-drying involved quickly freezing the sample followed by a slow warming under a vacuum to remove residual water. Cell damage was often a major concern of freeze-drying as a consequence of water sublimation if the temperature of the sample was increased too quickly.

Rearrangement of the molecules in the cell and/or losses of some components may have also occurred during the freeze-drying process. Therefore, this sample preparation procedure was not appropriate when the aim of the analysis is mapping elements and other components in the cell. In freeze-fracturing, a cell suspension is trapped between two shards of substrate in a sandwich form and then plunged into liquid propane. If the analysis was not performed immediately, the sample could be preserved in liquid nitrogen. However, the sandwich was separated at the moment of analysis, and the ice matrix within the sample was fractured as well, exposing the contained cells. Freeze-fracturing is a procedure that better preserves the sample components in the cell (Grovenor et al. 2006; Winograd and Bloom 2015; Chandra 2008a).

7.3.2 Applications of SIMS

Due to its high spatial resolution (50–100 nm), the NanoSIMS has been applied for the metallodrugs imaging in the cells as well as the organic ligands associated with the metal centre in the metallodrugs. However, the ligands could only be visualized by the NanoSIMS if they contained non-endogenously ubiquitous elements in their chemical structure such as Br or F or the ligands were isotopically labelled with ^{13}C or ^{15}N. For metal detection via the NanoSIMS, semi-thin sections must be analysed, and a long acquisition time (>10 h) is necessary to obtain a sufficient secondary ions yield. A long acquisition time would destroy thin sections, and therefore, they were not appropriate for analysis.

Detecting compounds with mass ranging from a few to 10 kDa with a spatial resolution of about 100 nm, as achieved by the ToF-SIMS, was challenging. Hence, focusing a narrow primary ion beam requires a compromised condition among sensitivity and mass resolution. Imagings of the cisplatin (Gulin et al. 2015) and the boron (Chandra et al. 2008) in the glioblastoma cells were examples in which the spatial resolution achieved was 500 nm for the species and the compounds had a mass in the range of 1–1000 Da, which was sensitive sufficient for Pt detection.

The excellent spatial resolution provided by the SIMS, a consequence of the very small spot sizes produced by focusing the primary ion beams on the sample surface, has demonstrated applicability to a large variety of biological samples. Specifically, any element and respective isotopes could be detected, enabling both qualitative and quantitative determination. However, given certain conditions, the secondary ion intensity of an element of interest depended not only on its local concentration in the sampling area but also on its ionization yield. Unfortunately, the ionization yield could vary greatly according to the chemical environment where the ions were ejected. The sample matrix may have prevented a direct quantification, as deduced from the ion signal intensity detected. The most common approach to circumvent the matrix effect was the use of relative sensitivity factors (RSFs) that could be obtained by analysing a reference matrix (chemically similar to that of the sample) with a known concentration of the element of interest. By simultaneously measuring the ion signal of the element of interest and the ion signals of the elements that constituted the reference matrix, a RSF could be established with the ratio of the respective ion signals. The RSF was then applied to the samples with a similar matrix to find the concentration of the element of interest, assuming the matrix effect was cancelled and the ionization yield remained unchanged. However, such an approach was applied to the samples with homogenous matrices and the element of interest was ideally a minor constituent. This was the case with the dopant determination or depth profiling of an implant in a semiconductor where there was very little change in the matrix composition. The successful quantification in this case was due to a uniform substrate with a fixed composition. The matrix signal remained unchanged as well as the ionization yields of the matrix element and the element of interest, allowing a robust applicability of the RSFs because they were not affected by the minor change in the local composition. The signals remained linear across a large dynamic range, from a few percentages down to parts per million or even parts per billion levels. In contrast, the biological materials generally had a heterogeneous local composition, and variations in ionization yields were expected from one element to another, which induced variation of the RSFs. In addition, local changes in the concentration of the element of interest could hardly be taken into account to apply a systematic correction of values (even the content of major constituents such as C, H, and O were not really constant across the different compartments). Therefore, such an approach would be feasible for quantification based on an average area that has been described by Dérue et al. (2006) in the determination of Na^+, K^+, Mg^{2+}, and Ca^{2+} in frozen-hydrated ionic solutions by the SIMS (Cunha et al. 2016).

In a study conducted by Becker et al. (2010), the SIMS was employed in conjunction with LA-ICP-MS for metals and biomolecules imaging in a mouse heart. In this study, the SIMS was used as an elemental and biomolecular mass spectrometry technique for imaging of Na, K, and cellular substances, such as phosphocholine, choline, and cholesterol, in adjacent tissue sections. The SIMS measurements were carried out using a ToF-SIMS instrument, which was equipped with a pulsed primary Bi cluster (Bi_n^+) ion source running at 25 keV. Mouse heart sections of 16 mm × 9 mm and 40 μm thickness were mounted on glass substrates and analysed using the SIMS, whereas the pixel distance was 33 μm, and 150 shots were applied on each pixel. Upon measurements using SIMS, both K and Na were detected in heart tissue.

The potential of SIMS as a metal imaging technique in biomedical research has also been demonstrated by mapping ^{12}C, ^{39}K, ^{23}Na, ^{40}Ca, and ^{10}B isotopes in T98G human glioblastoma cells (human brain cancer). The technique was used to better understand the biological effect of boron neutron capture therapy (BNCT). The elements were detected at subcellular levels in cell samples prepared using the cryogenic sandwich-fracture procedure (Chandra et al. 2008). In another work (Nygren et al. 2014), arsenate and Hg species in metal-resistant oral bacteria were imaged through ToF-SIMS to elucidate how this bacteria takes up and transforms toxic metals inside the cells. Strains of bacteria were isolated from the oral cavity of healthy volunteers and grown in the presence of Hg (1–10 μg mL^{-1}) or As (0.1–1.0% m/v).

Crude biofilms of air-dried bacteria on Al foil surfaces were analysed, enabling subcellular lateral resolution. To this end, the bacterial strains (*S. oralis* and *F. nucleatum*), after exposure to As and Hg, were washed in sterile water, centrifuged, smeared onto the Al foil, and then dried at room temperature. For As and Hg measurements, a ToF-SIMS instrument equipped with a 25 keV Bi_3 LMIG and 10 keV C60 sputter gun was employed. Free Hg ions were found at the cell surface, whereas in the periplasmic space, Hg was found as methylmercury (in-depth profiles revealed higher levels of the methylated isotopes ^{200}Hg and ^{202}Hg). The arsenate ions were converted to arsenite ions and found inside the cells, close to the cell membrane. Maia et al. (2006) applied ToF-SIMS for Cd, Cu, Cr, Hg, and Zn imaging in liver tumour cells (human HepG2 cell line) doped with these elements. The authors developed a protocol for sample preparation by combining rapid freezing, freeze-fracturing, and imprinting for transferring the cells to a silicon wafer. They observed that only Cr and Cu diffused into the cell following the simulated intoxication. The other elements (Cd, Hg, and Zn) were either too low to be detected in the cells or were indistinguishable from the background.

Alzheimer's disease (AD) is a complex neurological disorder caused by a number of interconnected genetic and lifestyle variables that culminate in multiple changes in brain structure and function (Braidy et al. 2014). These changes can be microscopically observed on a gross anatomical level, such as brain atrophy, extracellular amyloid plaque (Am), and neurofibrillary tangle formation, as well as alterations of metabolic activity. At a molecular level, metal dyshomeostasis is frequently observed in AD due to anomalous binding of metals such as Fe, Cu, and Zn or impaired regulation of redox-active metals that can induce the formation of cytotoxic reactive oxygen species and neuronal damage. Though metals likely play a significant role in AD, little is known about their sources, mechanisms of transport, chelation, biochemistry functions, and interactions with proteins. In an attempt to understand the role of metals in AD, studies have been conducted using imaging techniques in order to find metal distribution in parts of the brain, especially in cellular structures (Quintana et al. 2007). Analysis of the hippocampus of AD patients and the thalamus of transgenic APP/PS1 mice by SIMS enabled imaging of the morphological and chemical modifications that took place in well-characterized pathological brain regions. It was also possible to image the pathological iron-ferritin-hemosiderin distribution at the subcellular level in the hippocampus of the brain of the AD patients, as well as Ca-Fe mineralization in thalamus amyloid deposits (AmD) in the transgenic mice brains. A NanoSIMS instrument equipped with Cs^+ and O^- sources was employed for Fe and Ca detection with resolutions of 50–100 nm and 150–200 nm, respectively. Semi-thin (300–400 nm thickness) and thin (60 nm thickness) sections directly mounted on silicon plates or copper grids were analysed. To allow straightforward elements mapping on the thin section, TEM copper grids covered with a thin Formvar (a thermoplastic resin) film followed by a thin carbon layer were used, combining mechanical strength with electrical conductivity and preventing damage during ion bombardment. Images of CN^-, P^-, and S^- distribution were acquired simultaneously by using Cs^+ LMIG as a primary ion beam source. For simultaneous detection of Ca^+ and Fe^+, an O^- beam was employed (Quintana et al. 2006, 2007).

NanoSIMS was employed in studies dealing with Au (Wedlock et al. 2011b), Pt (Legin et al. 2014; Wedlock et al. 2013), and Ru (Lee et al. 2015) metal-based anticancer drug imaging. In vitro inter- and intracellular distribution of an isotopically labelled Ru(II)–arene (RAPTA) anti-metastatic compound for human ovarian cancer cells was imaged with ultrahigh resolution. Images of ^{13}C, ^{15}N, and Ru indicated that the phosphine ligand remains coordinated with the Ru(II) ion, whereas the arene detaches. It was demonstrated that Ru complexes were localized mainly on the membrane or at the interface of cells, which correlates with the anti-metastatic effects (Lee et al. 2015).

SIMS and energy-filtered transmission electron microscopy (EFTEM) techniques were employed to detect Au in situ in tumour cells of an individual submitted to treatment with an Au(I) phosphine-based anticancer agent. NanoSIMS ion maps of ^{12}C, $^{14}N^-$, $^{31}P^-$, $^{34}S^-$, and $^{197}Au^-$ rendered images of cellular morphology and distribution of Au at the subcellular level. By employing EFTEM, images of Au distribution and nuclear and mitochondrial morphology were obtained. The Au image was comparable to that obtained with NanoSIMS. The subcellular distribution of Au was associated with S-rich regions in the nucleus and cytoplasm, supporting evidence of the mechanism of action of Au(I) complexes based on inhibition of thiol-containing proteins (Wedlock et al. 2011b).

Isotope selective NanoSIMS combined with confocal laser scanning microscopy was employed for mapping ^{15}N-labelled ammine cisplatin complexes in human colon cancer cells (SW480 cells) (Legin et al. 2014). NanoSIMS proved to be a highly sensitive technique for subcellular imaging of Pt in cells exposed to cisplatin. By doping the cells with ^{15}N-labelled cisplatin, it was possible to monitor the N/Pt stoichiometry of the compound, which suggested partial dissociation of the Pt–N bonds, at least in nucleoli of cells treated with elevated amounts of cisplatin. The co-localization of Pt with S and P-rich structures observed is consistent with the known high affinity of Pt for S donors and binding to DNA. The cleavage of ligands and Pt co-localizing with S-rich nucleoli corroborate the role of thiol-bearing molecules (affecting Pt–N bond stability) in cisplatin metabolism. The partial dissociation of Pt–N bonds raises the question of possible implications on the mechanism of action with respect to nuclear targets. The accumulation of Pt within acidic organelles, as detected by the combined application of NanoSIMS and fluorescence microscopy, might be relevant for the detoxification and/or mode of action of Pt drugs. In the case of the investigated cells, the cisplatin aggregation in S-rich acidic cytoplasmic organelles may contribute to their intrinsic resistance mechanisms. NanoSIMS has provided information not only about internalization and subcellular localization of different Pt anticancer drugs but also Pt and ligands imaging, as well as ligand exchange in platinum-amine antitumour drugs labelled with ^{15}N. Clear differences in localization and time course among cisplatin and a polynuclear Pt drug (TriplatinNC) could be observed. The metabolism of the drug was also evidenced by the presence of ^{15}N in the absence of ^{195}Pt and vice versa (Wedlock et al. 2013).

SIMS equipped with a C_{60}^+ primary ion source has been employed for Au nanoparticle (AuNP) imaging in macrophage-like RAW 264.7 cells. The AuNPs and two pharmaceutical compounds, amiodarone and elacridar, were successfully imaged within the cells. To verify if SIMS could detect functionalized and non-functionalized NPs simultaneously, fluorophore-functionalized AuNPs were evaluated as a model system. The fluorescent characteristics of the functionalized NPs enabled their detection and localization within the cell, without interference from the other biologically relevant components of the cell. This unique capability of SIMS is useful in studies about targeted drug delivery through AuNPs (Bloom et al. 2016).

In summary, SIMS is a very important tool in investigations dealing with metals' involvement in neurodegenerative diseases and for understanding the mechanisms of action of metal-based anticancer drugs. Imaging of element distribution in biological tissues and cells is another important capability of SIMS. This imaging technique may be one of the key tools for the identification of possible biomarkers when seeking treat disease (Bourassa and Miller 2012).

7.4 Bioimaging Using μXRF and SXRF

Micro X-ray fluorescence (μXRF) is an elemental analysis technique that relies on the same principles as X-ray fluorescence (XRF). In traditional XRF, X-ray photons produced by X-ray tubes irradiate the sample atoms and eject a core electron with a binding energy lower than the incident photon energy. The resulting hole is then filled by an outer electron, thus resulting in X-ray fluorescence emission. The sensitivity increases with atomic number. Therefore, X-ray fluorescence is particularly appropriate for the study of heavy elements. The μXRF has a spatial resolution with a diameter many orders of magnitude smaller than conventional XRF. In μXRF, the X-ray beam is restricted through a pinhole aperture that blocks much of the X-ray flux, which has an adverse effect on the sensitivity of trace elemental analysis. In synchrotron XRF (SXRF), X-ray photons produced by synchrotron radiation provide X-rays with intensity that is about one order of magnitude higher than conventional X-ray tubes, thus providing higher sensitivity. Synchrotron is an ideal source for high-precision XRF analysis of heterogeneous and complex materials due to its small source size, low divergence, high photon flux, and linearly polarized nature. Because the electron binding energy is characteristic of the atom, the emitted X-ray fluorescence energy is typical for each element, and scanning a sample at a given energy with a focused beam yields two-dimensional (2D) or three-dimensional (3D) elemental distribution maps. The pioneering work in SXRF spectrometry by Horowitz and Howell (1972), Sparks et al. (1977) and Sparks Jr (1980) focused on the spot analysis of heterogeneous samples. These works led to μ-SXRF development, which enables elemental imaging with nanometre spatial resolution using advanced synchrotron sources, state-of-the-art X-ray optics, and novel detectors. Spatial resolution for practical analysis is now approaching a few tens of nanometres; nanoimaging and nanoanalysis (NINA) beam lines provide imaging with a 20 nm lateral resolution (Martínez-Criado et al. 2012).

Micro-SXRF spectrometry is routinely combined with micro-XAFS (X-ray absorption fine structure) techniques as well as micro-diffraction and micro-computed tomography (CT). SXRF microprobes are composed of several key elements: the magnet source, the monochromator, the focusing optics, the detection system, and the ancillary environment (e.g. vacuum, cryostat). None of the existing synchrotron XRF microprobe end stations is similar in terms of energy, lateral resolution, level of detection, or conditions of measurements, and the beam line should be carefully selected depending on each case and purpose. Soft X-ray (energy ranging from 0.12 to 12 keV) is suited for biological materials analysis as it gives access to distribution of low-Z elements such as C, N, O, F, Fe, Zn, Mg, and others with fundamental importance for metabolism in biological systems on cellular or subcellular levels. It is more difficult to focus photons with high energy than low energy, so hard X-ray (energy > 12 keV) imaging beam lines have operated for a long time with a micro-focused beam. Due to the high penetration of X-rays in matter, three-dimensional maps can be obtained by translating and rotating the sample. After mathematical reconstructions, the internal structure and elemental distribution of the sample are visualized without any sectioning of the sample and can be performed with a micro- or nano-focused beam (De Jonge and Vogt 2010).

7.4.1 Sample Preparation and Measurements

SXRF spectrometry has a high sensitivity for most of the biologically important elements and does not destroy the sample, though special sample preparation is sometimes required for high-resolution and accurate imaging. The conservation of the cell structure and metal distribution during the sample preparation and during the measurements has to be considered. SXRF enables investigation of the samples in their frozen hydrated state using a cryostat operating with helium or liquid nitrogen at a very low temperature (−263 and −196 °C, respectively). As such, the sample preparation is minimal and does not involve chemical reagents and resin embedding, hindering metal redistribution and speciation changes, mainly soluble species and low stable complexes (Matsuyama et al. 2010). Current beam lines are equipped with cryoenvironments. To preserve the integrity of cell membranes and organelles as well as elemental and chemical species redistribution, samples are frozen in liquid nitrogen-cooled isopentane or liquid ethane, transferred to the microscope in this frozen state, and analysed at a low temperature (Sarret et al. 2013). When necessary, the frozen sample can be cut into thin sections using a cryo-microtome. Sample degradation related to intense beam exposure can also be limited using fast detectors and appropriate scans to reduce beam time exposure, without sample damage (Lombi et al. 2011). For instance, fast fluorescence detectors use electronic data collection with lower counting than traditional detectors, which reduces the time for mapping (Lombi et al. 2011; Ryan et al. 2009). In addition, large samples can be scanned with reduced sample degradation by increasing the speed of the scanning.

7.4.2 Applications

Micro-SXRF has proved to be a feasible tool to study distribution of a wide range of biologically relevant elements and markers in tissues and cells. Images of the location and structural details of Au-labelled organelles were obtained, which correlated well with the subcellular distribution visualized by means of optical fluorescence microscopy (McRae et al. 2006). In this study, adherent mouse fibroblast cells were grown on silicon nitride serving as biocompatible XRF support substrate and labelled with organic fluorophore-Au clusters in combination with primary antibodies specific for mitochondria or the Golgi apparatus, respectively. Raster scanning of in-air dried cells with incident X-ray energy of 11.95 keV, sufficient to ensure excitation of the Au L line, provided 2D maps with submicron resolution for Au as well as for most biologically relevant elements (K, Cl, Cu, Zn, Ca, P, S, and Mn). Calibration to elemental area densities (μg cm^{-2}) was carried out by comparison of X-ray fluorescence signal strength from the sample with fluorescence from thin film standards. An image of mitochondria together with various elements was also obtained by labelling the adenosine triphosphate (ATP) with Au colloidal particles (Matsuyama et al. 2009). Chemical nanoimaging of human cells was also performed using functionalized CdSe/ZnS quantum dots (QDs) (Corezzi et al. 2009), which were used to label cancer markers. It was found that QDs were suitable for nano-SXRF analysis because of their unique Se emission band. Another example of application was silver NPs imaging in animal tissues (Smulders et al. 2015) combined to XAS (X-ray absorption spectroscopy).

Because analysis time at synchrotron facilities is limited, μXRF as a benchtop instrument represents an alternative for elemental mapping under laboratory conditions (Qin et al. 2011). However, μXRF has lower resolution and sensitivity in comparison to μ-SXRF. Therefore, benchtop μXRF can be utilized for a coarse scan to identify regions of interest (ROI) for further analysis by μ-SXRF. Applications for elemental imaging in biological samples by means of benchtop μXRF are, for example, the investigation of elemental distribution in teeth or bone tissue, where high analyte concentrations are present (Dias et al. 2015; Blaske et al. 2014). To investigate element distribution in the liver of a Wilson's disease (WD) patient, a liver biopsy specimen from the patient was analysed by means of XRF (Hachmöller et al. 2016). Firstly, benchtop μXRF was employed for a coarse scan of the sample under laboratory conditions. The resulting distribution maps of Cu and Fe enabled the determination of an ROI for further analysis. The results indicated the suitability of benchtop μXRF for the analysis of thin sections of biological materials such as liver tissue. Although Cu concentrations are elevated in WD, a sample with a thickness of 3 μm comprises a very small amount of analyte for determination by means of benchtop μXRF. Then, in order to obtain more detailed elemental information, the ROI was analysed by μ-SXRF with a beam size of 4 μm, reaching a resolution at the cellular level. Images of Cu, Fe, Zn, and Mn were obtained, whereas areas with high Cu intensities and areas with high Fe and Zn intensities showed an inverse correlation. In addition to μ-SXRF, X-ray absorption near edge structure spectroscopy (XANES) was employed to identify the oxidation states of Cu within the WD liver tissue, where a mixture of Cu(I) and Cu(II) was identified.

Advances in μ-SXRF instrumentation, such as improved detector sensitivity and reduced data collection times, have enabled 3D mapping (based on tomographic approaches) of trace metal. SXRF microtomography was employed by Bourassa et al. (2014) to image 3D distribution of Zn, Fe, and Cu in a zebrafish embryo at the pharyngula stage, 24-hours post-fertilization (hpf). In order to preserve the native elemental composition, the embryo was embedded at cryo-temperature in a methacrylate-based resin. Beam attenuation by the resin matrix was minimized by excising the specimen through fs laser sectioning prior to mounting on the tomographic sample stage. Based on 60 tomographic projections acquired over 100 h of beam time, it was possible to reconstruct the corresponding 3D elemental volumes using a maximum likelihood expectation maximization (MLEM) algorithm. The technique was further employed for 3D Zn and Fe imaging in a zebrafish embryo at the onset of the hatching period (48 hpf) (Bourassa et al. 2016). Calibration to elemental densities was performed by comparing the fluorescence emission of the sample with that of a thin film standard relative to the photon flux captured by two ion chambers positioned upstream and downstream of the sample. The reconstructed volumetric data revealed distinct differences in the elemental distributions, with Zn predominantly localized in the yolk and yolk extension and Fe in various regions of the brain as well as the myotome extending along the dorsal side of the embryo. The yolk syncytial layer exhibited higher concentration of Fe, likely due to enhanced expression of the Fe transporters ferroportin-1 and transferrin in that layer. The liver appeared as a prominent, Fe-rich anatomical feature of the embryo. As a major storage site of ferritin, the primary protein responsible for intracellular Fe storage and release, this organ was readily identified in the Fe distribution map. Higher levels of Fe in the retinal epithelium were associated with the presence of several Fe-dependent enzymes critical for cellular function. The Zn distribution map indicated higher levels of Zn in the outer layers of the retina, with concentrations ranging from 1 to 1.5 mM, probably due to metallothionein transcripts in the retina of zebrafish embryos.

It can be concluded that μ-SXRF became an advanced and essential analytical technique in life sciences. The increase of lateral resolution and sensitivity enables 3D imaging, offering new insights in metallomic studies.

7.5 Elemental Mapping in Electron Microscopy Techniques

Electron-ionizing radiation can provide images/mapping with resolution at the nanometre scale due to electron-specimen interactions (Pennycook 2012; Fiori 1988). The electron wavelength at 100 keV is about 0.004 nm, which is much lower than that of visible light in conventional optical microscopes (Williams and Carter 2009). As a consequence, visible light enables obtaining sub-micrometre spatial resolution images, while electron microscopy (EM) enables nanometre spatial resolution images.

The most common EM techniques employed for obtaining biological tissue images are transmission electron microscopy (TEM) and scanning transmission electron microscopy (STEM). For both techniques, a thin sample specimen (thickness in the nanometre range) is illuminated with an electron beam; for TEM analysis, the electrons transmitted through the sample are collected, and for STEM analysis, the specimen is scanned with electrons, which also pass through the sample to produce contrast image. In physical terms, the difference between TEM and STEM is that the latter represents a perfect incoherent image and the former represents a coherent one. The electron beam direction propagations are opposite for TEM and STEM, but the two techniques are optically equivalent when equivalent microscope apertures are used (Pennycook 2012; Cunha et al. 2016). In STEM, the electron beam is focused into a small probe that scans the sample to acquire spatial information (X,Y), whereas in TEM, broad parallel beams are focused over the sample to acquire spectral data to generate a plane-by-image plane (Williams and Carter 2009).

When the electron beam illuminates the sample specimen, it can interact with it differently, yielding transmitted electrons, elastically and inelastically scattered (energy-loss) electrons, secondary electrons, backscattered electrons, Auger electrons, and X-ray photons (Williams and Carter 2009). Electron energy loss spectroscopy (EELS) and energy-filtered (EF) images, approaches of STEM and TEM, respectively, are based on these phenomena, enabling elemental mapping at the nm scale (Cunha et al. 2016; Hofer et al. 2000; Aronova and Leapman 2012). A contrast image is produced by elastic electron scattering, which is deflected by a positively charged nucleus whose scattering angle depends on the atomic number (Zeta contrast information). Inelastic interactions between the incident beam and sample electrons provide elemental information and chemical mapping of the sample, as is the case with EELS and energy-dispersive X-ray spectrometry (EDX)(Fiori 1988; Hofer et al. 2000; Aronova and Leapman 2012). In EDX, an electron beam typically in the range of 10–20 keV strikes the surface of the sample and causes X-rays to be emitted from the irradiated material, which are then detected.

For EELS, the inelastic scattered electrons pass through a magnetic field and produce an energy-loss (EEL) spectrum that provides a lot of information. This spectrum can be divided into a low-loss energy range (up to 20 eV) and a high-energy range (higher than 100 eV) (Hofer et al. 2000). The part of the EEL spectrum corresponding to the low-loss energy range contains fine structures related to excitation of molecular levels, enabling the determination of frozen water, proteins, lipids, sugars, etc. The high-energy range of the EEL spectrum contains information that is related to the core edges associated with excitation of inner-shell electrons, thus providing elemental information (Cunha et al. 2016; Aronova and Leapman 2012).

Especially for TEM, the incorporation of a slit in the detection system enables obtaining an EF image, which removes the effects of chromatic aberration and creates unique contrast effects in the image. The retraction of the slit from the imaging filter system enables reverting EF to EELS mode. Currently, with the use of a field-emission gun (FEG), it is possible to operate the same microscope in EFTEM and STEM-EELS modes (Leapman 2017) using a fixed beam of EFTEM, or in STEM (Aronova and Leapman 2012). It is important to note that STEM-EELS and EFTEM enable mapping of elements ranging from Li to U (Hofer et al. 2000; Ortega 2005; Jurowski et al. 2015).

Data obtained using EELS mode provide information depicted in a cube, where two of the cube axes correspond to spatial information and the third dimension represents the EEL spectrum (Hofer et al. 2000; Aronova and Leapman 2012). For accurate element mapping in cellular structures, nanometre spatial resolution is required (Lobinski et al. 2006), as provided by TEM and STEM. In order to map whole organelles and cells (size ranging from 1 to 10 μm) and minimize the natural variability of composition of the organelles, it is necessary to analyse specimen areas up to 10 μm (Aronova and Leapman 2012). For more accurate analysis of large specimens, EFTEM is more appropriate than STEM-EELS. On the other hand, STEM-EELS is preferred for analysing samples rich in water and proteins, which can be damaged by the beam radiation. The STEM-EELS technique is also more appropriate for trace elements determination in biological specimens because it allows for a higher number of incident electrons per area, increasing the sensitivity (Cunha et al. 2016; Aronova and Leapman 2012).

In order to obtain accurate elemental mapping by means of EM techniques, the background signal must be subtracted from the ionization edge signal of the elements of interest. This is usually carried out using the three-window method, where two images of background are obtained, before (pre-edge) and after (post-edge), to generate the ionization edge signal of the element of interest (Hofer et al. 2000). Especially for mapping of metals present in trace concentrations (μg g^{-1} or lower) in biological specimens, STEM-EELS is preferred due to the high sensitivity, providing the production of higher numbers of electrons per pixel (Cunha et al. 2016).

Leapman et al. (1993) employed STEM-EELS to evaluate the distribution of Ca in frozen tissues of mouse cerebellar cortex with a thickness of 100 nm. Despite the calcium L_{23} signal being practically free of the sample matrix, production of a reference spectrum of $CaCl_2$ deposited on carbon film was necessary for correct selection of the element edge due to the small signal obtained for Ca from the specimen. Additionally, to provide satisfactory reduction of carbon background, filters were used to obtain the Ca mapping. The total sample area analysed was about 10^6 nm^2, whereas the electron beam energy was up to 100 keV. The Ca concentrations found in endoplasmic reticulum and mitochondria were 4.9 ± 0.4 and 1.4 ± 0.4 mmol kg^{-1} (dry weight), respectively. These results demonstrated that the endoplasmic reticulum is the organelle responsible for Ca regulation in neurons and other cell types. The same authors proposed STEM-EELS for Fe mapping in mouse brain cells to determine the number of Fe atoms in ferritin particles. Brain tissue samples with 200 nm thicknesses (cut using an ultramicrotome), perfused with 2% paraformaldehyde and 1% glutaraldehyde in a 0.1 M sodium cacodylate buffer and fixed with an osmium tetroxide solution, were analysed. To minimize mass loss of the tissues by the electron beam, the sample specimen was cooled to −160 °C during STEM-EELS analysis. Images were obtained using a scanning transmission microscope equipped with a nanometre-sized probe source operating up to 100 kV. To estimate the number of Fe atoms in each ferritin particle, the Fe and C signals were integrated over boxed areas. The Fe to C atoms ratio was given by:

$$\frac{\mathrm{NFe}}{\mathrm{NC}} = \frac{\mathrm{IFe}(\beta,\Delta 1)}{\mathrm{IC}(\beta,\Delta 2)} \times \frac{\sigma \mathrm{C}(\beta,\Delta 2)}{\sigma \mathrm{Fe}(\beta,\Delta 1)}$$

where Nx is the number of x atoms (Fe and C), Ix is the core edge signal integrated over scattering angle β and energy window Δ, and σx is the partial ionization cross section.

The number of C atoms in each boxed area was about 2.25×10^5, while the number of Fe atoms in each ferritin particle ranged from 1200 to 3600, which was in agreement with the maximum number of Fe atoms per ferritin particle (4500) previously proposed (Zhang et al. 2005).

Morello et al. (2002) employed STEM-EELS for Mn mapping in mitochondria of brain cells of healthy rats, where Mn is normally present in very small concentrations (up to 2 $\mu g\ g^{-1}$). Samples were prepared by conventional perfusion and fragmentation in small pieces (<40 nm) by means of a stereomicroscope, followed by fixation with 1% osmium tetroxide. Manganese subcellular deposits were quantitatively evaluated based on the relative optical density of Mn particles in mitochondria. The lowest level of bright intensity matched the background signals and was equal to 0 (zero) optical density units (ODU), while the highest level of bright intensity was equal to 20,000 ODU, corresponding to the maximum concentration of Mn. For Mn localization in the mitochondria and mapping construction, the signals from the STEM-EELS spectrum ranging from 637 to 665 eV (corresponding to the $MnL_{2,3}$ edge) were considered. The images obtained showed Mn particles localized mainly in mitochondria, where the smallest Mn particles were about 0.2 nm in diameter.

It is possible to produce a 3D map of an element (elemental tomography) in a given specimen by means of computational methods after collecting STEM-EELS images at energy losses above and below the core edge of the element (Aronova and Leapman 2012; Sousa and Leapman 2012; Collins and Midgley 2017). Using this approach, 3D maps of Fe bound to melanin and ferritin in fungus cells and degenerating neurons in mice, respectively, were obtained (Leapman et al. 1993; Zhang et al. 2005).

New STEM-EELS and EFTEM applications are being devoted to identification of markers and nanomaterials used in human diagnostics or therapeutic applications. Elemental mapping in luminescent semiconductor nanocrystals (also called QDs) used in immunoanalysis is cited as an example. The STEM-EELS and EFTEM spectra of QDs present peaks corresponding to specific element ionization edges, enabling the localization of specific biomolecules in cells (Nisman et al. 2004; Van Schooneveld et al. 2010; Chen et al. 2017). Labelled heavy atoms, such as Au, in clusters attached to antibody fragments can be detected by STEM-EELS tomography, which enables mapping of specific proteins in cells (Sousa et al. 2007). Protein mapping in cells is also possible via metal-tagging TEM (METTEM). In this case, the protein identification is based on the fusion of metallothionein with a specific protein, followed by treatment with Au salts that produces Au-thiolate clusters with 1 nm in diameter (Risco et al. 2012).

The interaction of the electron beam produced by electron microscopy with the atoms of a sample generates X-ray photons that can be used for X-ray microanalysis. The X-ray spectrum of an element contains characteristic peaks that correspond to transitions of electrons from the inner shells of the atom. The X-ray signal can be measured using energy-dispersive spectrometry (EDS) or wavelength-dispersive spectrometry (WDS). EDX is widely employed in electron microscopes, but the spatial resolution of about 500 nm of EDS is low when compared to STEM-EELS and EFTEM (Ortega 2005; Fiori et al. 1988).

The X-ray signal intensity is dependent on beam size and specimen thickness. In order to provide a suitable X-ray signal for elemental mapping, a relatively large beam (10–50 nm) is required for the thermionic source. On the other hand, for a FEG l, a 2 nm beam size ensures enough current at the EDX detector (Friel and Lyman 2006). The signal-to-noise ratio (SNR) increases when the specimen thickness increases, improving the EDS mapping. This is different from EFTEM mapping, where better mapping is obtained when thinner specimens are analysed (Watanabe et al. 2003).

With respect to X-ray collection, the EDX analysis can be divided into spot analysis and elemental mapping. In spot analysis, an analogue scanning beam is focused over a specific cellular region, and an X-ray spectrum is collected over a period of time. In elemental mapping, 2D arrays of X-ray data are collected across a specimen using a digital scanning raster; that is, at each point in a preset matrix, the beam is fixed for a given amount of time, and a spectrum is collected (Wong et al. 1989). Data acquisition ranging from 1 h to 10 h or more is required in order to obtain useful maps by EDX (Friel and Lyman 2006).

The EDS application for mapping is limited by relatively low sensitivity due to the poor collection of X-rays, which is geometrically unfavourable. In order to improve the EDX sensitivity, a dual energy-dispersive spectroscopy detector (mounted on opposite sides of the electron column) has been proposed, especially for Zn, which is not detected by STEM-EELS due to the poor edge definition (Wu et al. 2013b). For elemental mapping, EDX is combined with EM techniques, which provide an electron beam for generation of specific X-ray photons from the elements present in the specimen (Penen et al. 2016).

The most common application of TEM or STEM combined with EDX for elemental mapping in biological samples has dealt with Ca, K, Mg, Na, and Fe, probably due to the high concentration of these elements in most samples (Fiori et al. 1988; Penen et al. 2016; Kowarski 1984; Somlyo 1984). By using such a combination, Somlyo (1984) analysed muscle tissue cryo-sections, where Ca and K were mapped with subcellular resolution. It was observed that a high probe current is necessary to achieve spatial resolution, at the expense of damaging the specimen. Therefore, a FEG probe is mandatory for analysis using EDX associated with STEM to provide elemental maps with subcellular resolution. According to Somlyo (1984), the boundaries of diffusible elements in cells are preserved in cryo-sections. The author compared STEM-EELS and EDX and concluded that STEM-EELS should be the first choice for elemental mapping in biological samples due to the higher resolution achieved.

Although EDX enables elemental mapping, most biological samples analyses have been carried out based on spot analysis without interest in getting a map, only a X-ray spectrum. The main reason might be the low sensitivity of EDX for the elemental mapping in subcellular structures (Kowarski 1984). However, this mapping is possible when EDX is combined with EM techniques, as is the case with Al involved in neurodegenerative disorders (Yumoto et al. 2009; Komine et al. 2000);

distribution of Ca, Mg, and Zn in organelle cells (Ruiz et al. 2001; Komine et al. 2000); and metal in QDs or NP bound to biomolecules labelled with an antibody that enables molecule identification (Scotuzzi et al. 2017).

Combination of chemical images in biological samples obtained by different techniques is a current trend. An example of this multimodal application is the combination of EFTEM and NanoSIMS. With this multimodal image approach, the same biological cell/organelle can be analysed in parallel by these techniques; while NanoSIMS enables isotope ratio determination, EFTEM does the morphological and structural characterization (Cunha et al. 2016; Penen et al. 2016).

7.5.1 Sample Preparation for STEM-EELS, EFTEM, and EDX

Chemical fixation and freezing are the main sample preparation procedures for cells and biological tissue analyses by EELS, EFTEM, or EDX, approaches that are equal or similar to SIMS. In order to fix the elements bound to insoluble macromolecules (proteins, nucleic acids, carbohydrates, and lipids), the samples are usually treated with glutaraldehyde and osmium tetroxide solutions, followed by dehydration with polar solvents and embedding in epoxy resins. However, the abovementioned procedure is inconvenient for soluble macromolecules because the soluble compounds and elements are washed out from the cells and tissues. In this case, precipitation and cryofixation are more suitable for sample preparation (Thompson et al. 2016). The precipitation method is based on the use of specific reagents that produce insoluble compounds of the element of interest. One example is the precipitation of Ca with ammonium oxalate, producing calcium oxalate (Nagata 2004). In the case of cryofixation, the specimen is submitted to very low temperature, usually by immersion in cooled fluid pentane or liquid nitrogen or helium, followed by sample cutting with a cryomicrotome (Thompson et al. 2016).

Although chemical fixation is widely used for elemental mapping in cells, it is not suitable when observation of detailed structures at the molecular resolution level is desired. Cryofixation methods are preferred in this case, involving freezing the sample quickly, usually in liquid nitrogen. Because water is the major component, the biological specimen is immobilized as vitrified water. Moreover, cryofixation preserves the hydrated state of the molecules and reduces possible electron radiation damages (Thompson et al. 2016; Mitsuoka 2011).

7.6 Conclusion

Elemental imaging (2D and 3D) techniques largely used in metallomic bioimaging were addressed in this chapter. Each technique has its specificity, strengths, and limitations, being complementary to each other. Relatively low spatial resolution

(>5 μm) is achieved with LA-ICP-MS, but the technique provides quantitative and isotopic analysis. On the other hand, quantitative analysis by SIMS is more difficult, but it enables very high spatial resolution (>50 nm) and isotopic analysis. Electron microscopy and X-ray provide very high resolution (>2 nm), whereas X-ray also enables speciation analysis. The main characteristics of the techniques discussed in the present chapter are summarized in Table 7.1.

Table 7.1 Techniques employed for imaging of elements in metallomics and respective characteristics

Technique	LOD, μg g^{-1}	Condition	Spatial resolution, μm	Selectivity	Remarks
LA-ICP-MS	0.01	Ambient	Down to 1	Multielemental and isotopic analysis	High sensitivity for transition metals Operates at atmospheric pressures Low spatial resolution Qualitative and quantitative
ToF-SIMS	0.1	UHV	Down to 0.1	Molecular analysis	Low sensitivity for transition metals Small sampling area Sample must resist high vacuum Samples must be flat Quantitative
NanoSIMS	0.1	UHV	Down to 0.05	Multielemental and isotopic analysis	Low sensitivity for transition metals Small sampling area Sample must resist high vacuum Samples must be flat Qualitative
SRXRF	0.1	Ambient	Down to 0.02	Multielemental analysis ($Z \geq 6$)	Good sensitivity for transition metals High spatial resolution Operates at atmospheric pressures Samples must be flat Qualitative and quantitative
EM-EELS	1000	UHV	Down to 0.001	Multielemental analysis ($Z \geq 6$)	Complex procedures for specimen preparation FEG source is preferred Identification of water, lipid, protein, sugar, and proteins linked to metals Allows 3D chemical map
EM-EDX	1000	UHV	Down to 0.5	Multielemental analysis ($Z \geq 6$)	Complex procedures for specimen preparation FEG source is preferred Suitable for mapping of endogenous elements (Ca, Fe, K, Mg, and Na)

References

Amstalden van Hove ER, Smith DF, Heeren RMA (2010) A concise review of mass spectrometry imaging. J Chromatogr A 1217:3946–3954

Aronova MA, Leapman RD (2012) Development of electron energy loss spectroscopy in the biological sciences. MRS Bull 37(1):53–62

Arora M, Hare D, Austin C et al (2011) Spatial distribution of manganese in enamel and coronal dentine of human primary teeth. Sci Total Environ 409(7):1315–1319

Austin C, Hare D, Rozelle AL et al (2009) Elemental bio-imaging of calcium phosphate crystal deposits in knee samples from arthritic patients. Metallomics 1:142–147

Austin C, Fryer F, Lear J et al (2011) Factors affecting internal standard selection for quantitative elemental bio-imaging of soft tissues by LA–ICP–MS. J Anal At Spectrom 26:1494–1501

Bandura DR, Baranov VI, Ornatsky OI et al (2009) Mass cytometry: technique for real time single cell multitarget immunoassay based on inductively coupled plasma time-of-flight mass spectrometry. Anal Chem 81(16):6813–6822

Becker JS, Zoriy MV, Pickhardt C et al (2005a) Imaging of copper, zinc, and other elements in thin section of human brain samples (Hippocampus) by laser ablation inductively coupled plasma mass spectrometry. Anal Chem 77(10):3208–3216

Becker JS, Zoriy MV, Dehnhardt M (2005b) Copper, zinc, phosphorus and sulfur distribution in thin section of rat brain tissues measured by laser ablation inductively coupled plasma mass spectrometry: possibility for small-size tumor analysis. J Anal At Spectrom 20:912–917

Becker JS, Zoriy M, Becker JS et al (2007) Elemental imaging mass spectrometry of thin sections of tissues and analysis of brain proteins in gels by laser ablation inductively coupled plasma mass spectrometry. Phys Status Solidi C 4(6):1775–1784

Becker JS, Breuer U, Hsieh H-F et al (2010) Bioimaging of metals and biomolecules in mouse heart by laser ablation inductively coupled plasma mass spectrometry and secondary ion mass spectrometry. Anal Chem 82(22):9528–9533

Becker JS, Matusch A, Wu B (2014) Bioimaging mass spectrometry of trace elements– recent advance and applications of LA–ICP–MS: a review. Anal Chim Acta 835:1–18

Bellis DJ, Hetter KM, Jones J et al (2006) Calibration of laser ablation inductively coupled plasma mass spectrometry for quantitative measurements of lead in bone. J Anal At Spectrom 21:948–954

Bianga J, Touat-Hamici Z, Bierla K et al (2014) Speciation analysis for trace levels of selenoproteins in cultured human cells. J Proteome 108:316–324

Blaske F, Reifschneider O, Gosheger G et al (2014) Elemental bioimaging of nanosilver-coated prostheses using X-ray fluorescence spectroscopy and laser ablation–inductively coupled plasma–mass spectrometry. Anal Chem 86(1):615–620

Bloom AN, Tian H, Winograd N (2016) C60-SIMS imaging of nanoparticles within mammalian cells. Biointerphases 11(2):02A306-1–02A306-7

Böhme S, Stärk H-J, Kühnel D et al (2015) Exploring LA–ICP–MS as a quantitative imaging technique to study nanoparticle uptake in Daphnia magna and zebrafish (*Danio rerio*) embryos. Anal Bioanal Chem 407(18):5477–5485

Bonta M, Lohninger H, Laszlo V et al (2014) Quantitative LA–ICP–MS imaging of platinum in chemotherapy treated human malignant pleural mesothelioma samples using printed patterns as standard. J Anal At Spectrom 29:2159–2167

Bonta M, Limbeck A, Quarles CD Jr et al (2015) A metric for evaluation of the image quality of chemical maps derived from LA–ICP–MS experiments. J Anal At Spectrom 30:1809–1815

Bonta M, Hegedus B, Limbeck A (2016) Application of dried-droplets deposited on pre-cut filter paper disks for quantitative LA–ICP–MS imaging of biologically relevant minor and trace elements in tissue samples. Anal Chim Acta 908:54–62

Bourassa MW, Miller LM (2012) Metal imaging in neurodegenerative diseases. Metallomics 4(8):721–738

Bourassa D, Gleber S-C, Vogt S et al (2014) 3D imaging of transition metals in the zebrafish embryo by X-ray fluorescence microtomography. Metallomics 6(9):1648–1655
Bourassa D, Gleber S-C, Vogt S et al (2016) MicroXRF tomographic visualization of zinc and iron in the zebrafish embryo at the onset of the hatching period. Metallomics 8(10):1122–1130
Boxer SG, Kraft ML, Weber PK (2009) Advances in imaging secondary ion mass spectrometry for biological samples. Annu Rev Biophys 38:53–74
Braidy N, Poljak A, Marjo C et al (2014) Metal and complementary molecular bioimaging in Alzheimer's disease. Front Integr Neurosci 6:1–14
Castaing R, Slodzian G (1962) Microanalyse par emission ionique secondaire. J Microsc 1:395–410
Castro W, Hoogewerff J, Latkoczy C (2010) Application of laser ablation (LA–ICP–SF–MS) for the elemental analysis of bone and teeth samples for discrimination purposes. Forensic Sci Int 195(1–3):17–27
Chandra S (2008) Challenges of biological sample preparation for SIMS imaging of elements and molecules at subcellular resolution. Appl Surf Sci 255:1273–1284
Chandra S, Tjarks W, Lorey DR et al (2008) Quantitative subcellular imaging of boron compounds in individual mitotic and interphase human glioblastoma cells with imaging secondary ion mass spectrometry (SIMS). J Microsc 229:92–103
Chen D, Monteiro-Riviere NA, Zhang LW (2017) Intracellular imaging of quantum dots, gold, and iron oxide nanoparticles with associated endocytic pathways. Wiley Interdiscip Rev Nanomed Nanobiotechnol 9(2):1–19
Claverie F, Pecheyran C, Mounicou S et al (2009) Characterization of the aerosol produced by infrared femtosecond laser ablation of polyacrylamide gels for the sensitive inductively coupled plasma mass spectrometry detection of selenoproteins. Spectrochim Acta B 64(7):649–658
Collins SM, Midgley PA (2017) Progress and opportunities in EELS and EDS tomography. Ultramicroscopy 180:133–141
Corezzi S, Urbanelli L, Cloetens P et al (2009) Synchrotron based X-ray fluorescence imaging of human cells labeled with CdSe quantum dots. Anal Biochem 388:33–39
Cunha MML, Trepout S, Messaoudi C et al (2016) Overview of chemical imaging methods to address biological questions. Micron 84:23–26
De Jonge MD, Vogt S (2010) Hard X-ray fluorescence tomographyan emerging tool for structural visualization. Curr Opin Struct Biol 20:606–614
Dérue C, Gibouin D, Lefebvre F et al (2006) Relative sensitivity factors of inorganic cations in frozen-hydrated standards in secondary ion MS analysis. Anal Chem 78(8):2471–2477
Dias AA, Carvalho M, Carvalho ML et al (2015) Quantitative evaluation of ante-mortem lead in human remains of the 18th century by triaxial geometry and bench top micro X-ray fluorescence spectrometry. J Anal At Spectrom 30:2488–2495
Douglas DN, O'Reilly J, O'Connor C et al (2016) Quantitation of the Fe spatial distribution in biological tissue by online double isotope dilution analysis with LA–ICP–MS: a strategy for estimating measurement uncertainty. J Anal At Spectrom 31:270–279
Drescher D, Giesen C, Traub H et al (2012) Quantitative imaging of gold and silver nanoparticles in single eukaryotic cells by laser ablation ICP–MS. Anal Chem 84(22):9684–9688
Egger AE, Theiner S, Kornauth C et al (2014) Quantitative bioimaging by LA–ICP–MS: a methodological study on the distribution of Pt and Ru in viscera originating from cisplatin- and KP1339-treated mice. Metallomics 6(9):1616–1625
Farell J, Amarasiriwardena D, Goodman AH et al (2013) Bioimaging of trace metals in ancient Chilean mummies and contemporary Egyptian teeth by laser ablation–inductively coupled plasma–mass spectrometry (LA–ICP–MS). Microchem J 106:340–346
Feldmann J, Kindness A, Ek P (2002) Laser ablation of soft tissue using a cryogenically cooled ablation cell. J Anal At Spectrom 17:813–818
Fernandez B, Claverie F, Pecheyran C et al (2007) Direct analysis of solid samples by fs-LA–ICP–MS. TrAC-Trends Anal Chem 26(10):951–966
Fiori CE, Leapman RD, Swyt CR (1988) Quantitative X-ray mapping of biological cryosections. Ultramicroscopy 24(2–3):237–250

Fiori CE (1988) The new electron microscopy: imaging the chemistry of nature. Anal Chem 60(12):86R–90R

Frick DA, Günther D (2012) Fundamental studies on the ablation behaviour of carbon in LA–ICP–MS with respect to the suitability as internal standard. J Anal At Spectrom 27:1294–1303

Frick DA, Giesen C, Hemmerle T et al (2015) An internal standardisation strategy for quantitative immunoassay tissue imaging using laser ablation inductively coupled plasma mass spectrometry. J Anal At Spectrom 30:254–259

Friel JJ, Lyman CE (2006) X-ray mapping in electron-beam instruments. Microsc Microanal 12(1):2–25

Giesen C, Wang HAO, Schapiro D (2014) Highly multiplexed imaging of tumor tissues with subcellular resolution by mass cytometry. Nat Methods 11:417–422

Gray AL (1985) Solid sample introduction by laser ablation for inductively coupled plasma source mass spectrometry. Analyst 110:551–556

Grovenor CRM, Smart KE, Kilburn M et al (2006) Specimen preparation and calibration for NanoSIMS analysis of biological materials. Appl Surf Sci 252:6917–6924

Gulin AA, Pavlyukov MS, Gularyan SK et al (2015) Visualization of the spatial distribution of Pt^+ ions in cisplatin-treated glioblastoma cells by time-of-flight secondary ion mass spectrometry. Biochem (Mosc) Suppl Ser A Membr Cell Biol 9(3):202–209

Gunther D, Hattendorf B (2005) Solid sample analysis using laser ablation inductively coupled plasma mass spectrometry. TrAC-Trends Anal Chem 24(3):255–265

Hachmöller O, GuilhermeBuzanich A, Aichler M et al (2016) Elemental bioimaging and speciation analysis for the investigation of Wilson's disease using μXRF and XANES. Metallomics 8(7):648–653

Hagège A, Huynh TNS, Hébrant M (2015) Separative techniques for metalloproteomics require balance between separation and perturbation. TrAC-Trends Anal Chem 64:64–74

Hamilton JS, Gorishek EL, Mach PM et al (2016) Evaluation of a custom single Peltier-cooled ablation cell for elemental imaging of biological samples in laser ablation–inductively coupled plasma–mass spectrometry (LA–ICP–MS). J Anal At Spectrom 31:1030–1033

Hare D, Reedy B, Grimm R et al (2009) Quantitative elemental bio-imaging of Mn, Fe, Cu and Zn in 6–hydroxydopamine induced Parkinsonism mouse models. Metallomics 1:53–58

Hare D, Austin C, Doble P et al (2011) Elemental bio-imaging of trace elements in teeth using laser ablation–inductively coupled plasma–mass spectrometry. J Dent 39(5):397–403

Hare DJ, Lee JK, Beavis AD et al (2012) Three-dimensional atlas of iron, copper, and zinc in the mouse cerebrum and brainstem. Anal Chem 84(9):3990–3997

Hare DJ, George JL, Bray L et al (2014) The effect of paraformaldehyde fixation and sucrose cryoprotection on metal concentration in murine neurological tissue. J Anal At Spectrom 29:565–570

He Y, Esteban-Fernández D, Neumann B et al (2016) Application of MeCAT-Click labeling for protein abundance characterization of E. coli after heat shock experiments. J Proteome 136:68–76

Hofer F, Grogger W, Warbichler P et al (2000) Quantitative energy-filtering transmission electron microscopy (EFTEM). Mikrochim Acta 132(2–4):273–288

Hoffmann E, Stephanowitz H, Ullrich E et al (2000) Investigation of mercury migration in human teeth using spatially resolved analysis by laser ablation–ICP–MS. J Anal At Spectrom 15:663–667

Horowitz P, Howell JA (1972) A scanning X-ray microscope using synchrotron radiation. Science 178:607–608

Hugo L-F, Pessôa GS, Arruda MAZ et al (2016) LA-iMageS: a software for elemental distribution bioimaging using LA–ICP–MS data. J Cheminformatics 8:65–74

Jackson B, Harper S, Smith L et al (2006) Elemental mapping and quantitative analysis of Cu, Zn, and Fe in rat brain sections by laser ablation ICP–MS. Anal Bioanal Chem 384(4):951–957

Jiménez MS, Rodriguez L, Bertolin JR et al (2013) Evaluation of gel electrophoresis techniques and laser ablation–inductively coupled plasma–mass spectrometry for screening analysis of Zn and Cu-binding proteins in plankton. Anal Bioanal Chem 405(1):359–368

Jimenez MS, Luque-Alled JM, Gomez T et al (2016) Evaluation of agarose gel electrophoresis for characterization of silver nanoparticles in industrial products. Electrophoresis 37(10):1376–1383

Jurowski K, Buszewski B, Piekoszewski W (2015) Bioanalytics in quantitative (bio)imaging/mapping of metallic elements in biological samples. Crit Rev Anal Chem 45(4):334–347

Kang D, Amarasiriwardena D, Goodman AH (2004) Application of laser ablation–inductively coupled plasma–mass spectrometry (LA–ICP–MS) to investigate trace metal spatial distributions in human tooth enamel and dentine growth layers and pulp. Anal Bioanal Chem 378(6):1608–1615

Katz W, Newman JG (1987) Fundamentals of secondary ion mass spectrometry. MRS Bull 7:40–46

Kindness A, Sekaran CN, Feldmann J (2003) Two-dimensional mapping of copper and zinc in liver sections by laser ablation-inductively coupled plasma mass spectrometry. Clin Chem 49(11):1916–1923

Komine Y, Eggink LL, Park H et al (2000) Vacuolar granules in *Chlamydomonas reinhardtii*: polyphosphate and a 70-kDa polypeptide as major components. Planta 210(6):897–905

Konz I, Fernandez B, Fernandez ML et al (2013) Gold internal standard correction for elemental imaging of soft tissue sections by LA–ICP–MS: element distribution in eye microstructures. Anal Bioanal Chem 405(10):3091–3096

Konz I, Fernandez B, Fernandez ML et al (2014a) Design and evaluation of a new Peltier-cooled laser ablation cell with on-sample temperature control. Anal Chim Acta 809:88–96

Konz I, Fernández B, Fernández ML et al (2014b) Quantitative bioimaging of trace elements in the human lens by LA–ICP–MS. Anal Bioanal Chem 406(9):2343–2348

Kowarski D (1984) Intelligent interface for a microprocessor controlled scanning transmission electron microscope with X-ray imaging. J Electron Micro Tech 1(2):175–184

Leapman RD, Hunt JA, Buchanan RA et al (1993) Measurement of low calcium concentrations in cryosectioned cells by parallel-EELS mapping. Ultramicroscopy 49(1–4):225–234

Leapman RD (2017) Application of EELS and EFTEM to the life sciences enabled by the contributions of Ondrej Krivanek. Ultramicroscopy 180:180–187

Lear J, Hare DJ, Fryer F et al (2012) High-resolution elemental bioimaging of Ca, Mn, Fe, Co, Cu, and Zn employing LA–ICP–MS and hydrogen reaction gas. Anal Chem 84(15):6707–6714

Lee RFS, Escrig S, Croisier M et al (2015) NanoSIMS analysis of an isotopically labeled organometallic ruthenium(II) drug to probe its distribution and state in vitro. Chem Commun 51:16486–16489

Legin AA, Schintlmeister A, Jakupec MA et al (2014) NanoSIMS combined with fluorescence microscopy as a tool for subcellular imaging of isotopically labeled platinum-based anticancer drugs. Chem Sci 5:3135–3143

Limbeck A, Galler P, Bonta M et al (2015) Recent advances in quantitative LA–ICP–MS analysis: challenges and solutions in the life sciences and environmental chemistry. Anal Bioanal Chem 407(22):6593–6617

Lobinski R, Moulin C, Ortega R (2006) Imaging and speciation of trace elements in biological environment. Biochimie 88:1591–1604

Lombi E, De Jonge MD, Donner E et al (2011) Trends in hard X-ray fluorescence mapping: environmental applications in the age of fast detectors. Anal Bioanal Chem 400:1637–1644

Lum TS, Ho CL, Tsoi YK et al (2016) Elemental bioimaging of platinum in mouse tissues by laser ablation–inductively coupled plasma–mass spectrometry for the study of localization behavior of structurally similar complexes. Int J Mass Spectrom 404:40–47

Maia F-D, Chen B-J, Wu L-C et al (2006) Imaging of single liver tumor cells intoxicated by heavy metals using ToF-SIMS. Appl Surf Sci 252:6809–6812

Managh AJ, Edwards S, Bushell A et al (2013) Single cell tracking of gadolinium labeled $CD4^+$ T cells by laser ablation inductively coupled plasma mass spectrometry. Anal Chem 85(22):10627–10634

Managh AJ, Hutchinson RW, Riquelme P et al (2014) Laser ablation–inductively coupled plasma mass spectrometry: an emerging technology for detecting rare cells in tissue sections. J Immunol 193(5):2600–2608

Marshall J, Franks J et al (1991) Determination of trace elements in solid plastic materials by laser ablation–inductively coupled plasma mass spectrometry. J Anal At Spectrom 6:145–150
Martínez-Criado G, Tucoulou R, Cloetens P et al (2012) Status of the hard X-ray microprobe beamline ID22 of the European synchrotron radiation facility. J Synchrotron Rad 19:10–18
Matsuyama S, Shimura M, Mimura H et al (2009) Trace element mapping of a single cell using a hard X-ray nanobeam focused by a Kirkpatrick–Baez mirror system. X-Ray Spectrom 38:89–94
Matsuyama S, Shimura M, Fujii M et al (2010) Elemental mapping of frozen-hydrated cells with cryo-scanning X-ray fluorescence microscopy. X-Ray Spectrom 39:260–266
McRae R, Lai B, Vogt S et al (2006) Correlative microXRF and optical immunofluorescence microscopy of adherent cells labeled with ultra small gold particles. J Struct Biol 155(1):22–29
Mitsuoka K (2011) Obtaining high-resolution images of biological macromolecules by using a cryo-electron microscope with a liquid-helium cooled stage. Micron 42(2):100–106
Moraleja I, Esteban-Fernández D, Lázaro A et al (2016) Printing metal-spiked inks for LA–ICP–MS bioimaging internal standardization: comparison of the different nephrotoxic behavior of cisplatin, carboplatin, and oxaliplatin. Anal Bional Chem 408(9):2309–2318
Morello M, Canini A, Caiola MG et al (2002) Manganese detected by electron spectroscopy imaging and electron energy loss spectroscopy in mitochondria of normal rat brain cells. J Trace Microprobe T 20(4):481–491
Nagata T (2004) X-ray microanalysis of biological specimens by high voltage electron microscopy. Prog Histochem Cytochem 39(4):185–319
Nisman R, Dellaire G, Ren Y et al (2004) Application of quantum dots as probes for correlative fluorescence, conventional, and energy-filtered transmission electron microscopy. J Histochem Cytochem 52(1):13–18
Nygren H, Dahlén G, Malmberg P (2014) Analysis of As- and Hg-species in metal-resistant oral bacteria, by imaging ToF-SIMS. Basic Clin Pharmacol Toxicol 115:129–133
Noël M, Spence J, Harris KA et al (2014) Grizzly bear hair reveals toxic exposure to mercury through salmon consumption. Environ Sci Technol 48(13):7560–7567
Ortega R (2005) Chemical elements distribution in cells. Nucl Instrum Meth B 231(1–4):218–223
Osterholt T, Salber D, Matusch A et al (2011) IMAGENA: image generation and analysis – an interactive software tool handling LA–ICP–MS data. Int J Mass Spectrom 307(1–3):232–239
O'Reilly J, Douglas D, Braybrook J et al (2014) A novel calibration strategy for the quantitative imaging of iron in biological tissues by LA–ICP–MS using matrix-matched standards and internal standardization. J Anal At Spectrom 29:1378–1384
Pozebon D, Dressler VL, Matusch A (2008) Monitoring of platinum in a single hair by laser ablation inductively coupled plasma mass spectrometry (LA–ICP–MS) after cisplatin treatment for cancer. Int J Mass Spectrom 272(1):57–62
Pozebon D, Scheffler GL, Dressler VL et al (2014) Review of the applications of laser ablation inductively coupled plasma mass spectrometry (LA–ICP–MS) to the analysis of biological samples. J Anal At Spectrom 29:2204–2228
Pacholski ML, Winograd N (1999) Imaging with mass spectrometry. Chem Rev 99:2977–3005
Paul B, Paton C, Norris A et al (2012) CellSpace: a module for creating spatially registered laser ablation images within the Iolite freeware environment. J Anal At Spectrom 27(4):700–706
Paul B, Hare DJ, Bishop DP et al (2015) Visualising mouse neuroanatomy and function by metal distribution using laser ablation-inductively coupled plasma-mass spectrometry imaging. Chem Sci 6:5383–5383
Penen F, Malherbe J, Isaure MP et al (2016) Chemical bioimaging for the subcellular localization of trace elements by high contrast TEM, TEM/X-EDS, and NanoSIMS. J Trace Elem Med Bio 37:62–68
Pennycook SJ (2012) Scanning transmission electron: Z-contrast imaging. In: Kaufmann EN (ed) Characterization of materials. John Wiley & Sons, Inc., Hoboken, pp 1736–1763
Pessôa GS, Capelo-Martínez JL, Fdez-Riverola F et al (2016) Laser ablation and inductively coupled plasma mass spectrometry focusing on bioimaging from elemental distribution using MatLab software: a practical guide. J Anal At Spectrom 31:832–840

Pornwilard MM, Merle U, Weiskirchen R et al (2013) Bioimaging of copper deposition in Wilson's diseases mouse liver by laser ablation inductively coupled plasma mass spectrometry imaging (LA–ICP–MSI). Int J Mass Spectrom 354:281–287

Pozebon D, Dressler VL, Mesko MF (2010) Bioimaging of metals in thin mouse brain section by laser ablation inductively coupled plasma mass spectrometry: novel online quantification strategy using aqueous standards. J Anal At Spectrom 25:1739–1744

Pozebon D, Scheffler GL, Dressler VL (2017) Recent applications of laser ablation inductively coupled plasma mass spectrometry (LA–ICP–MS) for biological sample analysis: a follow-up review. J Anal At Spectrom 32:890–919

Pugh JAT, Cox AG, McLeod CW et al (2011) A novel calibration strategy for analysis and imaging of biological thin sections by laser ablation inductively coupled plasma mass spectrometry. J Anal At Spectrom 26:1667–1673

Qin ZY, Caruso JA, Lai B et al (2011) Trace metal imaging with high spatial resolution: applications in biomedicine. Metallomics 3(1):28–37

Quintana C, Bellefqih S, Laval JY et al (2006) Study of the localization of iron, ferritin, and hemosiderin in Alzheimer's disease hippocampus by analytical microscopy at the subcellular level. J Struct Biol 153:42–54

Quintana C, Wu T, Delatour B, Dhenain M et al (2007) Morphological and chemical studies of pathological human and mice brain at the subcellular level: correlation between light, electron, and NanoSIMS microscopies. Microsc Res Tech 70:281–289

Ranaldi MM, Gagnon MM (2009) Accumulation of cadmium in the otoliths and tissues of juvenile pink snapper (Pagrus auratus Forster) following dietary and waterborne exposure. Comp Biochem Phys C Toxicol Pharmacol 150(4):421–427

Ryan CG, Siddons DP, Moorhead G et al (2009) High-throughput X-ray fluorescence imaging using a massively parallel detector array, integrated scanning and real-time spectral deconvolution. 9th international conference on X-ray microscopy. J Phys Conf Ser 186(1):012013–012015

Reifschneider O, Wehe CA, Diebold K et al (2013a) Elemental bioimaging of haematoxylin and eosin-stained tissues by laser ablation ICP–MS. J Anal At Spectrom 28:989–993

Reifschneider O, Wehe CA, Raj I et al (2013b) Quantitative bioimaging of platinum in polymer embedded mouse organs using laser ablation ICP–MS. Metallomics 5:1440–1447

Reifschneider O, Wentker KS, Strobel K et al (2015) Elemental bioimaging of thulium in mouse tissues by laser ablation–ICPMS as a complementary method to heteronuclear proton magnetic resonance imaging for cell tracking experiments. Anal Chem 87(8):4225–4230

Risco C, Sanmartın-Conesa E, Tzeng WP et al (2012) Specific, sensitive, high-resolution detection of protein molecules in eukaryotic cells using metal-tagging transmission electron microscopy. Structure 20(5):759–766

Ruiz FA, Marchesini N, Seufferheld M et al (2001) The polyphosphate bodies of *Chlamydomonas reinhardtii* possess a proton-pumping pyrophosphatase and are similar to acidocalcisomes. J Biol Chem 276(49):46196–46203

Santos MC, Wagner M, Wu B (2009) Biomonitoring of metal contamination in a marine prosobranch snail (Nassarius reticulatus) by imaging laser ablation inductively coupled plasma mass spectrometry (LA–ICP–MS). Talanta 80(2):428–433

Sarret G, Smits EAHP, Michel HC et al (2013) Use of synchrotron-based techniques to elucidate metal uptake and metabolism in plants. Adv Agron 119:1–82

Scharlach C, Müller L, Wagner S et al (2016) LA–ICP–MS allows quantitative microscopy of europium-doped iron oxide nanoparticles and is a possible alternative to ambiguous prussian blue iron staining. J Biomed Nanotechnol 12:1001–1010

Scotuzzi M, Kuipers J, Wensveen DI et al (2017) Multi-color electron microscopy by element-guided identification of cells, organelles and molecules. Sci Rep 7:45970

Sela H, Karpas Z, Zoriy M et al (2007) Biomonitoring of hair samples by laser ablation inductively coupled plasma mass spectrometry (LA–ICP–MS). Int J Mass Spectrom 261(2–3):199–207

Sela H, Karpas Z, Cohen H et al (2011) Preparation of stable standards of biological tissues for laser ablation analysis. Int J Mass Spectrom 307(1–3):142–148

Shariatgorji M, Nilsson A, Bonta M et al (2016) Direct imaging of elemental distributions in tissue sections by laser ablation mass spectrometry. Methods 104:86–92
Smulders S, Larue C, Sarret G et al (2015) Lung distribution, quantification, co-localization and speciation of silver nanoparticles after lung exposure in mice. Toxicol Lett 238(1):1–6
Somlyo AP (1984) Compositional mapping in biology: X rays and electrons. J Ultra Mol Struct Res 88(2):135142
Sousa AA, Aronova MA, Kim YC et al (2007) On the feasibility of visualizing ultra small gold labels in biological specimens by STEM tomography. J Struct Biol 159(3):507–522
Sousa AA, Leapman RD (2012) Development and application of STEM for the biological sciences. Ultramicroscopy 123:38–49
Sparks CJ Jr, Raman S, Yakel HL et al (1977) Search with SR for super heavy elements in giant-halo inclusions. Phys Rev Lett 38:205–208
Sparks CJ Jr (1980) X-ray fluorescence microprobe for chemical analysis. In: Winick H, Doniach S (eds) Synchrotron radiation research. Springer, USA., New York. Chapter 14, pp 459–512
Thompson RF, Walker M, Siebert A et al (2016) An introduction to sample preparation and imaging by cryo-electron microscopy for structural biology. Methods 100:3–15
Todoli JL, Mermet JM (1998) Study of polymer ablation products obtained by ultraviolet laser ablation — inductively coupled plasma atomic emission spectrometry. Spectrochim Acta B 53(12):1645–1656
Uerlings R, Matusch A, Weiskirchen R (2016) Reconstruction of laser ablation inductively coupled plasma mass spectrometry (LA–ICP–MS) spatial distribution images in Microsoft Excel 2007. Int J Mass Spectrom 395:27–35
Van Malderen SJM, Managh AJ, Sharp BL et al (2016a) Recent developments in the design of rapid response cells for laser ablation–inductively coupled plasma–mass spectrometry and their impact on bioimaging applications. J Anal At Spectrom 31:423–439
Van Malderen SJM, Vergucht E, Rijcke M et al (2016b) Quantitative determination and subcellular imaging of cu in single cells via laser ablation–ICP–mass spectrometry using high-density microarray gelatin standards. Anal Chem 88(11):5783–5789
Van Schooneveld MM, Gloter A, Stephan O et al (2010) Imaging and quantifying the morphology of an organic-inorganic nanoparticle at the sub-nanometre level. Nature Nanotechnol 5(7):538–544
Wang HAO, Grolimund D, Giesen C et al (2013) Fast chemical imaging at high spatial resolution by laser ablation inductively coupled plasma mass spectrometry. Anal Chem 85(21):10107–10116
Wang S, Brown R, Gray DJ (1994) Application of laser ablation–ICPMS to the spatially resolved micro-analysis of biological tissue. Appl Spectrosc 48(11):1321–1325
Watanabe M, Williams DB, Tomokiyo Y (2003) Comparison of detectability limits for elemental mapping by EF-TEM and STEM-XEDS. Micron 34(2–5):173–183
Waugh AR, Bayly AR, Anderson K (1984) The application of liquid metal ion sources to SIMS. Vacuum 34:103–106
Wedlock LE, Berners-Price SJ (2011) Recent advances in mapping the sub-cellular distribution of metal-based anticancer drugs. Aust J Chem 64:692–704
Wedlock LE, Kilburn MR, Cliff JB et al (2011) Visualising gold inside tumour cells following treatment with an antitumour gold(I) complex. Metallomics 3(9):917–925
Wedlock LE, Kilburn MR, Liu R et al (2013) NanoSIMS multi-element imaging reveals internalisation and nucleolar targeting for a highly-charged polynuclear platinum compound. Chem Commun 49:6944–6946
Williams P (1985) Secondary ion mass spectrometry. Ann Rev Mater Sci 15:517–548
Williams DB, Carter CB (2009) The transmission electron microscope. In: Transmission electron microscopy. A textbook for materials science. Springer, USA., New York. Chapter 1, pp 3–22
Winograd N, Bloom A (2015) Sample preparation for 3D SIMS chemical imaging cells. In: He L (ed) Mass spectrometry imaging of small molecules. Springer, USA., New York. Chapter 2, pp 9–19
Wong JG, Wilkinson LE, Chen SW et al (1989) Quantitative elemental imaging in the analytical electron microscope with biological applications. Scanning 11:12–19

Wu B, Chen Y, Becker JS (2013a) Study of essential element accumulation in the leaves of a Cu-tolerant plant Elsholtzia splendens after Cu treatment by imaging laser ablation inductively coupled plasma mass spectrometry (LA–ICP–MS). Anal Chim Acta 633(2):165–172

Wu S, Kim AM, Bleher R et al (2013b) Imaging and elemental mapping of biological specimens with a dual-EDS dedicated scanning transmission electron microscope. Ultramicroscopy 128:24–31

Xu M, Bijoux H, Gonzalez P et al (2014a) Investigating the response of cuproproteins from oysters (Crassostrea gigas) after waterborne copper exposure by metallomic and proteomic approaches. Metallomics 6(2):338–346

Xu M, Frelon S, Simon O et al (2014b) Non-denaturating isoelectric focusing gel electrophoresis for uranium-protein complexes quantitative analysis with LA–ICP MS. Anal Bioanal Chem 406(4):1063–1072

Yumoto S, Kakimi S, Ohsaki A et al (2009) Demonstration of aluminum in amyloid fibers in the cores of senile plaques in the brains of patients with Alzheimer's disease. J Inorg Biochem 103(11):1579–1584

Zarco-Fernandez S, Coto-García AM, Munoz-Olivas R et al (2016) Bioconcentration of ionic cadmium and cadmium selenide quantum dots in zebrafish larvae. Chemosphere 148:328–335

Zhang P, Land W, Lee S et al (2005) Electron tomography of degenerating neurons in mice with abnormal regulation of iron metabolism. J Struct Biol 150(2):144–153

Zoriy MV, Kayser A, Izmer A et al (2005) Determination of uranium isotopic ratios in biological samples using laser ablation inductively coupled plasma double focusing sector field mass spectrometry with cooled ablation chamber. Int J Mass Spectrom 242(2–3):297–302

Zoriy MV, Dehnhardt M, Reifenberger G et al (2006) Imaging of Cu, Zn, Pb and U in human brain tumor resections by laser ablation inductively coupled plasma mass spectrometry. Int J Mass Spectrom 257(1–3):27–33

Zoriy MV, Dehnhardt M, Matusch A et al (2008) Comparative imaging of P, S, Fe, Cu, Zn and C in thin sections of rat brain tumor as well as control tissues by laser ablation inductively coupled plasma mass spectrometry. Spectrochim Acta B 63(3):375–382

Chapter 8
Chemical Speciation and Metallomics

Jemmyson Romário de Jesus, Luana Ferreira da Costa, Eraldo Luiz Lehmann, Rodrigo Moretto Galazzi, Katherine Chacón Madrid, and Marco Aurélio Zezzi Arruda

Abstract Chemical speciation approaches is an inherent part of metallomics, once metals/metalloids and organic structures need to be currently evaluated for attaining metallomics studies. Then, this chapter focuses on the applications of the chemical speciation applied to the human health risk, food and human diet, drugs, forensic, nanoscience, and geological metallomics, also pointing out the advances in such area. Some aspects regarding sample preparation is commented along this chapter, and some strategies for maintaining the integrity of the metallomics information are also emphasized.

Keywords Healthy · Food · Nutrition · Nanoscience · Forensic · Drugs · Geochemistry

Abbreviations

BIF	Banded iron formation
CE-ICP-MS	Capillary electrophoresis-inductively coupled plasma mass spectrometry
DNA	Deoxyribonucleic acid
ESI-FAIMS-IT-MS	Electrospray ionization-high-field asymmetric waveform ion mobility spectrometry-ion trap-mass spectrometry

J. R. de Jesus · L. F. da Costa · E. L. Lehmann · R. M. Galazzi · K. Chacón Madrid
M. A. Z. Arruda (✉)
Universidade Estadual de Campinas – Unicamp, Campinas, São Paulo, Brazil

Spectrometry, Sample Preparation and Mechanization Group (GEPAM), Department of Analytical Chemistry, Institute of Chemistry, University of Campinas (UNICAMP), Campinas, São Paulo, Brazil

National Institute of Science and Technology for Bioanalytics – INCTBio, Institute of Chemistry, University of Campinas (UNICAMP), Campinas, São Paulo, Brazil
e-mail: jemmyson_romario@yahoo.com; luana.costa@iqm.unicamp.br; eraldoluiz@iqm.unicamp.br; rodrigo_galazzi@hotmail.com; Katherine.madrid@iqm.unicamp.br; zezzi@iqm.unicamp.br

M. A. Z. Arruda (ed.), *Metallomics*, Advances in Experimental Medicine and Biology 1055, https://doi.org/10.1007/978-3-319-90143-5_8

ICP-MS	Inductively coupled plasma mass spectrometry
LC	Liquid chromatography
LC-ICP-MS	Liquid chromatography-inductively coupled plasma mass spectrometry
LC-MS/MS	Liquid chromatography-tandem mass spectrometry
nanoESI-Q-TOF	Nano-electrospray ionization quadrupole time-of-flight
nanoSIMS	Nanoscale secondary ion mass spectrometry
RP-HPLC-UV-ESI-MS	Reversed-phase high-performance liquid chromatography-ultraviolet-electrospray ionization spectrometry analysis
SEC-UV	Size-exclusion chromatography

8.1 Introduction

Metallomics area was coined recently by Prof. Hiroki Haraguchi, and it is defined as the science of biometals (Haraguchi 2004). In fact, this area congregates multidisciplinary concepts, and it is close not only to all omics (i.e., genomics, proteomics, and metabolomics) (Silva et al. 2010) but also to chemical speciation. As definition, chemical speciation comprises the identification and quantification of different inorganic and/or organic entities as well as the biomolecules involved in the given chemical species of interest (Arruda and Azevedo 2009).

In fact, the metallomics area involves not only the identification/quantification of the metal/metalloid present in a cell or tissue type but also the identification of what form it is present, the biomolecule to which they are bound, and the coordination groups involved (Mounicou et al. 2009). Then, it is easy to rationalize that these two areas, metallomics and chemical speciation, are in fact, complementary. Additionally, a crucial aspect is to decipher the exact mechanism involved in the selection of metals by biomolecules. This is not an easy task, once proteins compete with other molecules for metals rather than metals compete each other for proteins (Szpunar 2004). Additionally, in any given organism, the metallomes are dynamic, depending on both intrinsic (cellular and subcellular specificity, development) and extrinsic (nutrients, toxicants, drugs, stress, disease) factors.

Due to the complexity of the objectives when focusing on metallomics, as well as of metalloproteins (for instance, present stable, metals, and label, organic compounds; characteristics), the use of emerging techniques, as largely demonstrated along this book, can and should be used to increase our understanding regarding biological, environmental, and medicinal, among others, problems, where metalloproteins play an important role.

Then, inside the context of chemical speciation and metallomics, different approaches and strategies will be emphasized in this chapter, which are employed for elucidating some aspects of human health, food, drugs, forensic, nanoscience, and geology.

8.2 Elemental Speciation: An Approach for Human Health Risk Assessment

Metals and semi-metals have multiple effects on biological processes, playing crucial roles to human health (Hu et al. 2013). To understand the complex mechanism between biological processes and metal-associated human disease, one needs to look to interaction of metallic elements with the major classes of biomolecules, such as proteins, lipids, carbohydrates, and nucleic acids (Kulkarni et al. 2006). For example, among the known structurally characterized proteins, about one-third of these proteins are associated with metal (Shi and Chance 2011). The function of these proteins, termed metalloproteins, depends on the interaction between the proteins and those bound metals.

The ability of many metals to exist in multiple oxidation states and belong to molecules with different geometries allows to promote complex biochemical reactions and participate in highly specialized biological functions, such as electron transport, oxygen storage, metal transport, chemical bond hydrolysis, redox processes, and synthesis of biochemical compounds (Kulkarni et al. 2006; Zoorob et al. 1998; Arita and Costa, 2009). Zinc, for example, is the most abundant metal in cell and plays a vital role not only in the function of more than 300 enzyme classes but also in stabilizing the DNA double helix or in the gene expression control (Kulkarni et al. 2006; Shi and Chance 2011). Other trace metals, such as selenium, tungsten, and molybdenum, are also essential in the human health maintenance. Selenium, for example, is an essential micronutrient for humans, usually incorporated into antioxidant biostructures such as selenocysteine, and plays a key role in host oxidative defense (Shi and Chance 2008, 2011). However, in spite of recent efforts to understand the components of intracellular metal transport and its disposition, the overall blueprint of metal metabolism remains elusive (Kulkarni et al. 2006). In addition, although metals and metalloids are important in the full functioning of the human body as well as in maintaining human health, the majority metals, including essential metals, may be potentially toxic (Kannamkumarath et al. 2002).

Toxicity may vary significantly according to the element oxidation state, complex formation, and the elemental species biotransformation. Thus, the speciation can allow a better understanding of the toxicity mechanism of an element, as well as a refinement of the risk assessment, focusing the consequences of exposure upon the most relevant species.

8.2.1 Oxidation State and its Pattern of Toxicity

The relevance of speciation to health effects in human has been demonstrated, including acute toxicity, allergy, lung toxicity, neurotoxicity, reproductive toxicity, genotoxicity, and carcinogenicity. Most metal-associated disease may be related in

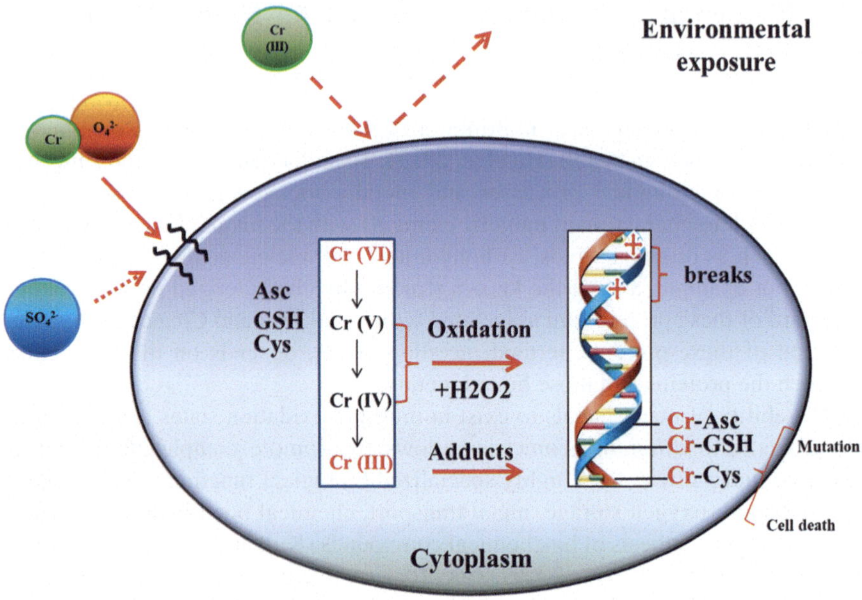

Fig. 8.1 Formation mechanism and biological effects of Cr-DNA. Asc, ascorbate; Cys, cysteine; GSH, glutathione

terms of the oxidation state type. Oxidation state can affect absorption, membrane mutagenic activity, epigenetic changes, or metal transport proteins (Shi and Chance 2011; Arita and Costa 2009). The valence state and/or associated ligands dictate availability for binding and hence the pattern of toxicity. Chromium serves as a good example of the importance of the oxidation state. Cr(III) is considered as an essential element, but Cr(VI) is genotoxic and carcinogenic. It does not appear to bind strongly to DNA but is reduced inside the cell to Cr(III), which binds strongly to DNA (Arita and Costa 2009; Zhitkovich 2005) (Fig. 8.1).

In fact, such behavior reflects the dependence between the degree of toxicity and the oxidation state. For example, Mn(II) chloride ($MnCl_2$) and Mn(IV) oxide (MnO_2) are both less toxic in vitro than Mn(III) pyrophosphate. One mechanism of manganese toxicity is by disruption of iron-sulfur clusters in mitochondrial enzymes, such as complex I and mitochondrial aconitase (Chen et al. 2001). The higher oxidative behavior of Mn(III) and its similarity with the ionic radius of Fe(III) have been suggested as reasons for its greater ability to inhibit iron-sulfur enzymes (Chen et al. 2001).

Numerous studies have implicated in structural alterations of chromatin and epigenetic changes as the primary events in nickel (Ni) carcinogenesis (Salnikow and Zhitkovich 2008; Oller et al. 1997). Nickel is a nonessential metal widely used in high scale in the industrial production of coins, jewelry, and batteries (Salnikow and Zhitkovich 2008). Several epidemiological studies have found Ni compounds to be carcinogenic (Salnikow and Zhitkovich 2008; Oller et al. 1997; Arita and Costa 2009). Occupational exposures to nickel usually involve multiple species, such as nickel(II) sulfate ($NiSO_4$), nickel chloride ($NiCl_2$), nickel oxide (NiO), and nickel(II) carbonate ($NiCO_3$) (Lee et al. 1995). For example, nickel sulfide particles have shown to selectively target heterochromatin (Arita and Costa 2009; Lee et al. 1995). These effects of nickel compound on heterochromatin led to the discovery of silencing of genes, including DNA methylation, by Ni compounds. Changes in DNA have been observed in vivo in nickel-induced tumors of wild type C57BL/6 mice and mice heterozygous for the tumor suppressor p53 gene injected with nickel sulfide (Oller et al. 1997; Arita and Costa 2009; Lee et al. 1995). The promoter of the tumor suppressor gene p16 was found to be hypermethylated in all tumors of these mice. Although the mechanisms by which nickel induces DNA hypermethylation are presently unknown, a proposed model includes the ability of nickel to substitute for magnesium (Mg) in the phosphate backbone of DNA. Ni(II) may be better at condensing heterochromatin than Mg(II) ions, increasing chromatin condensation, and triggering DNA methylation of critical tumor suppressor or senescence genes that can become incorporated into heterochromatin because of their proximity to this type of chromatin. In addition to gene silencing by DNA methylation, other nickel-induced epigenetic changes include the loss of histone acetylation in H2A, H2B, H3, and H4, increase in H3K9 dimethylation, and increases in the ubiquitylation of H2A and H2B at a global level. Ni(II) binding to H3 and H2A may cause several lesions including oxidative DNA damages via Fenton-like mechanisms and disrupt gene expression results in carcinogenicity (Arita and Costa 2009; Lee et al. 1995; Oller et al. 1997; Salnikow and Zhitkovich 2008).

Environmental exposure likely affects millions of people drinking metal-contaminated water that lives in the neighbor of numerous toxic sites or use products containing them. For example, arsenic (As) is a metal associated with serious human disease. Same studies have shown that chronic, low-dose exposure to arsenic is associated with skin, bladder, lung, kidney, and liver cancer. Exposure to As occurs currently either in the form of arsenite [As(III)] or arsenate [As(VI)]. The increased cancer risk is attributed to arsenite rather than the less toxic arsenate. Although proposed mechanisms for the carcinogenicity of arsenic include oxidative stress, inhibition of DNA repair, perturbation of signal transduction pathways, and chromosomal aberrations, the mechanism for carcinogenicity of arsenite remains unclear (Arita and Costa 2009; Salnikow and Zhitkovich 2008; Hei et al. 1998; Huang et al. 2004).

Wilson and Menkes diseases are promoted by Cu transport. The Wilson ATPase gene encodes a Cu-transporting P-type ATPase (ATP7B). ATP7B binds six Cu atoms at the N-terminus of the molecule. Defects in this protein result in the toxic accumulation of Cu in various tissues (Kulkarni et al. 2006). Menkes disease is another genetic disorder of Cu metabolism characterized by an impairment of the absorption of dietary Cu and severe disturbance in the intracellular transport of Cu (Kulkarni et al. 2006; de Bie et al. 2007). The Menkes ATPase gene encodes a Cu-transporting P-type ATPase (ATP7A) with a high degree of homology with ATP7B. This protein also binds Cu in a 1:6 protein to Cu ratio. The defects in this protein lead to the clinical symptoms of classic Menkes disease. As a result, the

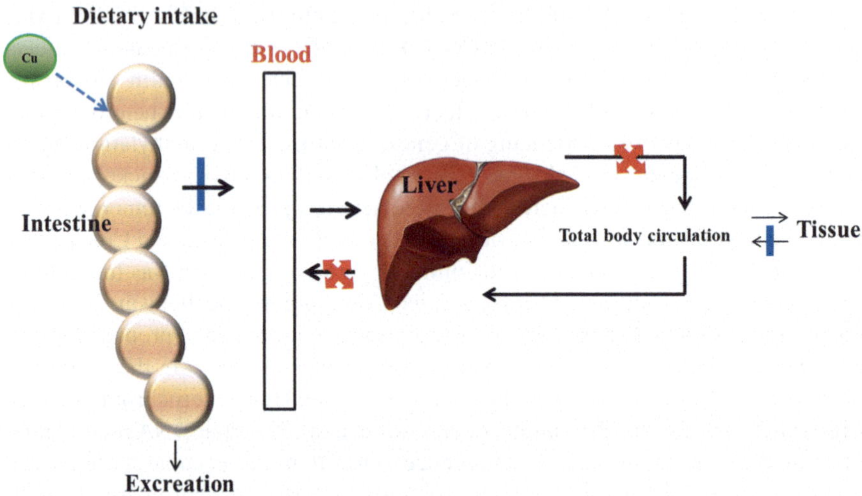

Fig. 8.2 Schematic representation of transport failure of copper into biological system causing Wilson and Menkes diseases

supply of copper is prevented from reaching developmentally important Cu enzymes, such as dopamine β-hydroxylase, superoxide dismutase, and amine oxidase (Kulkarni et al. 2006; de Bie et al. 2007; La Fontaine et al. 2016; Lee et al. 1995).

Oxidation state is critical for ion transport which is exemplified by different classes of Fe(II) and Fe(III) transporters, for example, hereditary hemochromatosis is human disease associated with metabolism aberration of metal in the organism. Hereditary hemochromatosis is a disorder of Fe metabolism where the uptake of Fe from the digestive system is not downregulated, resulting in excessive Fe uptake (Kulkarni et al. 2006). The body continues to absorb high levels of dietary iron regardless of the levels of stored iron in the body. The main iron transport protein in plasma is transferrin, which has two ferric iron-binding sites. Some proteins are involved in the cellular uptake of iron, including transferrin receptors 1 and 2 as well as the divalent metal transporter (DMT-1) (Fig. 8.2).

The DMT-1 is important in the uptake of iron in the gut, in addition to intracellular iron trafficking following endocytosis of transferrin-bound iron. DMT-1 transports Fe(II), but not the Fe(III) (Shindo et al. 2006). Fe(II) is soluble under physiological conditions and can also diffuse across membranes. In contrast, Fe(III) is prone to hydrolysis in aqueous environmental, producing poorly soluble products (Shindo et al. 2006). The regulation of gene products involved in iron metabolism is influenced by two cytoplasmic iron-regulatory proteins with mRNA-binding ability (IRP-1 and IRP-2) (Kulkarni et al. 2006; Cairo & Recalcati 2007). Cellular export of iron is accomplished by ferroportin and the export from non-intestinal cells requires ceruloplasmin (Kulkarni et al. 2006; Cairo and Recalcati 2007; Shindo et al. 2006).

Epidemiological risk assessment studies revealed a high incidence of lung cancer, for example, following occupational exposure to Cr(VI) (Zhitkovich 2005). However, the precise mechanism(s) of Cr(VI) carcinogenicity is not well understood. It has been suggested that Cr(VI) mediates a majority of its cytotoxic and genotoxic effects by inducing oxidative stress, forming stable-Cr-DNA adducts, protein-DNA cross-links, and DNA single- and double-strand breaks. Cr(VI)-induced DNA damage can affect DNA replication, transcription, and translation, resulting in altered gene expression. Until recently, changes in gene expression mediated by DNA damage were considered the key mechanism underlying the genotoxic and carcinogenic activities of Cr(VI) (Zhitkovich 2005; Salnikow and Zhitkovich 2008).

Cadmium (Cd) is another toxic nonessential transition metal classified as a human carcinogen by the National Toxicology Program (Arita and Costa 2009). Sources of human exposure to cadmium include employment in metal industries, production of certain batteries, some electroplating processes, and consumption of tobacco products. Cadmium poses a great health risk to humans because the body has limited capacity to respond to cadmium exposure, as the metal cannot undergo metabolic degradation to less toxic species, being poorly excreted. Multiple studies have linked occupational exposure to cadmium with pulmonary cancer, as well as prostate, renal, liver, hematopoietic system, urinary bladder, pancreatic, and stomach cancers. Although the potential mechanism of cadmium carcinogenesis is unknown, suggested mechanisms for Cd-carcinogenesis include aberrant gene activation and signal transduction, suppressed apoptosis and disruption of E-cadherin-mediated cell-cell adhesion, and/or altered DNA repair (Joseph 2009). Studies suggest sequestration of Cd(II) by proteins as an important factor for Cd-carcinogenesis (Arita and Costa 2009; Joseph 2009). Such species bond to a high-molecular-mass protein, as albumin in the blood, being absorbed by the liver. In the liver, cadmium induces and binds to the small proteins, as thioneine, thus forming a metallothionein. If the Cd-thioneine complex is then subsequently released from the liver and reaches the kidney, it causes more damage to the kidney (Arita and Costa 2009; Joseph 2009).

8.2.2 Sample Preparation and Analytical Techniques Used for Elemental Speciation in Human Sample

In order to assess the risk to human health from exposure to elemental species, several challenges are found to determine for a widening variety of species. One example is the species identification. This requires reliable procedure for material sampling and species isolation without changing their composition. The detection can be based on the measurement of their element in the species or the molecule in which the element is incorporated.

The clinical samples (blood and urine) are the main sources used in elemental speciation and fractionation for human health risk assessment (Oller et al. 1997; Sariego Muniz et al. 2001). However, such samples are highly complex due to the presence of amino acids, lipids, hydrocarbons, and polysaccharides, among others. Thus, cleanup procedures are necessary to remove all those compounds that are of no interest for the purpose of the analysis and whose presence may even compromise the detection sensitivity of the analytical method. Different extraction procedure types are described in the literature to remove interferents, varying from simple aqueous or solvent extraction to more complex methods, such as enzymatic extraction, solid-phase extraction, solid-phase microextraction, steam distillation, supercritical fluid extraction, liquid-gas extraction (purge and trap), accelerated solvent extraction, and microwave-assisted extraction (Gómez-Ariza et al. 2001; Rezende et al. 2015).

Regarding techniques used to elucidate the biological significance and toxicity of the metal species on the molecular basis, some of them have been employed to detect, identify, and quantify metals in biological sample. However, there are two main approaches that have been commonly applied in this type of study. The first is the widely used analytical techniques of separation, in particular, liquid chromatography (LC) and gel electrophoresis (GE) in combination with inductively coupled plasma mass spectrometry (ICP-MS). These techniques are ideal partners in the comprehensive structural and functional characterization of metalloproteins. The second approach is high-throughput X-ray absorption spectroscopy (HT-XAS) to provide a direct metal analysis of proteins and proteomic metal distribution in tissues and cells. The computational bioinformatics is another approach that play crucial roles to functional characterization of metalloproteins, complementing such experimental methods (Shi and Chance 2008, 2011; Zoorob et al. 1998; Easter et al. 2010).

When species can be converted to hydrides, such as is routinely done for selenium, arsenic, and antimony, hydride generation using atomic absorption spectrometry (AAS) may be very interesting because AAS is a cheap detection technique. An online method was developed for the speciation of arsenic in human serum, including methylarsonic acid (MMA), dimethylarsinic acid (DMA), arsenobetaine, and arsenocholine. It has been applied for the speciation of arsenic in persons with abnormally high arsenic concentrations in serum, such as dialysis patients (Zhang et al. 1998). The method is based on cation-exchange LC separation and continuous hydride generation AAS for the measurement of arsenic in the LC eluent. By developing, a substantial improvement in chromatographic resolution for the separation of these four arsenic species was obtained. The response is different for the different species. The detection limits (as arsenic) were 1.0, 1.3, 1.5, and 1.4 μg L^{-1} for MMA, DMA, arsenobetaine, and arsenocholine, respectively, in serum. The concentration of the four species was determined in serum samples of six patients with chronic renal insufficiency.

8.3 Food and Human Diet

There are a great amount of metals, which are important to the human health. Some of them, found at trace levels, are known to be deficient in human diet. On the other hand, there are metals that can be harmful, and these must be present at low concentration in food. Among the metals known for their toxicity or connections with diseases, arsenic, nickel, aluminum, and mercury can be mentioned. However, the toxicity of a given element is not limited by its total concentration in food. For some elements, the toxicity of a given food depends on the organometallic species present and also its distribution and concentration, besides the inorganic forms (Peeters et al. 2017; Brombach et al. 2017; Cheyns et al. 2017; Yim et al. 2017). For mercury, the inorganic form presents lower toxicity compared to the organomercury species (Brombach et al. 2017). The opposite occurs for organic arsenic species, which are considered less toxic than the inorganic forms arsenite and arsenate (Sadee et al. 2016; Jitaru et al. 2016; Yim et al. 2017; Cheyns et al. 2017). In this way, chemical speciation of metals together with the total determination is important to evaluate the health risk from such metals consumption in food, especially because the food toxicity cannot be estimated by their total content only.

Until recently, there was a growing concern about which toxic metals are present in food, and such toxicity was related to the concentration of the inorganic species present. More recently, studies showed that not only the inorganic species should be considered but also organic species bonded to a metal, which can be also a potential source of the inorganic specie, and being more or less toxic, depending on the form considered.

Some food like rice, seaweed, wheat, and broad bean are the base of the human diet in some countries (Yim et al. 2017). There are also beverages, cocoa, or tea infusions, for example, which can be source of metals such as aluminum or nickel (Peeters et al. 2017; Ščančar et al. 2013 a,b; Kralj et al. 2005). In fact, depending on their concentrations or if these foods are contaminated with toxic metals, the presence of their inorganic or organic forms can be potentially dangerous, increasing health problems.

More recently, arsenic was found in rice, broad bean, or seaweed, because such cultures were grown and harvested in contaminated soil and/or water (Yim et al. 2017). The species of arsenic found in broad bean and rice are the carcinogenic inorganic species arsenite, As(III), and arsenate, As(V), as well as the less toxic but still classified as cancer promoters monomethylarsonic acid (MMA) and dimethylarsinic acid (DMA) (Sadee et al. 2016; Taylor and Jackson 2016). In rice, others nontoxic arsenic species are present, namely, arsenobetaine (AsB), arsenocholine (AsC), trimethylarsine oxide (TMAO), and arsenosugars (Jitaru et al. 2016). Another food presenting high concentrations of arsenic is seaweed, with more than 85% of the soluble As as arsenosugars and with smaller quantities of methylated As compounds as DMA. However, taxonomy plays an important role in the As storage, and, therefore, the amount of arsenic and its species depends on the type of seaweed, brown, red, or green (Taylor and Jackson 2016). Inorganic arsenic (iAs) and some of the species mentioned, MMA, DMA, and AsB, can also be found in meat due to the use of drugs during the lifespan of the animal, being accumulated in the muscles, fat, and skin. In poultry and swine production, phenylarsenic compounds were widely used before withdrawing. For example, roxarsone (ROX) and *p*-arsanilic acid (*p*-ASA) were used to enhance weight gain and pigmentation of the meat; another purpose is control of parasites (Hu et al. 2017).

Similar to arsenic, mercury enters in the human food chain through consumption of rice, grown in contaminated soil and seafood. The two most important mercury species present in rice are inorganic mercury (iHg) and methylmercury (MeHg). As mentioned before, the last is more toxic than the former. This organomercury compound presents higher absorptivity in the human gastrointestinal tract when compared to the iHg, being neurotoxic (Brombach et al. 2017).

Some nickel compounds are harmful to humans, namely, insoluble particulate species, which are carcinogenic, and other water-soluble nickel compounds are mutagenic and genotoxic. This element can be found in cocoa, foodstuffs with the highest concentration of nickel, and it is usually consumed as beverage or as a component in chocolate-based products. Other food containing nickel are nuts, mussels, pea nodules, and tea infusions (Cacho et al. 2010; Ščančar et al. 2013a, 2013b). Besides the free nickel present in cocoa, responsible for the allergic reactions, it is found other two major components, namely, Ni-citrate and Ni-gluconate (Peeters et al. 2017). For black, green, white, red, and chamomile tea infusions, the major compound is Ni-quinate, which comprehends more than 84% of total nickel concentration (Ščančar et al. 2013b).

Another study evaluated the speciation of nickel in pea nodules, and the main components found were nickel malate and nickel citrate (Cacho et al. 2010).

Despite of many studies determining total concentration of toxic metals and less frequent the speciation in food, mostly after decomposition or extraction the analyzed solution is derived from the raw food. Besides that, food processing can also change the concentration of metals and its organometallic compounds in food. One process is the degree of polishing rice which decreases the concentration of As, suggesting that arsenic is mostly on the bran or the surface of the grain and the higher percentage of polishing the more is reduced the amount of arsenic (Jitaru et al. 2016; Yim et al. 2017). However, the same is not observed for MeHg; the polished and whole grain show similar total Hg and methylmercury concentrations (Brombach et al. 2017). Another food processing is washing of rice or soaking of seaweed; this processes reduces the amount of arsenic; however, the decrease is not as effective as the cooking (Cheyns et al. 2017). To nickel, different types of food processing consist in forms of contamination, namely, drying, canning, catalytic hydrogenation, and cooking in stainless steel recipients (Peeters et al. 2017; Ščančar et al. 2013b). If kitchen utensils are non-corroded, the concentration of released nickel is low and therefore does not pose a risk to human health (Ščančar et al. 2013b).

From the few studies evaluating food preparation, it is known that cooking rice to dryness does not decrease the As species concentration, but a decrease of total As, iAs, and DMA is seen when boiled in excess of water. In general, the As is released upon food preparation, and the observed reduction is due to the removal of the broth (Jitaru et al. 2016; Cheyns et al. 2017). When the broth is considered in the analysis, for example, as soup of seaweeds, the concentration of As species does not decrease. When food preparation consisted in steaming or boiling, for carrots, leeks, and mussels, a decrease in concentrations of As species is observed. Frying food, for example, onions, mushrooms, fish, scallops, and scampi, presents no significant effect on As species concentrations. Heating wine and preparing fish or crustacean also did not present differences in As concentrations. Other aspects on food processing worth mentioning are high temperatures, increased contact time, and water movement which tends to increase the release of As (Cheyns et al. 2017).

However, for precooked rice baby food, the cooking process does not seem to affect the concentration of MeHg, but removes considerable amounts of iHg (Brombach et al. 2017). In the case of leaves that are used to make tea, boiling water is enough to extracts 52–87% of the nickel present those leaves, classifying nickel as a high extractable element (Ščančar et al. 2013a, 2013b). In black tea, Al-citrate was found as extractable aluminum. When such tea is prepared with milk, besides the mentioned compound, aluminum associated with the milk proteins was also found (Kralj et al. 2005). In the case of other elements, not all are highly extractable by hot water, which means that total concentration in the plant should not be considered present in the tea infusion. Some elements, like nickel, the speciation in tea infusions are pH dependent (Ščančar et al. 2013b).

The matrix of the food is also important. Comparing rice and mussels, the first releases inorganic arsenic easier than DMA, and the second releases all quantified species, organic and inorganic species. Considering the types of arsenic species, the inorganic arsenic is released upon food preparation more easily than small organometallic species followed by larger molecules. When studying bioavailability and bioaccessibility of metals, the type of food and its composition are important factors that change its absorption by the body (Cheyns et al. 2017). For example, nickel species at low molecular mass are, in general, more available for absorption by intestinal cells when compared to high-molecular-mass nickel species (Ščančar et al. 2013a).

As other plants, the content of nickel in cocoa or tea leaves reflects the environmental conditions in which the food was grown (Peeters et al. 2017; Yim et al. 2017; Ščančar et al. 2013b). The concentration of nickel in leaves depends on the time of harvesting for white, green, oolong, and black tea, which the first contains the youngest leaves and sprouts (Ščančar et al. 2013b).

Studying the metals present in food, the samples were currently digested using a microwave oven or heated in open flasks. For seaweed, only nitric acid was used; to rice, a mixture of nitric acid and hydrogen peroxide was used (7:1 ratio), and broad beans were digested using nitric acid and hydrogen peroxide (Taylor and Jackson 2016; Yim et al. 2017; Sadee et al. 2016). Depending on the type of the sample, it may be necessary to freeze, dry, and/or ground with a pestle and mortar or with a stainless steel coffee grinder, before acid digestion. In this case, the objective was to determine the total metal present in the sample.

To evaluate the organic species bound to the metal, present in the food sample, a gentler preparation is needed. It is possible to use mixtures of methanol and water (1:1) for seaweed and rice, followed by mild heating and sonicated in room temperature, respectively. For turkey meat, an alkaline solution was used to digest the frozen samples. The digestion of broad bean used a 1% (v/v) nitric acid solution followed by a gentle heating in a microwave. There are cases in which centrifuge and filtering may be necessary after these procedures. Total metal is usually determined by inductively coupled plasma mass spectrometry (ICP-MS) , depending on the analyzed metal; a collision cell is used to eliminate interferences (Taylor and Jackson 2016; Sadee et al. 2016). In some cases, indium (In) or/and iridium (Ir) are used as internal standard (Sadee et al. 2016). The organic species containing the metal of interest are usually analyzed by a HPLC-ICP-MS.

8.4 Drugs and Diagnostic Agents

Paraphrasing Paracelsus, *Poison is in everything, and no thing is without poison. The dosage makes it either a poison or a remedy*, seems to be that the doubt still remains when considering chemical elements for drug design, once some of them (essentials, nonessential and radionuclides) present in the periodical table have the potential to be used as therapeutic drugs or as diagnostic agents (Chellan and Sadler 2016). In fact, from those 60 detectable elements in the human body, only 25 are believed to participate in the healthy functioning of the human body (Nielsen 1999). Additionally, there are many examples of where a biological function requires synergistic action of two or more elements, such as cytoplasmic superoxide dismutase that uses both copper and zinc to convert superoxide into oxygen and water (Takahashi et al. 1984). Even the noble gases have an important role as drugs. For example, xenon is used as general anesthetic and presents a high affinity for the glycine site on the *N*-methyl-D-aspartate (NMDA) receptor, and unlike most other clinical NMDA receptor antagonists, it is not neurotoxic (Harris et al. 2013). Then, it is clear that future advances in the design of drugs, also for those presenting metals in their constitutions, require knowing their mechanism of action, including metabolism and target sites, and the temporal speciation of elements in their biological environments, at atomic levels, is the major challenge (Chellan and Sadler 2016).

Inside this context, chemical speciation and metallomics are important auxiliary areas, and for demonstrating the ambiguity of the functions of some non-usual elements, some of them will be pointed out here, belonging to a diversity of drugs.

An important example of cancer drugs is related to the antitumor properties of cisplatin, being its clinical use approved by the FDA in 1978. Recently, the consecutive stages involved in the inhibition of the transcription by Pt-containing drugs was summarized as (1) cellular accumulation by both passive and active uptake, (2) activation of the Pt(II) complex, (3) binding to nucleic acids to form a variety of Pt-DNA adducts, and (4) the cellular response to DNA damage (Todd and Lippard 2009).

Cisplatin and the successive generations of platinum-based anticancer drugs have demonstrated that metal complexes are highly promising for anticancer treatment. Platinum drugs attack nuclear DNA, causing DNA distortions. DNA damage caused by cisplatin is recognized by proteins that activate downstream events, leading to a predominantly apoptotic response (Tan et al. 2014).

As there are different chemical structures of Pt-anticancer agents, such as cisplatin, carboplatin, oxaliplatin, and satraplatin, among others, then metallomics studies clearly offer a way to improve the knowledge about the behavior of these drugs in the organism. Such aspect was explored by using analytical methodologies for metallomics studies of antitumor Pt-containing drugs (Esteban-Fernández et al. 2010). The authors present a diversity of possibilities, but they emphasizes the use of multidimensional hyphenated techniques (i.e., LC-ICP-MS, RP-HPLC-UV-ESI-MS, ESI-FAIMS-IT-MS) for a better understanding of the Pt-based drug interactions in the organism. Through such approaches, some examples of interactions between Pt-based drugs and albumin, transferrin, and hemoglobin were commented. For example, cisplatin may also lead to albumin dimerization. Experiments using SEC-UV (Ivanov et al. 1998) allowed detecting Pt species of around 140 kDa, and probably the disulfide bond cleavages could be related with this aggregation process. Additionally, albumin seems to enhance the chemotherapeutic action of carboplatin. Such conclusion was obtained after the administration of this drug and a significant amount of albumin. In fact, there was a great accumulation of the carboplatin in several organs, suggesting a possible role in the transport and delivery of Pt drugs by this enzyme.

Regarding hemoglobin, it seems to have a reactivity order between the Pt-based and this enzyme, which follow the sequence: oxaliplatin, cisplatin, and carboplatin (Mandal et al. 2004). For each drug, at least three Pt-containing species were detected using SEC-ICP-MS from in vitro experiments in human blood. Additionally, some adduct forms were identified, such as [(ahbh)2-oxaliplatin] and [(ahbh)2-Pt-DACH] by nanoESI-Q-TOF (Peng et al. 2005).

In another example, ^{82}Rb, which decays by electron capture, is presented as an attractive radionuclide for positron emission tomography (PET), and it has been used clinically for PET myocardial perfusion imaging (MPI) (Beller and Bergmann 2004). Using the ^{82}Rb high-count-rate PET, a protocol was developed for simultaneous measurement of myocardial blood flow (MBF) and MPI. This procedure shows low selectivity for individual patient anatomy or injection site, making it a promising tool for obtaining accurate MBF data without loss of image quality (Tout et al. 2012).

Metallomics and chemical speciation have also been used for drug development. In fact, Aleksenko et al. (2013) demonstrated an integrated CE-ICP-MS and ICP-MS approach for speciation of Ru(III) in Ru-drug, which it was bound to holo-transferrin in simulated cancer cytosol, but not with albumin. The authors conclude that the speciation changes taking place through a redox mechanism rather than due to ligand-exchange transformations. Additionally, the authors noted, by the first time in the literature, the evolution of novel Ru species at simulated intracellular conditions. Apparently, there is a change between Fe and Ru from the transferrin adduct, hypothesizing that Ru is involved in Fe metabolism.

The evaluation of recent progress in using metallomics strategies and techniques for advancing the preclinical development of anticancer ruthenium agents can be found in an interesting review proposed by Timerbaev (2016), which includes current challenges in studying the drug's action mechanism, transport to cancer cell, the cellular activation, and subcellular distribution and targeting.

Focusing on gallium, Ossipov et al. (2013) demonstrate another interesting example, quantifying the in vitro binding of a Ga-based anticancer drug – tris(8-quinolinolato)gallium(III) – to serum albumin and transferrin. The fractionation of Ga between protein-free and protein-rich fractions was carried out by ICP-MS. Comparative kinetic studies revealed that the drug exhibits a different reactivity toward individual proteins, because Ga bounds fast with albumin (ca. 10%), but it was needed ca. 48 h to bound with transferrin (at higher than 50%).

Vanadium in another element that is apparently essential for man, but its roles in the body (ca. 2.4 mg) are poorly understood. Vanadium complexes have been on clinical trials recently as antidiabetic agents. The similarities of vanadate and phosphate are likely to be central to its biological effects. For example, vanadate-dependent haloperoxidases mimic enzymes involved in the metabolism of phosphate, blocking the protein-binding domain of phosphate. This competitive binding may account for the insulin-mimetic/insulin-enhancing potential of vanadium compounds, which can be used to alleviate insufficient insulin response in diabetes mellitus (Thompson and Orvig 2006). In fact, such compounds not are able to completely make up for the lack of insulin, as type 1 diabetes, but they can lower the dependence on exogenous insulin (Thompson et al. 2009).

Another intriguing element is the tellurium, which has been regarded as toxic and nonessential element. However, some pharmacological studies have been revealing that both organic and inorganic compounds have good potential in medicine (Cunha et al. 2009). One of such drugs is the AS101, a nontoxic Te(IV) compound, and a study of the oral and intraperitoneal with such drug revealed a significant reduction of clinical manifestations of inflammatory bowel diseases (Halpert et al. 2014). Such drug also presents antiapoptotic activities through the downregulation of colonic cytokine levels (IL-17 and IL-1β) and by the blockade of leucocyte (neutrophil and macrophage) migration into the colon (Halpert et al. 2014).

8.5 Forensic Approaches

Forensic science is the branch of science that studies crimes, not necessarily related to murder, but any situation in which the laws can be unenforced. Forensic chemistry is the application of chemical identification and quantification of substances present in the samples obtained in the crime scene or suspect of one. Some examples are commented such as the blood of the victim containing drugs, thus explaining if the victim was drugged or if it contains some poisons, which may explain if the victim was unconscious. Other examples of samples are regarding a food that does not display the content of a product or the amount of proteins in powdered milk for feeding newborn below established by law. All of these samples, besides others, indicate a crime and can be investigated with the aid of forensic chemistry.

Regarding speciation analysis into account forensic chemistry, few studies are reported; however, for some cases, they are the key for the resolution of a crime, as well exemplified in the movie "Erin Brockovich." The movie presents a situation where residents of a nearby region of an industry were getting sick, and even dying due to contamination of Cr(VI) in the waters of the region, while respecting the maximum value allowed for disposal of total Cr in wastewaters. The movie was based on a real case, occurred in Hinkley (CA, USA), from 1952 to 1996, where a gas and electric company did contaminate water, ground water, and soil with Cr(VI), causing disease in animals and humans nearby the company.

Regarding to metallomics applied to forensic chemistry, not only the function of toxic elements to human and animal health can be evaluated, but also elements present in certain types of products used to supply deficiencies. For example, iodine is an element associated with the thyroid gland responsible for producing and releasing hormones related to metabolism. When not properly working, it can cause hypothyroidism, commonly due to privation of iodine, which slows the metabolism, causing fat accumulation in the body; however, the excessive consumption of some iodine species could also be harmful, causing hyperthyroidism, accelerating the metabolism and the weight loss, including muscle mass (Wilson et al. 2016).

Following these thoughts, the addition of iodine is currently made in foods and supplements (for humans or pets). Wilson et al. (2016) studied some dog's foods and treats, acquired at local shops or obtained with the manufacture, and evaluated some iodine species, such as 3-monoiodotyrosine (MIT) and 3,5-diiodotyrosine (DIT) and 3,5,3′,5′-tetraiodothyronine (T4, also known as thyroxine),3,5,3′-triiodothyronine (T3) and diiodothyronine (T2). These iodinated substances are responsible for thyroid function, combining themselves to forming thyroid hormones. For such task, the total iodine was also evaluated for mass balance through an alkaline extraction with tetramethylammonium hydroxide (TMAH). Employing an ICP-MS, certified reference materials and spiked samples, the recoveries ranged from 84% to 113% (Wilson et al. 2016).

For the iodine speciation, an enzymatic extraction with protease pronase E, obtained from *St. griseus*, was then carried out. The extract obtained from each sample was divided into three aliquots: 1st, total iodine analysis by ICP-MS; 2nd, iodine speciation by LC-ICP-MS; and 3rd, species identification and quantification by LC-MS/MS. The results were in agreement with those obtained in the total iodine determination as well as in the total iodine enzymatic extracted for ICP-MS and the sum of the species obtained in enzymatic extraction for both LC-ICP-MS and LC-MS/MS. In this way, those dog's foods and treats could be helpful to regulate the thyroid from dogs (Wilson et al. 2016).

On the other hand, besides being able to verify legacies of food, one of the main objectives of forensic chemistry is to elucidate cases of death, for example, murder, suicide, accidental, or natural. One of the most famous deaths in world's history was that of the Emperor Napoleon Bonaparte. There are many speculations about murder or suicide, both by arsenic poising, even the autopsy indicating the cause of death from stomach cancer (Kintz et al. 2007). In fact, arsenic is present at different species in nature: H_3AsO_3 (arsenous acid) and its anions, also known as inorganic As(III), iAs(III), or just As(III), and H_3AsO_4 (arsenic acid) and its anions, also known as inorganic As(V), iAs(V), or just As(V). Additionally, some methylated species derived from inorganic species, as monomethylarsenic (MMA) and dimethylarsenic (DMA), both containing As(III) and As(V), arsenobetaine (AsB), arsenocholine (AsC), and some arsenosugars. Concerning the toxicity of As species, the As(III) is the most toxic specie, and AsB the less one. In fact, AsB is commonly found in edible fish and seafood, being considered inert specie. The toxicity of the arsenosugars was not tested (Kintz et al. 2007; Lehmann et al. 2013; Raab et al. 2005; Mandal and Suzuki 2002).

For investigating the Napoleon's death, Kintz et al. (2007) used two Napoleon's hair samples, obtained after his death. One of these samples was delivered to his servant, "Abraham Noverraz," and the other to his marshal, "Grand Maréchal Bertrand," who had been deported together with Napoleon to St. Helena Island. These samples would have been cut on the day of Napoleon's death (May 6, 1821). In fact, arsenic has a strong interaction with the keratin present in the hair. This element tends to greatly accumulate in nails and hair, which are considered the two main routes of elimination of As. Additionally, the change on As concentration over the hair may also indicate the exposure time or its poisoning in the body or, even if it was acute or chronic, the exposure or intoxication (Mandal and Suzuki, 2002).

Initially, total arsenic was determined through ICP-MS by acid decomposition with nitric acid. The amount of As found in both samples (42 ng mg^{-1} from Noverraz's and 37 ng mg^{-1} from Bertrand's samples) is higher than the acceptable (1 ng mg^{-1}). To evaluate whether these concentrations of arsenic belong to the hair or it comes from the contamination on its surface, the samples washed with acetone. The arsenic was then determined in acetone phase, and negligible values were found (Kintz et al. 2007). For As speciation, the samples were incubated in water for 6 h at 90°C. As(III), As(V), MMA, and DMA were evaluated through LC-ICP-MS, employing an anion-exchange column. The results showed higher amounts of As(V) (66.3% from Noverraz's and 53.2% from Bertrand's samples), followed by As(III) (31.1% from Noverraz's and 44.7% from Bertrand's samples). MMA and DMA were lower than 1% (Kintz et al. 2007). Arsenobetaine (AsB) was evaluated employing a cation-exchange column and LC-ICP-MS. This specie could indicate that the arsenic became from ingestion of seafood and fishes, because St. Helena Island is located at the middle of the sea. AsB was not found in both samples. However, this molecule is poorly incorporated in the hair (Kintz et al. 2007). Through these results, the hypothesis that Napoleon's death occurred by poising by arsenic became plausible. However, murder or suicide remains as doubt of his death.

Currently, blood, saliva, hair, nails, urine, and feces are the most employed human samples for forensic chemistry studies involving living suspects or victims. For dead bodies, liver, kidney, and brain tissues, stomach contents, and vitreous humor are those most used samples. As example, vitreous humor has become an interesting sample because it is a body fluid, which presents ca. 99% water and the remaining 1% as salts, sugars, and proteins. In addition, due to the anatomy of the eyes, the degradation or contamination of this fluid hardly occurs, maintaining the integrity of the information (Santos Junior et al. 2014).

Several studies involving postmortem period, also called thanatochemistry (Thanatos, Greek god of death), were made employing vitreous humor, which include toxicological (Jenkis and Oblock 2008; Costa et al. 2014; Carvalho et al. 2013), estimation of postmortem interval (PMI) (Santos Junior et al. 2014; Swain et al. 2015; Lendoiro et al. 2012), and postmortem biochemistry. For such task, Santos Junior et al. (2014) employed vitreous humor for searching metallomics biomarkers to estimate postmortem interval. Samples were collected and the interval of the estimated death, by evaluating the decomposition of the corpse, by coroner. Samples were decomposed in a microwave with nitric acid to carry out a semiquantitative elementary evaluation through ICP-MS. It was noted that the concentration of Fe, Mg, and Mo presents significant difference when the PMI increases. However, during a quantitative analyzes, Mg and Mo did not show a significant correlation between their concentration and PMI, presenting random behaviors; Fe, on the other hand, presents a significant correlation, increasing its concentration when increases the PMI.

For elucidating this phenomenon, samples were diluted, and metalloproteins were evaluated using a SEC-LC-ICP-MS. It was found Fe- and Mg-binding proteins. The fraction corresponding to these protein (ca. 77 and 67 kDa) was collected, and the proteins were identified by ESI-MS-MS as serotransferrin (77 kDa) and serum albumin (67 kDa) (Santos Junior et al. 2014). Serum albumin is the most abundant protein in plasma, binding several constituent, including ions, and it regulates the osmotic pressure of the blood. In the described study, this protein could indicate passive exudation due blood-ocular barrier failure (Santos Junior et al. 2014). Serotransferrin is synthetized and secreted in lens and retinal pigmented cells. The presence of this protein in vitreous humor is because, after death, tissues behind the blood-ocular barrier produce and secrete this protein to capture iron ions to maintain physiological functions; an inflammatory or decomposition process disrupts the barrier, causing the presence of transferrin in vitreous humor. In this way, the concentrations of Fe and serotransferrin presented to be an interesting metallomics biomarkers to estimate the postmortem interval, increasing their amount by increasing the PMI.

8.6 Nanoscience

The industry of manufactured products composed by nanoparticles (NPs) has grown fast in the last years. A sort of NPs, which the size ranges from 1 to 100 nm, has been synthesized and widely employed in different areas for a diversity of purposes (Ma et al. 2015; Arruda et al. 2015). Taking into account this information, a variety of NPs in the environmental matrices has been expected to lead to the appearance, including plants and animals (Anjum et al. 2013). Once the interactions between NPs and living organisms are not well understood, studies evaluating the effects of

NPs are prone to be performed for expanding the information concerning the possible toxicity due to NP exposition (Ma et al. 2015; Arruda et al. 2015). For this task, techniques based on characterization, separation, and quantification are required, and multidisciplinary areas, including the omics such as metallomics, are welcome to obtain additional results and expand the understanding regarding this issue (Li et al. 2014).

As metallomics refers to metal-containing biomolecules such as metalloproteins and metalloenzymes (Haraguchi 2004), the coupling of liquid chromatography (LC) to both inductively coupled plasma mass spectrometry (ICP-MS) and high-resolution mass spectrometry is then required to measure the elements and metallo-species, as well as to identify which biomolecules are important to the system evaluated. Indeed, this approach has been employed to obtain additional information concerning the distribution and function of the elements and evaluate the changes in the metallo-biomolecules in the biological systems upon a stress condition (Arruda et al. 2015; Li et al. 2014). Additionally, when ICP-MS is coupled to field flow fractionation (FFF) or used in the single particle mode (SP-ICP-MS), it is possible to obtain information concerning size of NPs as well to detect and quantify dissolved versus particulate forms of the element with respect to NP size in the same sample (Abad-Alvaro et al. 2017). Despite few studies concerning NP toxicity through a metallomics evaluation, the established metallomics approaches as well as innovations in such area are considered in this topic to expand the understanding concerning the effects of NPs in different organisms. Then, to consider the nanoscaled metal-related materials and since NPs may exhibit different properties when compared with those materials at the microscale, the nanometallomics field has been emerging (Li et al. 2014).

Some studies reveal the toxicity of the NPs at macroscopic level once, in specific cases, the LC_{50} was observed. This parameter should be considered in a first step, because some external factor can led to a stress condition, as well as assess the toxicity of any contaminant (Mehennaoui et al. 2016; Li et al. 2010). However, it is utmost important to consider other strategies in order to get additional information concerning any alteration in the organisms. Then, approaches involving different analytical techniques must be considered to evaluate the molecules that can change upon a stress condition. Regarding metallomics, the coupling between LC and ICP-MS is the most strategy employed, once it is possible to evaluate the elements, which can be related to any biomolecule (Haraguchi 2004). Additionally, the LC-ICP-MS with different strategies can also be used to separate the NPs according to size or distinguish them of their free ions. This information may be valuable for assess the NPs and their stabilities in the environmental matrices (Malysheva et al. 2015; Soto-Alvaredo et al. 2013).

Inside this context, Peng et al. (2015) used LC synchronously coupled to ICP-MS and electrospray ionization quadrupole time-of-flight mass spectrometry (ESI-QTOF-MS) to assess the toxicity of CdSe/ZnS quantum dots (QDs) in human hepatocellular carcinoma cells (HepG2) through a metallomics point of view. In total, four CdSe/ZnS QDs were used, and differences between them were both surface functional group (NH_2 or COOH) and the maximum emission wavelength. The first experimental step was the cellular culture treatment followed by QDs incubation for 24 h. The collected cells were then divided into two equal parts. One of them was directly digested by concentrated nitric acid, and the total amount of QDs in the cells was determined by ICP-MS. The other part was ultrasonicated, being the supernatant separated through a size exclusion chromatography (SEC) column. Then, two fractions, QD-1 and QD-2, presenting distinct retention times were collected and mixed together for subsequent digestion with nitric acid, and the total amount of QD-1 and QD-2 in the cells determined by ICP-MS. The objective of this experiment was to access the total amount of QD-1 and QD-2 in the cells. Additionally, the retention time for two kinds of metallothionein was fixed by SEC-ICP-MS. Using such strategy, two distinct peaks, QD-1 and QD-2, were observed for all evaluated QDs, while no signal was found for the control group. According to the authors, this result suggests that QD-1 and QD-2 were formed as the result of cellular uptake of CdSe/ZnS QDs. In fact, the amount of Cd/Se in both QD-1 and QD-2 in the cells were higher than 70%, indicating that the QDs ingested by the cells mainly existed as QD-1 and QD-2. The characterization of such fractions was carried out through a combination of sort techniques. For all QDs-1, the retention time and morphology obtained by SEC-ICP-MS and transmission electron microscopy (TEM), respectively, were similar to those visualized for original QDs. However, the molar ratio for QD-1 was less than the original QDs, indicating that QD-1 species are a kind of QD-like NPs, which might contain both original and degraded QDs. For QDs-2, the retention performance by SEC-ICP-MS was similar to those obtained for two kinds of Cd-metallothioneins. Through a combined hyphenated coupling and reverse-phase liquid chromatography (RPLC) with ICP-MS and ESI-QTOF-MS, the QDs-2 was confirmed as being a kind of metallothionein. In the speciation analysis, after the incubation of HepG2 cells with four CdSe/ZnS QDs, only two species, QD-1 and QD-2, were found in the cells. According to the authors, the transformation and species of QDs in biological systems cannot be neglected, once the toxicity of QDs will depend of its stability and transformation, as well the elemental release after cell uptake. Employing this combination between inorganic and organic mass spectrometry-based hyphenated techniques was possible provided valuable information regarding the metabolism and toxicity of QDs in living cells. Despite the authors assessed QDs uptake, stability, and transformation, the evaluation of the NPs/QDs toxicity is still a challenge.

Regarding imaging techniques, e.g., laser ablation coupled to ICP-MS (LA-ICP-MS), despite NPs analysis is still a challenge due to NP heterogeneity with respect to size, very low concentration, and lack of standards, in addition to evaluate distinct species from NPs interaction with any organism, assess NPs or released ions distribution, after a fractionation step, through imaging techniques can bring valuable information concerning NPs behavior when in contact with a cell (Pozebon et al. 2017). In this way, techniques based on imaging have been developed in the last years and properly discussed in Chap. 7 of this book, such as LA-ICP-MS (Pozebon et al. 2017; Pessôa et al. 2016; Oliveira and Arruda 2015), X-ray fluorescence microscopy (XFM) (Ganio et al. 2016), and secondary ion mass spectrometry (SIMS), among others (Gimenez et al. 2016; Shi et al. 2016). In fact, such strategy is useful once it can quantify and provide spatial maps to assess NPs or elemental distribution in different tissues (Pozebon et al. 2017; Ganio et al. 2016; Mehennaoui et al. 2016; Gimenez et al. 2016). Taking into this fact, some studies were already published using imaging techniques to expand the information regarding to the effects by NP exposure (Ganio et al. 2016; Mehennaoui et al. 2016; Gimenez et al. 2016, Shi et al. 2016). An important factor that should be considered is the analytical issue of such approach in order to get reliable results to provide additional understanding concerning any biological system. In this way, Ganio et al. (2016) evaluated the biometals quantification and elemental distribution by XFM imaging in order to compare the results with those obtained by ICP-MS, an established and robust technique for elemental measurement. The main goal of this work was compare the accuracy and precision of the ICP-MS method for metal determination with the synchrotron-based XFM to determine the variation between isogenic, developmentally synchronous *C. elegans* adults, a kind of roundworm. According to the authors, the XFM method can be used as alternative to the ICP-MS. Additionally, through XFM technique is also possible assess the metal concentration in any organism by a spatial map. Thus, such strategy can also be adapted to evaluate any change in the metal distribution through an external factor, e.g., NP exposure. In fact, if NP interaction with the cells of any organism is different considering of element ones, NP/released ion quantification and distribution would be achieved using imaging strategy, e.g., LA-ICP-MS. For example, accumulations/aggregations and the increase of NP concentrations in a specific cell region are not necessarily expected to be similar than the chemical elements that compose the NPs (Pozebon et al. 2017). Regarding NP evaluation, some techniques have been employed, since those already established to those recently developed. Shi et al. (2016) used the TEM to study the 10 nm gold (Au) NP distribution in rats after intervaginal space injection (ISI) in the tarsal tunnel. As a control for Au distribution, an intravenous injection (IVI) of AuNPs was considered. Quantitative analysis of Au in the blood and Au distribution in the organs was carried out by ICP-MS. In such evaluation, some differences in terms of ratios of AuNP content and relative organ AuNP distribution proportions were seen. The AuNP content was higher in the skin and muscle of the rats after ISI than that after IVI. For the lungs, heart, and intestines, similar results are observed for both kinds of injections. Otherwise, the AuNP content in the blood was much lower after ISI than IVI. Once that AuNP amounts were relatively higher in some organs at the earlier time points after ISI compared to those after IVI, the authors obtained TEM images from the heart of rats to evaluate whether the AuNPs were transferred for the organs. One hour after ISI in the tarsal tunnel, some AuNPs were found distributed within the loose connective tissues. According to the authors, the AuNP distribution is not mediated by the circulation, but rather via loose connective tissues, which is present in almost every part of the body, allowing AuNP transfer to the organs.

In the same way, Mehennaoui et al. (2016) used nanoscale SIMS (nanoSIMS) to assess Ag distribution in two populations of a crustacean (*Gammarus fossarum*), after exposure to different AgNPs and $AgNO_3$, which were used as a positive control to compare AgNP effects and silver ion effects. The nanotoxicology, through LC_{50}, physiological, and behavioral responses, was also evaluated in this work.

According to acute toxicity test, $AgNO_3$ and 23 nm AgNPs led to a significant decrease in survival rates, osmoregulation, and locomotor activity. Otherwise, the physiological and behavioral responses were not affected by $AgNO_3$ and AgNP treatments, once antioxidant responses, defense mechanisms, cellular damage, energy reserves, and ventilatory activity were not impacted. However, the locomotion was significantly decreased by $AgNO_3$ and AgNP (23 nm) expositions. The Ag uptake as well as cellular regions was assessed through a nanoSIMS analysis. The results showed no presence of Ag in the gill tissues of control crustaceans. For those crustaceans exposed to 23 and 27 nm AgNPs, an internalization of Ag within gill tissues was seen, while for $AgNO_3$ exposition, the Ag seems to be present only in the external membrane. An important aspect to be considered is the influence of the AgNP synthesis method on the ion release. The authors suggested that both AgNPs can have different or additional uptake route compared to $AgNO_3$ and 23 nm AgNPs may readily release the ions in the cells after uptake, being more toxic for crustaceans.

8.7 Geochemistry Studies

The transdisciplinarity of metallomics has a great impact also considering geochemistry studies. The metals/metalloids essentiality to all living organisms and the consideration of the existence a biological selection of metals in biomolecules since long time ago, in very different circumstances, have a great influence on the advances of geochemistry research (Anbar 2008). For example, in the history evidences throughout geologic time, both the physical Earth and its biota were far different from today. Elemental cycle and the life have been strongly influencing each other along the way, consequently inferring the existence of a bioinorganic bridge between biological and environmental evolved (Williams 2002; Battistuzzi et al. 2004; Dupont et al. 2010). Chemical characteristics of rocks formed from ancient seafloor sediments and genomic researches have suggested changes in the distribution of the metals in *Archaea*, *Bacteria*, and *Eukarya*, which have been essential for understanding the evolution of life on Earth (Battistuzzi et al. 2004; Dupont et al. 2006; Parnell et al. 2012; Ralser 2014).

In geochemistry, the history of atmospheric composition is vital for understanding the origin and evolution of life. In the primary stages of the Earth, the atmosphere and oceans were anoxic, possibly with the ocean chemistry being dominated by iron and then by sulfide. Then, fluctuations in the atmospheric oxygen levels and gradual oxygenation of the oceans created changes in environmental availability of the elements (Williams 2002; Farquhar et al. 2007; Dupont et al. 2010), with a consequent switch on metals used in the enzymatic pathways and the composition of biomacromolecules (Canfield 1998; Saito et al. 2003; Holland 2006). These variations in the Earth redox chemistry changed the use of metals in the geochemical timeline, and, due to this reason, the metallo-biomolecules can be mapped for looking to a molecular biosignature (Thauer 1998; Kasting 2005) through the genetic, metallocompound, and elemental biodistribution in the environment.

Methanogenesis is characterized by several species of *Achaea* that use high levels of nickel and tungsten as a key metal cofactor in methane produced by enzymes, providing a biosignature for that Earth time (Thauer 1998; Kasting 2005). Determination of the concentration of Ni in simulated Precambrian seawater and diverse iron hydroxides showed decay from molar Ni to iron ratio registered in banded iron formations (BIFs), consequently by reducing the flux of Ni in the oceans (Konhauser et al. 2009; Wasylenki et al. 2015). Such Ni decrease may have severely limited the biogenic methane production, belike by decreasing on the methyl coenzyme M reductase activity, because this element is a component of its active site (Thauer 1998). Recently, by using chromatography and mass spectrometry, Wang and Wasylenki (2017) investigated the fractionation of stable isotopes of Ni during the Ni coprecipitation with ferrihydrite and, then, its transformation to hematite. The results of the Ni isotopes in BIFs reflected that the incorporated Ni is isotopically heavier than both surface-adsorbed and surface-dissolved Ni, with values following the order: δ60/58Ni incorporated > δ60/58Ni solution > δ60/58Ni surface. These results suggest that the Ni isotopic signatures in BIFs represent an important study for reconstruction of the Ni isotopic composition of ancient seawater over Precambrian geological history.

The geological events, biological evolution, and chemical reactions (including climate conditions) profoundly influenced the cycling of elements (Williams 2002; Saito 2009; Day et al. 2017; Wang and Wasylenki 2017). The hypothesis that the availability of metals influences on changes of metallic center in metallo-molecules has been generating by many researches. For example, the progressive rise of the atmospheric oxygen increased the use of copper and molybdenum, once Cu- and Mo-enzymes catalyze reactions of oxygen transformation (Scott et al. 2008; Scott and Lyons 2012). To understand the sedimentary Mo accumulation is necessary to know its biogeochemical cycling. Some work regarding to Mo speciation from both modern and ancient environments, using spectroscopic techniques, have proposed hypothesized by its fixation in sediments through different pathways involving Mo reduction, Mo bonding with organic matter, and Fe phases, among others (Scott and Lyons 2012; Wagner et al. 2017). The Mo speciation may elucidate geochemical mechanisms insight into sedimentary Mo precipitation, consequently inferring about changing on the marine redox chemistry throughout Earth's evolution.

Techniques used in multidisciplinary areas have been useful for establishing the paleoenvironmental, consequently, increasing the comprehension of the Earth life's history. For such task, the studies currently involve techniques such as X-ray absorption/fluorescence, radionuclide, inductively coupled plasma (ICP-MS), and bioinformatics (Zerkle et al. 2005; Sevcenco et al. 2009, 2011; Dickson et al. 2014). The modeling and simulation methods characterizing the bioavailability of trace elements have expectations of success of alternative metabolic engineering strategies. Model metallomes developed considering fluctuations through the geologic time, as environmental conditions (i.e., temperature and redox state of the oceans and atmosphere), indicate that changes in the biomolecules expression could show a metal biosignature in the expressed cellular metal content (Dupont et al. 2006; Sevcenco et al. 2009; Ralser 2014). The microbial metalloproteomics using structural bioinformatics provide the information on the evolution of Eukaria.

Some studies of metal-binding biomolecule structures for the Archaea, Bacteria, and Eukarya indicate the use of metals hierarchy in prokaryotes as Fe, Zn > Mn, Mo, Co, Cu, Ni > W, and V, although model metallomes can vary with the oxygen tolerance, optimum growth temperature, and phylogeny (Zerkle et al. 2005; Sevcenco et al. 2009).

Studies based on the determination of elemental isotope ratios, currently by mass spectrometric techniques, are reported as a tool to build the history of the Earth's mantle (Dickson et al. 2014; Day et al. 2017; Wagner et al. 2017). Results of the Os isotope data for abyssal peridotites from three ocean basins (Indian, Atlantic, Arctic), as well as Mesozoic to Proterozoic-aged Os-rich alloys from ophiolites, were used to investigate melt depletion processes in the mantle and, consequently, estimate the primitive mantle composition for highly siderophile elements (Day et al. 2017). The variations in the molybdenum, vanadium, uranium, chromium, and iron stable isotope compositions, redox-sensitive elements, in the seawater and marine sedimentary, both past and present, have been crucial for understanding functions of rise atmospheric oxygen in the Earth evolution history (Emerson & Huested 1991; Crusius et al. 1996; Montoya-Pino et al. 2010; Brenneckaa et al. 2011). The Fe isotope fractionations, occurred during redox and non-redox processes, and Fe isotope composition are potentials to use as a biosignature and as a tracer of the redox state of ancient oceans (Partin et al. 2013; Azrieli-Tal et al. 2014). Furthermore, the determination of the proportions of reactive and unreactive iron species has been attracted great interest in recent years (Partin et al. 2013; Azrieli-Tal et al. 2014; Wasylenki et al. 2015). Banded iron formations may have the ability to sorb and record many of dissolved metals isotopes, which has the capacity of informing the isotope compositions of Archaean seawater (Konhauser et al. 2009; Wasylenki et al. 2015). These isotopes data may represent relationship between the chemistry of the atmosphere and oceans, from which life evolved, and they can be linked to the evolution of ancient environments on Earth.

Then, the identification of metallomes for modern and ancient biochemistry generates geochemical signatures, and the chemical speciation for specific determination of metallomes has been necessary for understanding biological evolution, consequently, to infer geochemical aspects of the biosphere throughout the time and the evolution of ancient environment on Earth.

8.8 Conclusions

It is clear the enormous potential of both areas: metallomics and chemical speciation. In fact, the synergic effect provoked when both areas are explored together makes their borders diffuse, being difficult to recognize each other, and, then, seems to be that both are almost the same.

Inside this context, it is ease to rationalize that the progress of metallomics area needs necessarily of chemical speciation approaches. Only through this way, the puzzle can be complete, making the metallomics an essential area for drug discovery, nutrition, toxicology, and forensic, among others.

Acknowledgments The authors gratefully acknowledges the financial support of Fundação de Amparo à Pesquisa do Estado de São Paulo, FAPESP; the Conselho Nacional de Desenvolvimento Científico e Tecnológico, CNPq; and Financiadora de Estudos e Projetos, FINEP.

References

Abad-Alvaro I, Bolea E, Laborda F, Castillo JR (2017) An ICP-MS-based platform for release studies on silver-based nanomaterials. J Anal At Spectrom 32(6):1101–1108

Aleksenko SS, Matczuk M, Lu X, Foteeva LS, Pawlak K, Timerbaev AR, Jarosz M (2013) Metallomics for drug development: an integrated CE-ICP-MS and ICP-MS approach reveals the speciation changes for an investigational ruthenium(III) drug bound to holo-transferrin in simulated cancer cytosol. Metallomics 5(8):955-963

Anbar AD (2008) Elements and Evolution. Science 322:1481–1483

Anjum NA, Gill SS, Duarte AC, Pereira E, Ahmad I (2013) Silver nanoparticles in soil–plant systems. J Nanopart Res 15:1896–1922

Arita A, Costa M (2009) Epigenetics in metal carcinogenesis: nickel, arsenic, chromium and cadmium. Metallomics 1(845):222–228

Arruda MAZ, Azevedo RA (2009) Metallomics and chemical speciation: towards a better understanding of metal-induced stress in plants. Ann Appl Biol 155:301–307

Arruda SCC, Silva ALD, Galazzi RM, Azevedo RA, Arruda MAZ (2015) Nanoparticles applied to plant science: a review. Talanta 131:693–705

Azrieli-Tal I, Matthews A, Bar-Matthews M, Almogi-abin A, Vance D, Archer C, Teutsch N (2014) Evidence from molybdenum and iron isotopes and molybdenum–uranium covariation for sulphidic bottom waters during Eastern Mediterranean sapropel S1 formation. Earth Planet Sci Lett 393:231–242

Battistuzzi FU, Feijao A, Hedges B (2004) A genomic timescale of prokaryote evolution: insights into the origin of methanogenesis, phototrophy, and the colonization of land. BMC Evol Biol 4(1):1–14

Beller GA, Bergmann SR (2004) Myocardial perfusion imaging agents: SPECT and PET. J Nucl Cardiol 11:71–86

Brenneckaa GA, Herrmanna AD, Algeoc TJ, Anbar AD (2011) Rapid expansion of oceanic anoxia immediately before the end-Permian mass extinction. Proc Natl Acad Sci U S A 108(43):17631–17634

Brombach C-C, Manorut P, Kolambage-Dona PPP, Ezzeldin MF, Chen B, Corns WT, Feldmann J, Krupp EM (2017) Methylmercury varies more than one order of magnitude in commercial European rice. Food Chem 214:360–365

Cacho C, Brito B, Palacios J, Pérez-Conde C, Cámara C (2010) Speciation of nickel by HPLC-UV/MS in pea nodules. Talanta 83:78–83

Cairo G, Recalcati S (2007) Iron-regulatory proteins: molecular biology and pathophysiological implications. Expert Rev Mol Med 9(33):1–13

Canfield DE (1998) A new model for Proterozoic ocean chemistry. Nature 396(1):450–452

Carvalho WM, Fukushima AR, Fontes LR, Fuzinato DV, Florio JC, Chasin AAM (2013) Cocaine postmortem distribution in three brain structure: a comparison with whole blood and vitreous humos. J Forensic Legal Med 20:143–145

Chellan P, Sadler PJ (2016) The elements of life and medicines. Philos Trans A Math Phys Eng Sci 373:20140182

Chen JY, Tsao GC, Zhao Q, Zheng W (2001) Differential cytotoxicity of Mn(II) and Mn(III): special referenceto mitochondrial [Fe-S] containing enzymes. Toxicol Appl Pharmacol 175(2):160–168

Cheyns K, Waegeneers N, de Wiele TV, Ruttens A (2017) Arsenic release from foodstuffs upon food preparation. J Agric Food Chem 65:2443–2453

Costa JL, Morrone AR, Resende RR, Chasin AAM, Tavares MFM (2014) Development of a method for the analysis of drugs of abuse in vitreous humor by capillary electrophoresis with diode array detection (CE–DAD). J Chromatogr B 945–946:84–91

Crusius J, Calve S, Pedersen T, Sage D (1996) Rhenium and molybdenum enrichments in sediments as indicators of oxic, suboxic and sulfidic conditions of deposition. Earth Planet Sci Lett 145:65–78

Cunha RLOR, Gouvea IE, Juliano L (2009) A glimpse on biological activities of tellurium compounds. An Acad Bras Cienc 81:393–407

Day JMD, Walker RJ, Warren JM (2017) 186Os–187Os and highly siderophile element abundance systematics of the mantle revealed by abyssal peridotites and Os-rich alloys. Geochim Cosmochim Acta 200(1):232–254

de Bie P, Muller P, Wijmenga C, Klomp LWJ (2007) Molecular pathogenesis of Wilson and Menkes disease: correlation of mutations with molecular defects and disease phenotypes. J Med Genet 44(11):673–688

Dickson AJ, Cohen AS, Coe AL (2014) Continental margin molybdenum isotope signatures from the early Eocene. Earth Planet Sci Lett 404:389–395

Dupont CL, Yang S, Palenik B, Bourne PE (2006) Modern proteomes contain putative imprints of ancient shifts in trace metal geochemistry. Proc Natl Acad Sci U S A 103(47):17822–17827

Dupont CL, Butcher A, Valas RE, Bourned PE, Caetano-Anollés G (2010) History of biological metal utilization inferred through phylogenomic analysis of protein structures. Proc Natl Acad Sci U S A 107(23):10567–10572

Easter RN, Qilin C, Lai B, Ritman EL, Caruso JA, Zhenyu Q (2010) Vascular metallomics: copper in the vasculature. Vasc Med 15(1):61–69

Emerson SR, Huested SS (1991) Ocean anoxia and the concentrations of molybdenum and vanadium in seawater. Mar Chem 34:177–197

Esteban-Fernández D, Moreno-Gordaliza E, Cañas B, Palaciosa MA, Gómez-Gómez MM (2010) Analytical methodologies for metallomics studies of antitumor Pt-containing drugs. Metallomics 2:19–38

Farquhar J, Peters M, Johnston DT, Strauss H, Masterson A, Wiechert U, Kaufman AJ (2007) Isotopic evidence for Mesoarchaean anoxia and changing atmospheric sulphur chemistry. Nature 449(1):706–709

Ganio K, James SA, Hare DJ, Roberts BR, McColl G (2016) Accurate biometal quantification per individual *Caenorhabditis elegans*. Analyst 141:1434–1439

Gimenez Y, Busser B, Trichard F, Kulesza A, Laurent JM, Zaun V, Lux F, Benoit JM, Panczer G, Dugourd P, Tillement O, Pelascini F, Sancey L, Motto-Ros V (2016) 3D imaging of nanoparticle distribution in biological tissue by laser-induced breakdown spectroscopy. Sci Rep 6(29936):1–9

Gómez-Ariza JL, Morales E, Giráldez I, Sánchez-Rodas D, Velasco A (2001) Sample treatment in chromatography-based speciation of organometallic pollutants. J Chromatogr A 938(1–2):211–224

Halpert G, Eitan T, Voronov E, Apte RN, Rath-Wolfson L, Albeck M, Kalechman Y, Sredni B (2014) Multifunctional activity of a small tellurium redox immunomodulator compound, AS101, on dextran sodium sulfate-induced murine colitis. J Biol Chem 289:17215–17227

Haraguchi H (2004) Metallomics as integrated biometal science. J Anal At Spectrom 19:5–14

Harris K, Armstrong SP, Campos-Pires R, Kiru L, Franks NP, Dickinson R (2013) Neuroprotection against traumatic brain injury by xenon, but not argon, is mediated by inhibition at the N-methyl-d-aspartate receptor glycine site. Anesthesiology 119:1137–1148

Hei TK, Liu SX, Waldren C (1998) Mutagenicity of arsenic in mammalian cells: role of reactive oxygen species. Proc Natl Acad Sci U S A 95(14):8103–8107

Holland HD (2006) The oxygenation of the atmosphere and oceans. Philos Trans R Soc Lond B Biol Sci 361:903–915

Hu LG, He B, Wang Y, Jiang G, Sun H (2013) Metallomics in environmental and health related research: current status and perspectives. Chin Sci Bull 58(2):169–176

Hu Y, Zhang W, Cheng H, Tao S (2017) Public Health Risk of Arsenic Species in Chicken Tissues from Live Poultry Markets of Guangdong Province, China. Environ Sci Technol 51:3508–3517

Huang C, Ke Q, Costa M, Shi X (2004) Molecular mechanisms of arsenic carcinogenesis. Mol Cell Biochem 255(1–2):57–66

Ivanov AI, Christodoulou J, Parkinson JA, Barnham KJ, Tucker A, Woodrow J, Sadler PJ (1998) Cisplatin binding sites on human albumin. J Biol Chem 273:14721–14730

Jenkis AJ, Oblock J (2008) Phencyclidine and cannabinoids in vitreous humor. Legal Med 10:201–203

Jitaru P, Millour S, Roman M, Koulali KE, Noël L, Guérin T (2016) Exposure assessment of arsenic speciation in different rice types depending on the cooking mode. J Food Compos Anal 54:37–47

Joseph P (2009) Mechanisms of cadmium carcinogenesis. Toxicol Appl Pharmacol 238(3):272–279

Kannamkumarath SS, Wrobel K, Wrobel K, B'Hymer C, Caruso JA (2002) Capillary electrophoresis-inductively coupled plasma-mass spectrometry: an attractive complementary technique for elemental speciation analysis. J Chromatogr A 975(2):245–266

Kasting JF (2005) Methane and climate during the Precambrian era. Precambrian Res 137(1):119–129

Kintz P, Ginet M, Marques N, Cirimele V (2007) Arsenic speciation of two specimens of Napoleon's hair. Forensic Sci Int 170:204–206

Konhauser KO, Pecoits E, Lalonde SV, Papineau D, Nisbet EG, Barley ME, Arndt NT, Zahnle K, Kamber BS (2009) Oceanic nickel depletion and a methanogen famine before the Great Oxidation Event. Nature 458(1):750–753

Kralj B, Križaj I, Bukovec P (2005) Speciation of aluminium in tea infusions by use of SEC and FPLC with ICP-OES and ES-MS-MS detection. Anal Bioanal Chem 383:467–475

Kulkarni PP, She YM, Smith SD, Roberts EA, Sarkar B (2006) Proteomics of metal transport and metal-associated diseases. Chem Eur J 12(9):2410–2422

La Fontaine S, Burke R, Giedroc DP (2016) Mammalian copper biology: hitting the pause button in celebration of three pioneers and four decades of discovery. Metallomics 8(9):810–812

Lee YW, Klein CB, Kargacin B, Salnikow K, Kitahara J, Dowiat K, Zhitkovich A, Christie NT, Costa M (1995) Carcinogenic nickel silences gene expression by chromatin condensation and DNA methylation: a new model for epigenetic carcinogens. Mol Cell Biol 15(5):2547–2557

Lehmann EL, Fostier AH, Arruda MAZ (2013) Hydride generation using a metallic atomizer after microwave-assisted extraction for inorganic arsenic speciation in biological samples. Talanta 104:187–192

Lendoiro E, Cordeiro C, Rodrıguéz-Calvo MS, Vieira DN, Suárez-Peñaranda MJ, López-Rivadulla M, Muñoz-Barús JI (2012) Applications of Tandem Mass Spectrometry (LC–MSMS) in estimating the post-mortem interval using the biochemistry of the vitreous humour. Forensic Sci Int 223(1–3):160–164

Li T, Albee B, Alemayehu M, Diaz R, Ingham L, Kamal S, Rodriguez M, Bishnoi SW (2010) Comparative toxicity study of Ag, Au, and Ag–Au bimetallic nanoparticles on *Daphnia magna*. Anal Bioanal Chem 398:689–700

Li Y-F, Gao Y, Chaiab Z, Chen C (2014) Nanometallomics: an emerging field studying the biological effects of metal-related nanomaterials. Metallomics 6:220–232

Ma C, White JC, Dhankher OP, Xing B (2015) Metal-based nanotoxicity and detoxification pathways in higher plants. Environ Sci Technol 49:7109–7122

Malysheva A, Lombi E, Voelcker NH (2015) Bridging the divide between human and environmental nanotoxicology. Nat Nanotechnol 10:835–844

Mandal BK, Suzuki KT (2002) Arsenic round the world: a review. Talanta 58:201–235

Mandal R, Kalke R, Li X-F (2004) Interaction of oxaliplatin, cisplatin, and carboplatin with hemoglobin and the resulting release of a heme group. Chem Res Toxicol 17:1391–1397

Mehennaoui K, Georgantzopoulou A, Felten V, Andreï J, Garaud M, Cambier S, Serchi T, Pain-Devin S, Guérold F, Audinot J-N, Giambérini L, Gutleb AC (2016) *Gammarus fossarum* (Crustacea, Amphipoda) as a model organism to study the effects of silver nanoparticles. Sci Total Environ 566-567:1649–1659

Montoya-Pino C, Weyer S, Anbar AD, Pross J, Oschmann W, Schootbrugge BV, Arz HW (2010) Global enhancement of ocean anoxia during Oceanic Anoxic Event 2: a quantitative approach using U isotopes. Geol Soc Am 38(4):315–318

Mounicou S, Szpunar J, Lobinski R (2009) Metallomics: the concept and methodology. Chem Soc Rev. 38(4):1119–1138

Nielsen FH (1999) Ultratrace minerals. In: Shils ME, Olsen JA, Shike M, Ross AC (eds) Modern nutrition in health and disease, 9th edn. Williams and Wilkins, Baltimore, MD, pp 283–303

Oliveira SR, Arruda MAZ (2015) Application of laser ablation (imaging) inductively coupled plasma mass spectrometry for mapping and quantifying Fe in transgenic and non-transgenic soybean leaves. J Anal At Spectrom 30(2):389–395

Oller AR, Costa M, Oberdörster G (1997) Carcinogenicity assessment of selected nickel compounds. Toxicol Appl Pharmacol 143(1):152–166

Ossipov K, Foteeva LS, Seregina IF, Perevalov SA, Timerbaev AR, Bolshov MA (2013) Metallomics for drug development: serum protein binding and analysis of an anticancer tris(8-quinolinolato)gallium(III) drug using inductively coupled plasma mass spectrometry. Anal Chim Acta 785:22–26

Parnell J, Hole M, Boyce AJ, Spinks S, Bowden S (2012) Heavy metal, sex and granites: crustal differentiation and bioavailability in the mid-Proterozoic. Geology 40(8):751–754

Partin CA, Lalonde SV, Planavsky NJ, Bekker A, Rouxel OJ, Lyons TW, Konhauser KO (2013) Uranium in iron formations and the rise of atmospheric oxygen. Chem Geol 362:82–90

Peeters K, Zuliani T, Žigon D, Milačič R, Ščančar J (2017) Nickel speciation in cocoa infusions using monolithic chromatography – post-column ID-ICP-MS and Q-TOF-MS. Food Chem 230:327–335

Peng L, He M, Chen B, Qiao Y, Hu B (2015) Metallomics study of CdSe/ZnS quantum dots in HepG2 cells. ACS Nano 9(10):10324–10334

Peng J, Mandal R, Sawyer M, Li X-F (2005) Characterization of intact hemoglobin and oxaliplatin interaction by nanoelectrospray ionization tandem mass spectrometry. Clin Chem 51:2274–2281

Pessôa GS, Capelo-Martínez JL, Fdez-Riverola F, López-Fernández H, Glez-Peña D, Reboiro-Jato M, Arruda MAZ (2016) Laser ablation and inductively coupled plasma mass spectrometry focusing on bioimaging from elemental distribution using MatLab software: a practical guide. J Anal At Spectrom 31(4):832–840

Pozebon D, Scheffler GL, Dressler VL (2017) Recent applications of laser ablation inductively coupled plasma mass spectrometry (LA-ICP-MS) for biological sample analysis: a follow-up review. J Anal At Spectrom 32(5):890–919

Raab A, Fecher P, Feldmann J (2005) Determination of Arsenic in Algae – Results of an Interlaboratory Trial: Determination of Arsenic Species in the Water-Soluble Fraction. Microchim Acta 151:153–166

Ralser M (2014) The RNA world and the origin of metabolic enzymes. Biochem Soc Trans 42(4):985–988

Rezende HC, Almeida ILS, Coelho LM, Coelho NMM, Marques TL (2015) Non-chromatographic methods focused on speciation of arsenic and selenium in food and environmental samples. Sample Prep 2:31–48

Sadee BA, Foulkes ME, Hill SJ (2016) A study of arsenic speciation in soil, irrigation water and plant tissue: a case study of the broad bean plant, *Vicia faba*. Food Chem 210:362–370

Saito MA (2009) Less nickel for more oxygen. Nature 458(1):714–715

Saito MA, Sigman DM, Morel FMM (2003) The bioinorganic chemistry of the ancient ocean: the co-evolution of cyanobacterial metal requirements and biogeochemical cycles at the Archean/Proterozoic boundary? Inorg Chim Acta 356(1):308–318

Salnikow K, Zhitkovich A (2008) Genetic and epigenetic mechanisms in metal carcinogenesis and cocarcinogenesis: nickel, arsenic and chromium. Chem Res Toxicol 21(1):28–44

Santos Junior JC, Mollo Filho PC, Guidugli RBF, Eberlin MN, Pessôa GS, da Silva EG, Arruda MAZ, Hoeh NF (2014) Metals and (metallo)proteins identification in vitreous humor focusing on post-mortem biochemistry. Metallomics 6(10):1801–1807

Sariego Muniz C, Gayón JMM, Alonso JIG, Sanz-Medel A (2001) Speciation of essential elements in human serum using anion-exchange chromatography coupled to post-column isotope dilution analysis with double focusing ICP-MS. J Anal At Spectrom 16(6):587–592

Ščančar J, Zuliani T, Milačič R (2013a) Study of nickel content in Ni-rich food products in Slovenia. J Food Compos Anal 32:83–89

Ščančar J, Zuliani T, Žigon D, Milačič R (2013b) Ni speciation in tea infusions by monolithic chromatography – ICP-MS and Q-TOF-MS. Anal Bioanal Chem 405:2041–2051

Scott C, Lyons TW, Bekker A, Shen Y, Poulton SW, Chu X, Anbar AD (2008) Tracing the stepwise oxygenation of the Proterozoic ocean. Nature 452:456–459

Scott C, Lyons TW (2012) Contrasting molybdenum cycling and isotopic properties in euxinic versus non-euxinic sediments and sedimentary rocks: refining the paleoproxies. Chem Geol 324-325:19–27

Sevcenco AM, Pinkse MWH, Bol E, Krijger GC, Wolterbeek HT, Verhaert PEDM, Hagedoorna PL, Hagen WR (2009) The tungsten metallome of Pyrococcus furiosus. Metallomics 1:395–402

Sevcenco AM, Pinkse MWH, Wolterbeek HT, Verhaert PEDM, Hagena WR, Hagedoorn PL (2011) Exploring the microbial metalloproteome using MIRAGE. Metallomics 3:1324–1330

Shi W, Chance MR (2008) Metallomics and metalloproteomics. Cell Mol Life Sci 65(19):3040–3048

Shi W, Chance MR (2011) Metalloproteomics: forward and reverse approaches in metalloprotein structural and functional characterization. Curr Opin Chem Biol 15(1):144–148

Shi X, Zhu Y, Hua W, Ji Y, Ha Q, Han X, Liu Y, Gao J, Zhang Q, Liu S, Ren K, Wu X, Li H, Han D (2016) An *in vivo* study of the biodistribution of gold nanoparticles after intervaginal space injection in the tarsal tunnel. Nano Res 9(7):2097–2109

Shindo M, Torimoto Y, Saito H, Motomura W, Ikuta K, Sato K, Fujimoto Y, Kohgo Y (2006) Functional role of DMT1 in transferrin-independent iron uptake by human hepatocyte and hepatocellular carcinoma cell, HLF. Hepatol Res 35(3):152–162

Silva MO, Sussulini A, Arruda MAZ (2010) Metalloproteomics as an interdisciplinary area involving proteins and metals. Expert Rev Proteomics 7:387–400

Soto-Alvaredo J, Montes-Bayón M, Bettmer J (2013) Speciation of silver nanoparticles and silver(I) by reversed-phase liquid chromatography coupled to ICPMS. Anal Chem 85:1316–1321

Swain R, Kumar A, Sahoo J, Lakshmy R, Gupta SK, Bhardwaj DN, Pandey RM (2015) Estimation of post-mortem interval: a comparison between cerebrospinal fluid and vitreous humour chemistry. J Forensic Legal Med 36:144–148

Szpunar J (2004) Metallomics: a new frontier in analytical chemistry. Anal Bioanal Chem 378:54–56

Takahashi N, Ortel TL, Putnam FW (1984) Single-chain structure of human ceruloplasmin: the complete amino acid sequence of the whole molecule. Proc Natl Acad Sci U S A 81:390–394

Tan C-P, Lu Y-Y, Ji L-N, Mao Z-W (2014) Metallomics insights into the programmed cell death induced by metal-based anticancer compounds. Metallomics 6:978–995

Taylor VF, Jackson BP (2016) Concentrations and speciation of arsenic in New England seaweed species harvested for food and agriculture. Chemosphere 163:6–13

Thauer RK (1998) Biochemistry of methanogenesis: a tribute to Marjory Stephenson. Microbiology 144(1):2377–2406

Thompson KH, Orvig C (2006) Metal complexes in medicinal chemistry: new vistas and challenges in drug design. Dalton Trans 6:761–764

Thompson KH, Lichter J, LeBel C, Scaife MC, McNeill JH, Orvig C (2009) Vanadium treatment of type 2 diabetes: a view to the future. J Inorg Biochem 103:554–558

Timerbaev AR (2016) Role of metallomic strategies in developing ruthenium anticancer drugs. Trends in Analytical Chemistry 80:547-554

Todd RC, Lippard SJ (2009) Inhibition of transcripton by platinum antitumor compounds. Metallomics 1:280–291

Tout D, Tonge CM, Muthu S, Arumugam P (2012) Assessment of a protocol for routine simultaneous myocardial blood flow measurement and standard myocardial perfusion imaging with rubidium-82 on a high count rate positron emission tomography system. Nucl Med Commun 33:1202–1211

Wagner M, Chappaz A, Lyons TW (2017) Molybdenum speciation and burial pathway in weakly sulfidic environments: insights from XAFS. Geochim Cosmochim Acta 206:18–29

Wang SJ, Wasylenki LE (2017) Experimental constraints on reconstruction of Archean seawater Ni isotopic composition from banded iron formations. Geochim Cosmochim Acta 206(1):137–150

Wasylenki LE, Howe HD, Spivak-Birndorf LJ, Bish DL (2015) Ni isotope fractionation during sorption to ferrihydrite: implications for Ni in banded iron formations. Chem Geol 400(1):56–64

Williams RJP (2002) The fundamental nature of life as a chemical system: the part played by inorganic elements. J Inorg Biochem 88:241–250

Wilson RA, Yanes EG, Kemppainen RJ (2016) Iodine speciation in dog foods and treats by high performance liquid chromatography with inductively coupled plasma mass spectrometry detection. J Chromatogr B 1022:183–190

Yim SR, Park GY, Lee KW, Chung M-S, Shim S-M (2017) Determination of total arsenic content and arsenic speciation in different types of rice. Food Sci Biotechnol 26(1):293–298

Zerkle AI, House CH, Brantley SI (2005) Biogeochemical signatures through time as inferred from whole microbial genomes. Am J Sci 305:467–502

Zhang X, Cornelis R, de Kimpe J, Mees L, Lameiere N (1998) Study of arsenic-protein binding in serum of patients on continuous ambulatory peritoneal dialysis. Clin Chem 44(1):141–147

Zhitkovich A (2005) Importance of chromium-DNA adducts in mutagenicity and toxicity of chromium(VI). Chem Res Toxicol 18(1):3–11

Zoorob GK, McKiernan JW, Caruso J (1998) ICP-MS for elemental speciation studies. Microchim Acta 128(3):145–168

Chapter 9
Advanced Nuclear and Related Techniques for Metallomics and Nanometallomics

Yu-Feng Li, Jiating Zhao, Yuxi Gao, Chunying Chen, and Zhifang Chai

Abstract Metallomics, focusing on the global and systematic understanding of the metal uptake, trafficking, role, and excretion in biological systems, has attracted more and more attention. Metal-related nanomaterials, including metallic and metal-containing nanomaterials, have unique properties compared to their macroscale counterparts and therefore require special attention. The absorption, distribution, metabolism, excretion (ADME) behavior of metal-related nanomaterials in the biological systems is influenced by their physicochemical properties, the exposure route, and the microenvironment of the deposition site. Nanomaterials not only may interact directly or indirectly with genes, proteins, and other molecules to bring genotoxicity, immunotoxicity, DNA damage, and cytotoxicity but may also stimulate the immune responses, circumvent tumor resistance, and inhibit tumor metastasis. Because of their advantages of absolute quantification, high sensitivity, excellent accuracy and precision, low matrix effects, and nondestructiveness, nuclear and related analytical techniques have been playing important roles in the study of metallomics and nanometallomics. In this chapter, we present a comprehensive overview of nuclear and related analytical techniques applied to the quantification of metallome and nanometallome, the biodistribution, bioaccumulation, and transformation of metallome and nanometallome in vivo, and the structural analysis. Besides, metallomics and nanometallomics need to cooperate with other -omics, like genomics, proteomics, and metabolomics, to obtain the knowledge of underlying

---

Part of this chapter was reproduced from (1) Li Y-F, Gao YX, Chai ZF, Chen CY. Nanometallomics: An emerging field studying the biological effects of metal-related nanomaterials. Metallomics, 2014, 6(2): 220–232; and adapted from (2) Li Y-F, Chen CY, Qu Y, Gao YX, Li B, Zhao YL and Chai ZF. Metallomics, elementomics and the analytical techniques. Pure Appl. Chem., 2008, 80(12): 2577–2594.

Y.-F. Li · J. Zhao · Y. Gao · Z. Chai (✉)
CAS Key Laboratory for Biomedical Effects of Nanomaterials and Nanosafety,
and Laboratory for Metallomic and Nanometallomics, Institute of High Energy Physics,
Chinese Academy of Sciences, Beijing, China
e-mail: chaizf@ihep.ac.cn

C. Chen
CAS Key Laboratory for Biomedical Effects of Nanomaterials and Nanosafety,
National Center for Nanoscience and Technology, Beijing, China

M. A. Z. Arruda (ed.), *Metallomics*, Advances in Experimental Medicine and Biology 1055, https://doi.org/10.1007/978-3-319-90143-5_9

mechanisms and therefore to improve the application performance and to reduce the potential risk of metallome and nanometallome.

Keywords Metallomics · Nanometallomics · Nuclear analytical techniques · Metals · Metal-related nanomaterials

Abbreviations

ADME	Absorption, distribution, metabolism, and excretion
ASE	Accelerated solvent extraction
CAE	Capillary array electrophoresis
CD	Dichroism spectroscopy
CE	Capillary electrophoresis
CEC	Capillary electrochromatography
CGE	Capillary gel electrophoresis
CNT	Carbon nanotubes
CT-SPECT	Computed tomography coregistered with single-photon emission computerized Tomography
CZE	Capillary zone electrophoresis
EC	Electrochromatography
EDX	Energy dispersive X-ray fluorescence
ESI-MS	Electrospray ionization mass spectrometer
EXAFS	Extended X-ray absorption fine structure
GC	Gas chromatography
GE	Gel electrophoresis
HPLC	High-performance liquid chromatography
ICP-AES	Inductively coupled plasma atomic emission spectrometry
ICP-MS	Inductively coupled plasma mass spectrometry
LA	Laser ablation
MALDI-TOF-MS	Matrix-assisted laser desorption/ionization time-of-flight mass spectroscopy
MECC	Micelle electrokinetic capillary chromatography
MEKC	Micelle electrokinetic capillary electrophoresis
MMP	Matrix metalloproteinases
MW	Multiwalled
NAA	Neutron activation analysis
nanoXRF	XRF mapping with the nano-sized spatial resolution
NATs	Advanced nuclear analytical techniques
NMR	Nuclear magnetic resonance spectroscopy
NPs	Nanoparticles

PAHs	Polycyclic aromatic hydrocarbons
PCBs	Polychlorinated biphenyls
PDDAC	Poly(diallydimethylammonium chloride)
PEI	Polyethyleneimine
PET	Positron emission tomography
PIXE	Proton-inducted X-ray emission spectrometry
PLE	Pressurized liquid extraction
POPs	Persistent Organic Pollutants
QDs	Quantum dots
RES	Reticuloendothelial systems
ROS	Reactive oxygen species
SANS	Small angle neutron scattering
SAXS	Small angle X-ray scattering
SCND	Single crystal neutron diffraction spectroscopy
SEM	Scanning electron microscopy
SFE	Supercritical fluid extraction
SIMS	Secondary ion mass spectroscopy
SPE	Solid-phase extraction
SPECT	Single-photon emission computed tomography
SPME	Solid-phase microextraction
SR	Synchrotron radiation
SR-μXRF	Synchrotron radiation-based microbeam X-ray fluorescence analysis
SW	Single-walled
SWE	Subcritical water extraction
TEM	Transmission electron microscopy
WDX	Wavelength dispersive x-ray fluorescence
XAS	X-ray absorption spectroscopy
XRD	X-ray diffraction
XRF	X-ray fluorescence analysis
μ-XRF	Microbeam X-ray fluorescence analysis

9.1 Introduction

9.1.1 Nuclear and Related Analytical Techniques

Advanced nuclear analytical techniques (NATs) are powerful tools for studying a number of basic issues in chemistry and biology. Because of their advantages of high sensitivity, excellent accuracy, low matrix effects, and nondestructiveness, NATs have been playing important roles in the study of metallomics and nanometallomics, especially where other analytical techniques might not be sufficient or where they cannot be simply applied at all.

NATs are techniques that deal with nuclear excitations, electron inner shell excitations, nuclear reactions, and/or radioactive decay (Chen et al. 2010; Gao et al. 2007). Nuclear mass, spin and magnetic moment, excited states and related parameters, and probability of nuclear reactions are nuclear parameters that serve as basis for nuclear analytical techniques. In principle, NATs are based on the properties of the nucleus itself, compared to nonnuclear techniques, which use properties of the atom as a whole, primarily governed by properties of the electrons arranged in shells. However, fundamentally as well as practically, no sharp borderline can be drawn between nuclear and atomic techniques. For example, mass spectrometry deals with ionized atoms and rarely with the bare nucleus, but the signal is determined by the mass differences of the nucleus, i.e., isotope ratios. Therefore, it can be also categorized as nuclear techniques in many textbooks and also in this chapter. Synchrotron radiation facilities can produce photons with high brilliance, high level of polarization, high collimation, low emittance, wide tunability in energy/wavelength, and pulsed light emission (Rowe et al. 1973), which was regarded as related techniques in this chapter.

9.1.2 Metallomics

In recent decades, -omics studies, especially genomics, proteomics, or metabolomics (metabonomics), aiming at the collective characterization and quantification of pools of biological molecules that translate into the structure, function, and dynamics of an organism or organisms have received great attention (Seemungal and Newton 2001; James 1997; Nicholson and Lindon 2008). Similarly, metallomics, aiming to provide a global and systematic understanding of the metal uptake, trafficking, role, and excretion in biological systems and potentially to be able to predict all of these in silico using bioinformatics, is attracting more and more attentions Mounicou et al. 2009; Haraguchi 2004). It focuses on the systematic study of metallomes as well as the interactions and functional connections of metal ions and their species with genes, proteins, metabolites, and other biomolecules within organisms and ecosystems (Ge and Sun 2009). Five successful meetings for metallomics have been held since 2007 and the journal *Metallomics* has been publishing since 2009 by the Royal Society of Chemistry, UK (Drickamer and Taylor 2002; http://pubs.rsc.org/en/journals/journalissues/mt#!recentarticles&all).

9.1.3 Nanometallomics

Nanomaterials are widely used in many fields due to their fascinating properties. However, safety concerns on the potential risk of nanomaterials when entering into biological systems directly during manufacturing processes or indirectly via the environment, food chains, etc. are raised (Brumfiel 2003; Colvin 2003; Service RF

2003; Hoet et al. 2004; Zhao et al. 2008). For example, some studies showed that nanoparticles, after deposition in the lungs, largely escape alveolar macrophage surveillance and gain access to the pulmonary interstitium with greater inflammatory effect than larger particles (Oberdörster 2000; Oberdoster et al. 2004). A fast translocation of nanoparticles from pulmonary and gastrointestinal epithelium into the systemic circulation through animal studies was also observed (Nemmar et al. 2002). Therefore, the study of the toxicological effect of nanomaterials is necessary and a new research field as nanotoxicology was formed in recent years (Zhao and Nalwa 2006; Oberdorster et al. 2005; Chen et al. 2013). Besides the toxicological effect of nanomaterials, the elucidation of the biological effects (including both toxicological and beneficial effects) of the nanomaterials is important, not only for the protection of human health and environmental integrity but also to aid industry and regulatory bodies in maximizing the application of nanomaterials.

The terminology "nanometallomics" was initiated as a branch of metallomics, and the application of advanced nuclear analytical techniques on the study of nanometallomics was also reviewed (Li et al. 2010, 2014). Besides, a session for nanometallomics was organized at the 4th international symposium on metallomics in Spain in July 2013 (http://www.metallomics2013.com/index.php/important-dates/conference-schedule). In this chapter, we will review on the systematically absorption, distribution, metabolism, and excretion (ADME) behavior and their interactions with genes, proteins, and other biomolecules of metallomes and metal-related nanomaterials in biological systems.

9.2 Advanced Nuclear and Related Techniques for Metallomics

9.2.1 Quantification of Metallomes

For -omics study, the most important requirement should be high throughput. Plasma-based instruments like inductively coupled plasma atomic emission spectrometry (ICP-AES) and inductively coupled plasma mass spectrometry (ICP-MS), which can quantify metallomes rapidly and simultaneously in one run, are ideal for this purpose. However, ICP-AES shows the low sensitivity and often the poor selectivity to accurately determine many of the less abundant but more important elements in biological systems. In contrast, ICP-MS is extremely sensitive, due to the efficient ionization from plasma coupled with the sensitive detection of the mass spectrometer. At its best, the parts per trillion detection limits are achievable that are generally 2–3 orders of magnitude lower than those in ICP-AES (Thompson and Walsh 1983). ICP-MS can detect most elements in biological systems, but sulfur, phosphorous, and halogens are not efficiently ionized by the ICP owing to their high ionization energies. Further, a number of polyatomic interferences also hinder the detection of S, P, and transition elements like Fe and V using ICP-MS (Szpunar 2005). Polyatomic interferences can be solved by either using high-resolution (sector field double focusing) mass spectrometer or, in more commonly available quadrupole mass analyzer, the collision/reaction cell techniques (Feldmann et al. 1999; Koppenaal et al. 2004). Therefore, S, P, Cl, Br, I, and even C have been detected by quadrupole ICP-MS (Szpunar 2005; Wang et al. 2007). Besides polyatomic interference effect, memory effect of Hg, U, Os, Hf, Pt, etc. is another important issue in precise ICP-MS detection. To solve this problem, complexing agents can be added to prevent the adherence of them to the walls of the spray chamber and the transfer tubing of the sample introduction system. For example, a series of complexing agents aiming to reduce the memory effect of Hg were compared, and –SH-containing reagents were found to be the best to eliminate it in the detection of Hg by ICP-MS (Li et al. 2006).

High-resolution mass spectrometer is superior to the common quadrupole detector since it can provide extremely low instrumental background and very good detection limits (0.1–1 ng mL^{-1}) for nearly all elements, but the high cost may hamper its wide application in metallomics study.

For ICP-MS, liquid and gaseous samples can be directly introduced into system, but solid samples need pretreatment like digestion or ashing to make them into liquid solutions. The application of laser-ablation (LA) technique in ICP-MS has made the direct and online analysis of solid samples possible which have enhanced greatly the application of ICP-MS (Jarvis and Williams 1993; Durrant and Ward 1994). Its fast data acquisition, minimal sample preparation, and high sensitivity offer low detection limits for a wide range of major and trace elements in complex biological, environmental, and geological samples.

Neutron activation analysis (NAA) is also a metallome quantification technique which can simultaneously measure more than 30 elements in one sample. The detection limits of NAA range from 10^{-6} to 10^{-13} g g^{-1} (Chai et al. 1992). In typically instrumental NAA, stable nuclides (^{A}Z, the target nucleus) in the sample undergo neutron capture reactions in a flux of (incident) neutrons. The radioactive nuclides (^{A+1}Z, the compound nucleus) produced in this activation process will, in most cases, decay through the emission of a beta particle ($ß^-$) and gamma ray(s) with an inherent half-life. A high-resolution gamma-ray spectrometer is used to detect these "delayed" gamma rays from the artificially induced radioactivity in the sample for both qualitative and quantitative analyses (Chai et al. 1992). One of the principal advantages of NAA is that it is nearly free of any matrix interference effects as the vast majority of samples are completely transparent to both the probe (the neutron) and the analytical signal (the gamma ray). Moreover, because NAA can most often be applied instrumentally (no need for sample digestion or dissolution), there is little, if any, opportunity for reagent or laboratory contamination (Chai and Zhu 1994).

X-ray fluorescence analysis (XRF) is another metallome analytical technique which is based on the detection of characteristic fluorescence after excited by primary X-ray with sufficient energy (Jenkins 1999). If the exciting source is replaced by a high-energy proton beam, it will be called proton-inducted X-ray emission spectrometry (PIXE) (Johansson 1989), which can provide better analytical sensitivity than XRF. If synchrotron radiation (SR) is used as exciting source, the

SR-XRF will achieve an absolute detection limit of 10^{-12}–10^{-15} g and relative detection limit of several μg/g, even in ng/g levels (Van Langevelde and Vis 1991). XRF is basically a qualitative technique, but quantification can also be achieved (Karanatsios et al. 1988; Tertian and Claisse 1982).

In general, ICP-MS, NAA, XRF, and PIXE are all excellent techniques for metallome quantification. NAA, XRF, and PIXE can achieve nondestructive and in situ metallomes analysis for solid samples, while ICP-MS needs laser ablation for this purpose.

After getting the concentrations of different elements in samples, correlation studies using mathematical methods may lead to further understanding on the roles and their cross-talk of different elements. For example, positive correlation between Se and Hg contents has been found in fish (Skerfving 1974; Zhang et al. 2004), marine mammals and birds (Bowerman et al. 1994; Caurant et al. 1996; Koeman et al. 1973; Scheuhammer et al. 1998), and animals and workers exposed to inorganic mercury (Alexander et al. 1983; Falnoga et al. 2000; Kosta et al. 1975; Chen et al. 2006), which leads to the study of Hg and Se antagonism or synergism mechanism (Burk et al. 1974; Gailer et al. 2000; Yoneda and Suzuki 1997). Other studies which measure metallomes in human autopsy tissue (Beneš et al. 2000) and cancerous lung (Xiong et al. 2006) also found significant relationship among some elements.

9.2.2 The Distribution of Metallomes

Besides total concentration of metallomes, the spatial distribution of them in samples is also very important in understanding their bioavailability, trophic transfer, and environmental risk. A number of complementary analytical techniques exist for the mapping of elemental distributions in biological tissues including SR-μXRF, microscopic EDX (energy dispersive X-ray fluorescence), microscopic WDX (wavelength dispersive X-ray fluorescence), microscopic PIXE, LA-ICP-MS, and microscopic SIMS (secondary ion mass spectrometry).

SR-XRF with microbeam (SR-μXRF) has been used widely in elemental distribution study. For example, cellular distribution of arsenic and other elements in arsenic hyperaccumulator *Pteris nervosa* and their relations to arsenic accumulation have been studied using line-scanning of XRF, and the transport route for arsenic in *Pteris nervosa* was found (Chen et al. 2003). By regulating with a slit or focusing system, such as Kirkpatrick-Baez mirror system (Ortega et al. 2004), refractive lenses (Snigirev et al. 1996), or a Fresnel zone plate (Snigirev et al. 1996), the beam size can be made in micron or even submicron level. Raster scanning of the specimen and acquisition of the entire X-ray spectrum yields quantitative topographical maps for a wide range of elements (Fahrni 2007). Different cells (Kemner et al. 2004; Yang et al. 2005), tissues (Paunesku et al. 2007; Ide-Ektessabi et al. 2002), and organ slices (Iida and Noma 1993) have been successfully studied using this microbeam two-dimensional (2D) SR-μXRF (Paunesku et al. 2006; Iida 1997).

However, either line-scanning or two-dimensional (2D) elemental mapping is performed along the sample surface. Therefore, the information retrieved is a product of the incident X-rays, the penetration capability, and the self-absorption correction depending on both the energy of the exciting radiation and the energy of the fluorescence radiation and has the disadvantage that it is not explicitly depth sensitive. To perform depth analysis, the confocal arrangement, which consists of X-ray optics in the excitation as well as in the detection channel, has been set up by several groups with a depth resolution from 10 to 40 μm and minimum detection limits are sub-mg/kg level (Kanngießer et al. 2003; Muradin 2000; Vincze et al. 2004; Kouichi Tsuji 2007; Janssens et al. 2004). Three-dimensional (3D) micro-XRF investigations have been successful in biological samples like rice grain (Nakano and Tsuji 2006), duckweed (Kanngießer et al. 2007a, b), and human joint bone (Zoeger et al. 2007).

Besides the abovementioned 3D SR-XRF, the SR-XRF tomography measurement is also possible to perform three-dimensional elemental analysis by measuring a series of projected distributions under various angles and back projected using the appropriate mathematical algorithms (Vincze et al. 2001; Gan et al. 2006). Since this method involves rotation of the sample over 180° or 360° relative to the primary beam, it is limited to the investigation of relatively small objects. The spatial resolutions for XRF tomography are situated at the 1–2 μm level while routinely a resolution of 5 μm is being employed (Hansel et al. 2002).

Other XRF-based techniques like EDX has been coupled to electron microscopes like TEM (transmission electron microscopy) or SEM (scanning electron microscopy) to map element in roots and leaves of *Arabidopsis thaliana* (Isaure et al. 2006), mice liver tissue (Kametani and Nagata 2006), and pyramidal neurons (Jiang and Sato 1999), among others. Although both SEM-EDX and TEM-EDX can provide very good spatial resolution at about 10 nm, the detection limit is at about g/kg level (Motelica-Heino et al. 1998) which may hold back their application in trace element detection in biological samples.

Just like SR-XRF, PIXE with proton microprobe has also been developed (Ryan 1995, 2000; Przybylowicz et al. 2001) and has been applied to elemental distribution in plant and animal tissues (Tylko et al. 2007; Kramer et al. 1997; Sie and Thresher 1992) and human blood cells and tumors (Johansson et al. 1987). Further, 3D micro-PIXE also has been developed to perform depth analysis recently (Karydas et al. 2007; Kanngießer et al. 2007a, b) and has the spatial resolution of 4 μm by using characteristic Ti-K-X-rays (4.558 keV) produced by 3 MeV protons with beam spot size of ~1 μm (Ishii et al. 2007).

With the microprobe technique, LA-ICP-MS has become one of more powerful tools in the quantitative analysis of major and minor element in situ owing to the very high sensitivity of ICP-MS and direct laser sampling to obtain more information from small samples in micro area. LA-ICP-MS is well established for elemental mapping in the geological sciences (Meurer and Claeson 2002; Devos et al. 2000) and has been applied to biological tissues such as plant leaves (Punshon et al. 2004), snake tails (Jackson et al. 2003), tree rings (Watmough et al. 1997), rat brains (Jackson et al. 2006), pig liver (Kindness et al. 2003), human brains

(Becker et al. 2007), and human teeth (Kang et al. 2004). The spatial resolution of LA-ICP-MS is about 10 μm with a detection limit of sub-mg/kg level (Motelica-Heino et al. 1998). Accurate quantitative analysis by LA-ICP-MS is a challenge because of the lack of suitable certified solid standards. Various calibration procedures have been used, and the most accurate approach is to have true matrix matching of standards and samples. Commercially available certified reference materials have also been used for calibration (Kindness et al. 2003; Raith et al. 1996).

SIMS operates on the principle that bombardment of a material with a beam of ions with high energy (1–30 keV) results in the ejection or sputtering of atoms from the material. A small percentage of these ejected atoms leave as either positively or negatively charged ions, which are referred to as secondary ions. The collection of these sputtered secondary ions and their analysis by mass-to-charge spectrometry give information on the composition of the sample, with the elements present identified through their atomic mass values. Counting the number of secondary ions collected can also give quantitative data on the sample's composition (Benninghoven et al. 1987). SIMS can be used for practically all elements of the periodic table (only the noble gases are difficult to measure because they don't ionize easily) with a detection limit of ng kg^{-1} level and lateral resolution at 10 μm (Briggs et al. 1988). Since ions of different mass are measured separately, SIMS is ideally suited for the study of isotopic compositions of small samples (Wilson et al. 1989). SIMS works by analyzing material removed from the sample by sputtering, and is therefore a locally destructive technique, which may not be a suitable technique for live biological samples but still has been used in soybean root (Lazof et al. 1994), grape seeds (Freitas et al. 1998), animal tissues, and cells (Chandra and Morrison 1992). During a SIMS measurement, the sample is slowly sputtered (eroded) away and depth profiles can be obtained, which is the 3-D measurements (Chandra 2004).

Taken together, SR-μXRF, PIXE, LA-ICP-MS and SIMS are all very good metallomes imaging analysis techniques with spatial resolution of at least μm level and detection limit of at least mg kg^{-1} level. Although TEM-EDX and SEM-EDX have much better spatial resolution, the sensitivity is much lower for trace elements than SR-μXRF, PIXE, LA-ICP-MS, and SIMS. When comparing with LA-ICP-MS and SIMS, the beam time limitation for SR-μXRF and PIXE may hinder their wide application.

Besides determination of the distribution of elements, monitoring the absorption, distribution, transportation, storage, retention, metabolism, excretion, and toxicity of metallome is very important. Isotopic tracing is such kind of techniques that may realize the above goals (Fu et al. 2005; Yin et al. 2017). The commercialized PET (positron emission tomography) and SPECT (single-photon emission computed tomography) are indeed such kind of isotopic tracing techniques since ^{18}F and ^{99m}Tc are generally used. The major concern for isotopic tracing is the radioactivity of these isotopes. Therefore, application of stable isotope tracing has been proposed combined with the application of NAA and ICP-MS (Hulle et al. 2005; Heumann et al. 1998). ICP-MS measurement is being exploited for the use of enriched stable isotopes for multi-tracer investigations, which can also help to avoid spectral and background interference problems (Lambertsson et al. 2001).

9.2.3 Speciation and Structural Analysis of Metallomes

The scheme for metallomes speciation and structural analysis includes extraction, separation, identification, and structural analysis of metallomes. The general route for metallomes speciation are as follows: different species in samples are first extracted using acid, base, or enzyme or just diluted; then, the extraction solution is separated by the powerful selective separation techniques like HPLC, gas chromatography (GC), gel electrophoresis (GE), or electrochromatography (EC) and monitored by a sensitive and element-specific detector like ICP-MS and NAA (Szpunar 2005; Montes-Bayón 2002; Chai et al. 1999, 2004; Shi et al. 2004).

In the point of metallomics speciation and structure characterization, the high throughput in extraction, separation, detection, identification, and structural analysis is needed. As aforementioned, ICP-MS and NAA are all metallomes detectors; therefore, the process of extraction, separation, identification, and structural analysis may be the bottlenecks for metallomics study.

For more details regarding speciation in metallomics, please, see also the Chap. 8 of this book.

9.2.4 Structural Analysis

A series of techniques can be used for metallomes structure characterization like ray-based techniques, magnetic resonance-based techniques, and spectroscopy-based techniques, among others (Yang and Gao 2002).

Ray-based techniques can character the structure on atomic level. The rays that can be used for structure analysis include X-ray, gamma ray, or neutron beam. In the X-ray-based techniques, X-ray crystallography is the most powerful tool for the determination of macromolecular three-dimensional structure at a resolution of 0.15–2 nm, but the requirement for single crystal will greatly limit its application to numerous biological samples. X-ray absorption spectroscopy (XAS), especially the extended X-ray absorption fine structure (EXAFS), may provide an alternative tool for the local structure around certain atoms at a resolution of 0.0001–0.001 nm without the requirement for crystalline samples (Wang and Liu 1994). For example, Hg in human hair and blood samples from long-term mercury-exposed populations has been studied using EXAFS, and the structural information like bond distances and coordination numbers of Hg were taken (Li et al. 2004, 2007). Further, EXAFS can give a refinement of the structure that were taken from X-ray crystallography since EXAFS have a higher spatial resolution than X-ray crystallography especially in local structures (Aksenov et al. 2001; Cheung et al. 2000; Hasnain and Strange 2003). Neutron beam-based techniques like single crystal neutron diffraction spectroscopy (SCND) can provide complementary structural information at about 20 nm taken from X-ray-based techniques since the light atoms like hydrogen can be easily detected even with the existence of heavier one (Hazemann et al. 2005). Both small angle X-ray scattering (SAXS) and small angle neutron scattering (SANS) are structural characterization tools of solid and fluid biological samples at relatively lower resolution (ca 100 nm) than single crystal X-ray or neutron diffraction spectroscopy but again does not need crystalline samples (Feigin and Svergun 1987).

Just like XAS, Mössbauer spectroscopy is also a kind of local structure characterization tools but is based on the gamma-ray absorption or emission rather than X-ray absorption (Dickson and Berry 1986). The major limitation for the application of Mössbauer spectroscopy in biological samples is that only a few isotopes can be used. The most commonly used isotope is ^{57}Fe, which has been used for the study of ferritins in biological samples, but the low content and natural abundance of Fe may hinder its application (Upadhyay et al. 2006).

Nuclear magnetic resonance (NMR) spectroscopy is a potentially powerful alternative to X-ray crystallography for the determination of macromolecular three-dimensional structure at the same resolution of 0.15–2 nm (Liang 1976). NMR has the advantage over crystallographic techniques in that experiments are performed in aqueous solution as opposed to a crystal lattice. Besides, NMR can be used to determine protein secondary structure content empirically (Wishart et al. 1992). The major difference of the structure of single crystals got through NMR or X-ray crystallography is that groups (10–50) of structures each satisfying the experimental constraints equally well for one unit cell from NMR while only one, or at most a few structures will be got from X-ray crystallography. Therefore, for NMR, the potential problem is either the entire ensemble of structures is evaluated or a mean conformation is produced and then evaluated (Savchenko et al. 2003). Mean structures from ill-defined portions of the polypeptide chain will have nonstandard geometries and may cause problems in analyses.

Spectroscopy-based techniques including ultraviolet-visible absorption spectroscopy; ultraviolet-visible luminescence spectroscopy such as molecular fluorescence spectroscopy, molecular phosphorescence spectroscopy, and chemiluminescence spectroscopy; circular dichroism spectroscopy (CD) and the aforementioned FTIR; and Raman spectroscopy can give structural information on molecular level. For example, CD spectra between 260 and approximately 180 nm can be analyzed for the different secondary structural types like alpha helix, parallel and antiparallel beta sheet, turn, and other (Provencher and Gloeckner 1981).

All these techniques used for structure characterization is generally low throughput considering the number of sample treated in one time and the time used for one sample, but attempts on high-throughput techniques like high-throughput XAS by automation of arraying small samples, rapid data collection of multiple low-volume low-concentration samples, and data reduction and analysis have been proposed (Scott et al. 2005).

Attempts on simultaneous speciation of species of different elements in biological samples have been performed. Since selenium plays important roles in detoxification of both arsenic and mercury, it has been paid much effort on the simultaneous speciation of both selenium and mercury species or both selenium and arsenic species by HPLC-ICP-MS (Pan et al. 2007; Li et al. 2007; Le et al. 1998). Solvent extraction with NAA detection for different organohalogen species in biological samples like milk, human tissues, and pine needles has been studied (Xu et al. 2003; Watanabe et al. 1987; Zhong et al. 2004; Zhang et al. 2006, 2007).

9.3 Advanced Nuclear and Related Techniques for Nanometallomics

9.3.1 Metal-Related Nanomaterials

Metal-related nanomaterials, including metallic and metal-containing nanomaterials, are frequently used as catalysts, sensors, or probes due to their unique crystalline forms and superior mechanical, electrical, magnetic, optical, and catalytic properties (Buzea et al. 2007). Among them, metal, metal crystals, metal nanoclusters, metal oxide, and other metal multicomponent core/shell nanomaterials, intermetallic or alloyed nanomaterials, metal fluorescent nanoclusters, and metal nanoparticle-based hybrid nanomaterials can be categorized as metallic nanomaterials. Nobel metal nanomaterials with interesting physical and chemical properties via controllable synthesis are ideal building blocks for enhanced functions and application potentials, e.g., single-component Pt, Pd, Ag, and Au nanomaterials. Quantum dots (QDs) are semiconductor nanomaterials (~2–100 nm), with their cores consisted of a variety of metal complexes such as semiconductors, noble metals, and magnetic transition metals, which can also be regarded as metallic nanomaterials. As a typical metal oxide, iron oxide nanoparticles are widely used in different medical applications. Metallofullerenes (fullerenes with metal atom(s) encapsulated) are novel forms of fullerene-based materials showing attractive applications in biomedicines (Wilson et al. 1999), which can be regarded as metal-containing nanomaterials. The endohedral metallofullerenes so far produced are centered on group 2 and 3 metallofullerenes such as Sc, Y, La, Ca, Sr, and Ba as well as lanthanide metallofullerenes (Ce–Lu). These metal atoms have been encapsulated in higher fullerenes, especially in C82. Besides metallofullerenes, metal-decorated multiwall carbon nanotubes (Espinosa et al. 2007) and metal-filled single-walled carbon nanotubes (Bendall et al. 2006) can also be regarded as metal-containing nanomaterials. In this paper, all these metallic or metal-containing nanomaterials are categorized as metal-related nanomaterials.

9.3.2 Absorption, Distribution, Metabolism, and Excretion of Metal-Related Nanomaterials in Biological Systems

Animals can be exposed to metal-related nanomaterials through different routes, mainly ingestion, inhalation, and dermal absorption. The absorption, distribution, metabolism, and excretion of these metal-related nanomaterials in the biological systems can be influenced by their physicochemical properties, the exposure route, and the environment of the deposition site.

9.3.2.1 Absorption and Excretion of Metal-Related Nanomaterials

Exposure through ingestion may lead to extended retaining of the nanomaterials in the body. Oughton et al. (2008) studied the dietary absorption (uptake) and excretion of a cobalt nanopowder (average particle size, 4 nm; surface area, 59 $m^2\ g^{-1}$) in earthworm *E. fetida*. It was found that cobalt nanoparticles were taken up to a high extent during 7 day of exposure (concentration ratios of 0.16–0.20 relative to the nanoparticles concentration in horse manure) and were largely retained within the worms for 8 weeks, with less than 20% of absorbed nanoparticles being excreted. Similarly, Meng et al. (2007) studied the copper content in mice after oral gavage of nano-copper particles at the dose of 70 mg kg^{-1} body weight. They found that massive copper enriches in renal tissue 24 h after the mice exposed to nano- and ion-copper. The copper content in renal tissue drops from 12.6 ± 2.2 μg g^{-1} to 6.5 ± 1.3 μg g^{-1} in ion-copper group at 72 h; however, in nano-copper group, it still maintains high copper content level in the kidney (11.5 ± 2.5 μg g^{-1}). These imply that the rate of elimination of nano-copper is very low in the kidney, with only 1.5 μg g^{-1} reduction within 48 h.

Exposure through inhalation showed different absorption and excretion behaviors. Kreyling et al. studied the absorption and excretion of ^{192}Ir nanoparticles (15 and 80 nm) after inhalation exposure by young adult rats (Kreyling et al. 2002). They found that particles were predominantly cleared via airways into the gastrointestinal tract and feces during week 1 after inhalation. Additionally, minute particle translocation of <1% of the deposited particles into secondary organs such as the liver, spleen, heart, and brain was found after systemic uptake from the lungs. The translocated fraction of the 80-nm particles was about an order of magnitude less than that of 15-nm particles.

The coating of nanomaterials was also found to influence their absorption and excretion behavior in rats after intravenous injection. Fischer et al. (2006) found that QDs coated with mercaptoundecanoic acid and crosslinked with lysine (denoted as QD-LM) are cleared from plasma with a clearance of 0.59 ± 0.16 mL $min^{-1}\ kg^{-1}$. A higher clearance (1.23 ± 0.22 mL $min^{-1}\ kg^{-1}$) exists when the QDs are conjugated to bovine serum albumin (denoted as QD-BSA, $p < 0.05$). QDs are not detected in feces or urine for up to 10 days after intravenous dosing. On the contrary, Chen et al. found 33.3% and 23.8% of the intravenously given silica-coated CdSe-core QDs in mice were cleared via feces and urine (Chen et al. 2008). This was attributed to the difference in coating materials of these CdSe QDs.

In general, vertebrate organisms tend to recognize nanoparticles as foreign objects, with elimination of the materials through the primary excretory organs/systems. It was found that the particles with sizes over 100 nm will be caught by the reticuloendothelial systems (RES) while particles with sizes below 5 nm can be removed by the kidneys (Li and Chen 2011).

9.3.2.2 Distribution of Metal-Related Nanomaterials in Biological Systems

It is necessary to find out where the nanoparticles accumulated after absorption for better understanding the biological effects of metal-related nanomaterials in cells, tissues, animals, and ecosystems.

Whole body quantification of exposed nanoparticles was generally achieved after sacrificing the animals. Liu et al. (Liu et al. 2009a; b) evaluated the distribution of nasal instilled copper nanoparticles (23.5 nm) in mice. After instillation of copper nanoparticles for 1 week, the copper level was significantly higher in the liver, kidneys, olfactory bulb, and blood in H-Nano group than the control, which are in agreement with the damages in the liver, kidneys, and olfactory bulb clearly seen in pathological examinations. Therefore, the nasal inhaled copper particles at very high dosage can translocate to other organs and tissues and induce certain lesions.

Wang et al. (2008a, b, c) studied the whole body distribution of 20 and 120 nm ZnO at doses of 1, 2, 3, 4, and 5 g/kg body weight on healthy adult mice after oral intake. After 5 g/kg 20 nm ZnO exposure, compared with the control, significant increase of Zn contents was found in the kidney, pancreas, and bone ($p < 0.05$) and slight increase in the liver and heart of the ZnO-treated mice. In contrast to the 120 nm ZnO-treated mice, the Zn content in the liver, kidney, and pancreas of the 20 nm ZnO group mice showed little higher, suggesting more Zn may excrete from the 20 nm ZnO administrated mice than 120 nm ZnO mice. Among the observed organs, the highest Zn content was found in the bone. The 120 nm ZnO mice retained significantly higher Zn in the bone than the 20 nm ZnO mice. The authors concluded that the liver, spleen, heart, pancreas, and bone are the target organs for 20 and 120 nm ZnO oral exposure.

Noninvasive study of the biodistribution of nanomaterials was also possible with specific tools. For example, the whole body distribution of silver labeled with ^{125}I after intravenous injection in Balb/c mice was studied using a computed tomography coregistered with single-photon emission computerized tomography (CT-SPECT). Predominant accumulation of the silver nanoparticles in the spleen (41.5% ID g^{-1}) and liver (24.5% ID g^{-1}) at 24 h was found. In vivo positron emission tomography (PET) imaging found that $^{64}Cu^{2+}$-labeled quantum dots (QD, 15.1 ± 7.6 nm) in mice were excreted via renal filtration shortly post-injection and accumulated in the liver (Tu et al. 2011).

Zhu et al. (2009) found the intratracheal-instilled nano-$^{59}Fe_2O_3$ could pass through the alveolar-capillary barrier into systemic circulation within 10 min after intratracheal instillation into the male SD rats at a dose of 4 mg/rat using isotopic tracing techniques. The nano-$^{59}Fe_2O_3$ in the lung was distributed to organs rich in mononuclear phagocytes, including the liver, spleen, kidney, and testicle. The plasma elimination half-life of nano-$^{59}Fe_2O_3$ was 22.8 days, and the lung clearance rate was 3.06 μg/day, indicating systemic accumulation and lung retention occurred. The extrapulmonary transported $^{59}Fe_2O_3$ was redistributed in many organs, which indicate that $^{59}Fe_2O_3$ can easily pass through a number of tissue compartments and accumulate in the extrapulmonary organs. The highest extrapulmonary ^{59}Fe levels were found in the liver, followed in the decreasing order: spleen, heart, kidney, pancreas, testicle, and brain. The ^{59}Fe in the liver and heart showed time-response of accumulation from post-instilled day 1 to day 21 and then decreased at day 50. The ^{59}Fe showed persistent high level until to post-instilled day 50 in spleen.

All these studies suggested an extensive distribution of metal-related nanomaterials in the tissues of the reticuloendothelial system, which requires further investigation of the interaction of nanoparticles with hepatic and splenic tissues at the cellular level. This is critical for evaluation of the in vivo effects and potential toxicity of metal-related nanomaterials.

For the distribution of metal-related nanomaterials in a specific animal organs or tissues, or in small-size animal models, even in cells, dedicated tools are necessary. Wang et al. (Wang et al. 2008c) studied the accumulation of two crystalline phases of TiO_2 nanoparticles (80 nm, rutile, and 155 nm, anatase; purity >99%) in murine brain using synchrotron radiation-based X-ray fluorescence (SR-XRF) after intranasally instilled with 500 μg of TiO_2 nanoparticle suspension every other day for 30 days. It was found that titanium accumulated mainly in the cerebral cortex, thalamus, and hippocampus, especially in the CA1 and CA3 regions of hippocampus at 30 days. The significantly increased Ti contents in the hippocampus result in the obviously irregular arrangement and loss of neurons in the hippocampus. Intranasal instillation of either rutile or anatase TiO_2 nanoparticles produced sustained accumulation in brain tissues especially in the hippocampus, which indicate that the TiO_2 nanoparticles can enter the brain via the olfactory bulb (Wang et al. 2008a; b).

μ-XRF mapping was applied to investigate the distribution of Cu nanoparticles in the whole body of *C. elegans* (Gao et al. 2008). It was found that the exposure to Cu nanoparticles can result in an obvious elevation of Cu and K levels and a change of biodistribution of Cu in nematodes. Accumulation of Cu occurs in the head and at a location 1/3 of the way up the body from the tail compared to the unexposed control. In contrast, a higher amount of Cu was detected in other portion of worm body, especially in its excretory cells and the intestine when exposed to Cu^{2+}. The nondestructive and multielemental μ-XRF provides an important tool for mapping the elemental distribution in the whole body of a single tiny nematode at low levels.

XRF mapping with the nano-sized spatial resolution (nanoXRF) has been applied to study the distribution of metal-related nanomaterials in cells. For example, Corezzi et al. (2009) studied the distribution of a commercially available QD-secondary antibody conjugate in SKOV3 cancer cells. Pixel-by-pixel analysis of the elements present in the core/shell of QDs (S, Zn, Cd, and Se) was performed to retrieve a topographical map of their intracellular distribution. It was found that the emission intensity of Zn and S, the elements that also constitute the shell of the QDs, is located mainly in the nuclear region, although a clear signal can also be detected in the cytoplasm. The relatively high concentration of these elements and their presence in both the test and control samples suggest that the main contribution to the detected signal comes from the elements naturally present in the cell. On the other hand, the Cd signal is almost undetectable, whereas the Se signal is clearly detected only in the cytoplasm of the labeled sample and is not present in either the cytoplasm or the nucleus of the control sample. Even if the microtubule structure cannot be seen because they are 25-nm-diameter polymers that course throughout the cytoplasm, this observation is consistent with a higher concentration of Se in the cytoplasm due to the specific binding of the secondary antibody conjugated to QDs to the primary antibody that recognizes the microtubule-associated protein β-tubulin.

The μ-XRF technique has also been applied to study cell-carbon nanotubes interactions. Bussy et al. (2008) studied the distribution of unpurified and purified single-walled (SW) and multiwalled (MW) carbon nanotubes (CNT) in macrophages by monitoring the catalyst metal particle employed in most synthesis technique and finally remaining attached to or contained in nanotubes. The μ-XRF technique allows CNT localization at the single-cell level with simultaneous analysis of the biological response through observation of changes in cell elemental composition (calcium in the present study).

Since the availability of synchrotron radiation-based bioimaging tools like SR-XRF is limited, commercially available laser-ablation inductively coupled plasma mass spectrometry (LA-ICP-MS) is a convenient alternative. LA-ICP-MS provides easy sample preparation, multielemental detection with high sensitivity, and high spatial resolution. The distribution of gold and silver nanoparticles in individual fibroblast cells was studied using LA-ICP-MS. Nanoparticles were visualized with respect to cellular substructures and were found to accumulate in the perinuclear region with increasing incubation time (Drescher et al. 2012).

9.3.2.3 Metabolic Responses of Metal-Related Nanomaterials in Biological Systems

After entering the body, metal-related nanomaterials may dissolve, be decomposed, be oxidized, or be reduced, and this may lead to the electronic and/or ionic transfer either within the nanoparticles lattice or on release to culture medium.

Nanomaterials may dissolve in physiologically relevant media to form partially soluble metal ions or metal-ligand complex. Franklin et al. (2007) compared the dissolution of both nano- and bulk ZnO in a freshwater system and found rapid dissolution of ZnO nanoparticles with a saturation solubility in the milligram per liter range, similar to that of bulk ZnO. Further toxicity test confirmed that comparable toxicity to *P. subcapitata* for nanoparticulate ZnO, bulk ZnO, and $ZnCl_2$, suggesting that the toxicity aroused solely from the dissolved zinc. Further, an enhanced dissolution of iron oxide nanoparticles in the acidic condition of lysosomes or in a microenvironment containing ligands with a strong affinity was also observed (Zhu et al. 2012; Wang et al. 2013).

The toxicity of nanomaterials may occur due to their decomposition. Derfus et al. (2003) observed the liberation of free Cd(II) from CdSe QDs and found that the cytotoxicity of CdSe QDs correlated with the free Cd(II) concentration in air-oxidized and UV-exposed samples. On the other hand, Qu et al. (2011) found that CdSe@ZnS core/shell QDs were degraded in *C. elegans* and Se^{2-} in the CdSe core was oxidized to Se(IV). Both results confirmed that QDs could be decomposed in the body and lead to the nanotoxicity of QDs.

X-ray absorption spectroscopy (XAS) is a powerful tool to investigate the electronic and/or ionic transfers of metal-related nanomaterials in biological systems and can provide structural details of the metabolites. The metabolism of gold nanorods in SD rats was studied using XAS, and it was found that long-term retention of gold nanorods in the liver and spleen did not induce changes in the oxidation states of gold, suggesting gold nanorods are inert in the body (Wang et al. 2010). For other elements, however, the oxidation state was found to have changed in the body. For example, Auffan et al. (2008) investigated the metabolism of three kinds of Fe nanoparticles in *E. coli*. According to Fe K edge XAS spectra, three compounds, including γFe_2O_3, Fe_3O_4, and $\gamma FeOOH$, had different pre-edge intensity, position of pre-edge and main edge, and ramped absorption position. The pre-edge information showed that Fe(II) in nMagnetite could be oxidated to Fe(III) in water and in contact with *E. coli*. All the Fe(0) atoms in nZVI were highly active to oxygen atoms under the same condition and transformed into $\gamma FeOOH$ and Fe_3O_4 supported by XRD and XAS results. The process of Fe(0) and Fe(II) oxidation resulted in ionic or electronic transfers on nanoparticle surface that might intervene with the metabolism of bacterial when cell membrane and component contacted with them. Consequently, the authors concluded that the toxicity of Fe nanoparticles containing ferrous and zerovalent iron resulted from the generation of reactive oxygen species or interference on electron/ion transport chains. This work suggested that surface properties of metal nanomaterials may affect their toxicity heavily by ROS generation, and suitable surface modification needs consideration in order to be more biocompatible and healthy.

The nanomaterials may induce other metabolic changes even if they are inert. The time-dependent metabolic changes in Au NRs-induced A549 and 16HBE were studied by ^1H NMR spectroscopy (Zhang et al. 2013). The dominant metabolites present in the cell extracts include a range of amino acids; organic acids, such as lactate, creatine, and citrate; membrane metabolites including choline, phosphocholine, and glycerophosphocholine; and a number of nucleosides and nucleotides, such as inosine, adenosine, uracil, inosine-5′-monophosphate, and nicotinamide adenine dinucleotide. A significant reduction in the levels of lactate in both 16HBE and A549 cells suggests that protein-coated Au NRs exposure inhibits the intracellular anaerobic glycolysis process.

9.3.3 Interactions with Genes, Proteins, and Other Biomolecules of Metal-Related Nanomaterials in Biological Systems

The metal-related nanomaterials and their metabolites can interact with genes, proteins, and other biomolecules in biological systems, and this may bring different biological effects.

Nanomaterials are engineered to have various unique material properties, which will affect their possible direct or indirect interaction with the DNA. Landsiedel et al. (2009) did not observe their genotoxicity of titanium dioxide and zinc oxide nanomaterials in vitro (Ames' Salmonella gene mutation test and V79 micronucleus chromosome mutation test) or in vivo (mouse bone marrow micronucleus test and Comet DNA damage assay in lung cells from rats exposed by inhalation). However, the study on silver nanoparticles in *A. cepa* found they could penetrate plant system and may impair stages of cell division causing chromatin bridge, stickiness, disturbed metaphase, multiple chromosomal breaks, and cell disintegration (Ng et al. 2010). For most nanomaterials, it is still unknown whether they directly interact with DNA or whether indirect effects such as inflammation-mediated oxidative stress may infer a threshold for the genotoxicity of some nanomaterials (Landsiedel et al. 2009).

As mentioned above, QDs have been used as fluorescent markers in biological application. However, it is essentials to know that QDs themselves do not induce adverse effects when they are used in in vivo studies or DNA-based assays. Green and Howman (2005) observed the DNA damage in plasmid nicking assays with water-soluble CdSe QDs by comparing the electrophoresis bands. Assays with DNA that had been incubated with quantum dots and exposed to UV light showed damage. Samples of DNA incubated with quantum dots in the dark also showed evidence of a damaged band not observed in assays using DNA stored alone in the dark. It is worth noting that the intensity of the bands in assays carried out with DNA that were incubated with quantum dots are significantly weaker than experiments run without quantum dots present. This was attributed to the DNA coordinating to the dots during incubation, resulting in smaller yields of DNA when isolated. Up to 70% of the DNA coordinated to the dots nonspecifically and was therefore unavailable for assay analysis. These results suggest that there may be serious issues to address concerning the use of such materials in DNA-based assays or in in vivo applications, as well as highlighting potential toxicological and environmental implications. For most nanomaterials, it is still unknown whether they directly interact with DNA or whether indirect effects such as inflammation-mediated generation of genotoxic ROS may infer a threshold for the genotoxicity of some nanomaterials.

Research shows that nanoparticles can stimulate or suppress the immune responses. Liu et al. studied the effect of $Gd@C_{82}(OH)_{22}$ nanoparticles on the release of Th1/Th2 cytokines and found markedly enhanced immune responses and stimulated immune cells to release more cytokines, helping eliminate abnormal cells (Liu et al. 2009b). Exosomes are extracellularly secreted membrane vesicles which act as Trojan horses for the dissemination and intercellular communication of natural nano-sized particles (like viruses). Upon exposure to magnetic iron oxide nanoparticles, maturation of dendritic cells and activation of splenic T cells were significantly induced by these exosomes. Furthermore, exosome-induced T-cell activation was more efficient toward sensitized T cells and in ovalbumin (OVA)-sensitized mice than in the unsensitized counterparts. The studies suggested that exosomes may act as conveyors for extrapulmonary signal transduction in nanoparticle-induced immune systemic responses (Zhu et al. 2012).

Since nanomaterials are foreign to the host, the penetration of nanomaterials into the organism results from the crossing of barriers like the skin, lungs, and gastrointestinal track, which are all patrolled and controlled by the immune system, and this will trigger the formation of a series of proteolytic enzymes and may cause the immunotoxicity (Dobrovolskaia and McNeil 2007). For example, the viability study of dendritic cells incubating in gold nanoparticles (NPs) showed that these NPs were not cytotoxic even at high concentration, but important amounts of gold NPs amassing in endocytic compartments were observed. Furthermore, the secretion of cytokines was significantly modified after such internalization, which suggests that a potential perturbation of the immune response may occur (Villiers et al. 2010).

When NPs enter a physiological environment, it can rapidly adsorb a layer of proteins to form the protein "corona" (Walkey and Chan 2012; Ge et al. 2011; Yang et al. 2013). Researchers found that size, shape, and surface characteristics of NPs could affect protein adsorption and also the structure of the adsorbed proteins. This will affect the reactivity of NPs with cells and determine the route and efficiency of NP uptake. For example, NPs were found to accelerate the fibrillation of proteins and peptides, a process that is associated with several diseases like Alzheimer's disease (Linse et al. 2007). The adsorbed proteins may also promote translocation of the NP across cellular barriers and clearance or accumulation in vital organs. Understanding the dynamics of this complex interaction can thus provide useful insights into cytotoxic, inflammatory potential and other key properties of these novel materials that can be explored for developing safer and value added nanomaterials for future applications. On the other hand, proteins in the corona were found to be conserved, and unique irrespective of the nanoparticles types and both size and surface properties of nanomaterials played a very significant role in determining the nanoparticles' coronas on different nanoparticles composed of identical materials (Lundqvist et al. 2008). The protein "corona" has the potential to alter the size and interfacial composition of nanomaterials, thus inducing changes between their biological identity and their synthetic identity. For example, the adsorption of proteins onto NPs was found to modify their melting point in a composition- and size-dependent manner (Laera et al. 2011). Thorough understanding of NP-protein interactions might lead to strategic manipulation of NP surfaces to adsorb specific functional proteins or small drug molecules intended for delivery in vivo. Furthermore this knowledge might also prove to be useful in predicting nanotoxicity related safety concerns.

NPs can also interfere with the biological process of cancer cells and this may help in designing drugs for cancer therapy. For example, it was found that metallofullerene nanoparticles, formulated as $[Gd@C_{82}(OH)_{22}]_n$, could penetrate plasma membrane of tumor cells and lead to the shrinkage of solid tumors and to decrease the activities of those enzymes responsible for catalyzing the production of reactive oxygen species in vivo (Yin et al. 2009; Chen et al. 2005; Wang et al. 2006). Besides, pretreatment of the cisplatin-resistant human prostate cancer (CP-r) cells with $[Gd@C_{82}(OH)_{22}]_n$ enhanced intracellular accumulation of cisplatin and formation of cisplatin-DNA adducts by restoring the defective endocytosis of the CP-r cancer cells. The results suggested that $[Gd@C_{82}(OH)_{22}]_n$ nanoparticles could overcome tumor resistance to cisplatin by increasing its intracellular accumulation through the mechanism of restoring defective endocytosis (Liang et al. 2010). More importantly, a recent study found that $[Gd@C_{82}(OH)_{22}]_n$ could inhibit the metastasis of pancreatic tumor through the inhibition of matrix metalloproteinases (MMP-9) activity mainly via an exosite interaction by interfering binding of the incoming ligands as remotely modulating the S1' loop (Kang et al. 2012). Through molecular-dynamics simulations, detailed inhibition dynamics and molecular mechanism behind the $[Gd@C_{82}(OH)_{22}]_n$–MMP-9 interaction was revealed, which provided insights for de novo design of nanomedicine for fatal diseases such as pancreatic cancer.

The interactions of DNA with nanomaterials also help for promising application in medical purposes. Hydroxylated C60 is actually a polyacid-like molecule, and each proton of the hydroxyl groups (C–OH) can dissociate in an aqueous solution, thus yielding a conjugated base C–O–. Fullerenols can encapsulate some Env plasmid DNA during the self-assembly by which the nanoadjuvant realizes dual functions as plasmid DNA carrier and activator of host immunity. Poly(diallydimethylammonium chloride) (PDDAC) and polyethyleneimine (PEI-modifed Au NRs) on the surface of the nanorod can significantly promote cellular and humoral immunity as well as T-cell proliferation through activating antigen-presenting cells if compared to naked HIV-1 Env plasmid DNA treatment in vivo. Artificially synthesized CpG oligodeoxynucleotides (CpG ODNs) can imitate bacterial DNA and effectively stimulate the mammalian immune system. Self-assembled polyvalent CpG-AuNP conjugates enhance the efficiency of cellular uptake and stimulate secretion of cytokines. These findings have shed light that the intentional design of the interaction of low-toxic nanomaterials with biomolecules can be used as a versatile platform for drug and gene delivery systems and biomedical applications.

9.4 Conclusions and Future Perspectives

As an integrated biometal science, metallomics has received more and more attention. The complexity of metallomics in living organisms results in significant analytical challenges. The comprehensive quantification, distribution, speciation, identification, and structural characterization of metallome require high-throughput and powerful analytical techniques. In general, high-throughput quantification of metallomes can be achieved by ICP-MS and NAA, while the high-throughput metallomes distribution mapping can be done by fluorescence-detecting techniques like SR-XRF, XRF tomography, EDX, and PIXE and ion-detecting-based SIMS. LA-ICP-MS can also realize metallomes distribution mapping. When it comes to identification and structural characterization of metallome, all the techniques are generally low-throughput. Therefore, this should be the bottlenecks in metallomics study. However, by automation of arraying small samples, rapid data collection of multiple low-volume low-concentration samples, and data reduction and analysis, high-throughput techniques may be and in fact have partially been achieved.

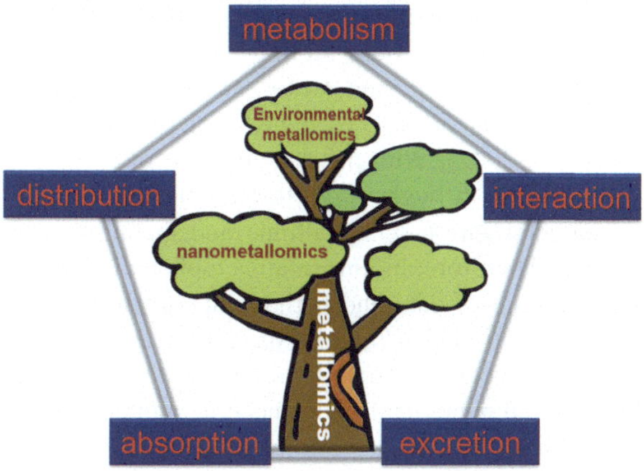

Fig. 9.1 Nanometallomics is a branch of metallomics studying the biological effects, such as absorption, distribution, metabolism, and excretion of metal-related nanomaterials and their interactions with genes, proteins, metabolites, and other biological molecules in the biological systems. Environmental metallomics is another branch of metallomics, in which nuclear and related analytical techniques could also play important roles

The relationship between metallomics and nanometallomics can be shown in Fig. 9.1, among which nanometallomics is a branch of metallomics for the study of the biological effects of metal-related nanomaterials. This includes the systematically study of the absorption, distribution, metabolism, and excretion of metallomes and metal-related materials in biological systems. Further, it also involves the systematically study on the interactions and functional connections of metallomes and metal-related nanomaterials with genes, proteins, metabolites, and other biomolecules within organisms. Other branches of metallomics also include environmental metallomics, which studies the biological effects of environmental related metals, including mercury, cadmium, lead, chromium, and arsenic, among others (Li et al. 2016).

The absorption, distribution, metabolism, and excretion of metal-related nanomaterials in the biological systems are influenced by their physicochemical properties, the exposure route, and the environment of the deposition site. Nanomaterials not only may interact directly or indirectly with genes, proteins, and other molecules to bring genotoxicity, immunotoxicity, DNA damage, and cytotoxicity but may also stimulate the immune responses, circumvent tumor resistance, and inhibit tumor metastasis.

Since nanometallomics is such a young research field, publications on risk evaluation and biological behavior of these metal-related nanomaterials are sometimes unreplicable and contradictory. Therefore, standardized procedure for studying of the biological effects of metal-related nanomaterials should be developed. Besides, due to the fascinating and unique properties of nanomaterials, dedicated analytical tools are necessary in this field. Further, nanometallomics need to develop with other -omics like genomics, proteomics, and metabolomics to improve the application performance and reduce the potential hazard of metal-related nanomaterials.

Bioinformatics uses computer software tools for database creation, data management, data warehousing, data mining, and global communication networking, which is the recording, annotation, storage, analysis, and searching/retrieval of nucleic acid sequence (genes and RNAs), protein sequence, and structural information (Lesk 2002). With the development of metallomics and nanometallomics, especially through high-throughput detection, distribution, speciation, and characterization of metallome and metallome, more and more data will be obtained. Therefore, bioinformatics tools will also be highly needed to treat such huge amount of data and perhaps the databases of metallome and nanometallome quantification, distribution, speciation, and structural information as well as methods to access, search, visualize, and retrieve the information. Besides, metallomics and nanometallomics need to cooperate with other -omics, like genomics, proteomics, and metabolomics, to obtain the knowledge of underlying mechanisms and therefore to improve the application performance and to reduce the potential risk of metallome and nanometallome.

Acknowledgments Y-F Li gratefully acknowledges the support of K. C. Wong Education Foundation, Hong Kong, and the CAS Youth Innovation Association, Chinese Academy of Sciences. This work was supported by the National Natural Science Foundation of China (11205168, 11405196, and U1432241) and the Ministry of Science and Technology of China (2011CB933401, 2012CB934003, and 2016YFA0201600). We thank staffs at Beijing Synchrotron Radiation Facility (BSRF) and Shanghai Synchrotron Radiation Facility (SSRF), who provided us beam time and technical assistance.

References

Aksenov VL, Kuzmin AY, Purans J, Tyutyunnikov SI (2001) EXAFS spectroscopy at synchrotron-radiation beams. Phys Part Nucl 32:1–33

Alexander J, Thomassen Y, Aaseth J (1983) Increased urinary excretion of selenium among workers exposed to elemental mercury vapor. J Appl Toxicol 3:143–145

Auffan M, Achouak W, Rose J, Roncato M-A, Chaneac C, Waite DT, Masion A, Woicik JC, Wiesner MR, Bottero J-Y (2008) Relation between the redox state of iron-based nanoparticles and their cytotoxicity toward *Escherichia coli*. Environ Sci Technol 42:6730–6735

Becker JS, Zoriy M, Przybylski M, Becker JS (2007) High resolution mass spectrometric brain proteomics by MALDI-FTICR-MS combined with determination of P, S, Cu, Zn and Fe by LA-ICP-MS. Int J Mass Spectrom 261:68–73

Bendall JS, Ilie A, Welland ME, Sloan J, Green MLH (2006) Thermal stability and reactivity of metal halide filled single-walled carbon nanotubes. J Phys Chem B 110:6569–6573

Beneš B, Jakubec K, Šmíd J, Spěváčková V (2000) Determination of thirty-two elements in human autopsy tissue. Biol Trace Elem Res 75:95–203

Benninghoven A, Rüdenauer FG, Werner HW (1987) Secondary ion mass spectrometry: basic concepts, instrumental aspects, applications and trends. Wiley, New York, pp 1–1227

Bowerman WW, Evans ED, Giesy JP, Postupalsky S (1994) Using feathers to assess risk of mercury and selenium to bald eagle reproduction in the Great Lakes region. Arch Environ Contam Toxicol 27:294–298

Briggs BYD, Brown A, Vickerman JC (1988) Handbook of static secondary ion mass spectrometry (SIMS). Anal Chem 60:1791–1799

Brumfiel G (2003) Nanotechnology: a little knowledge. Nature 424:246–248

Burk RF, Foster KA, Greenfield PM, Kiker KW, Hannon JP (1974) Binding of simultaneously administered inorganic selenium and mercury to a rat plasma protein. Proc Soc Exp Biol Med 145:782–785

Bussy C, Cambedouzou J, Lanone S, Leccia E, Heresanu V, Pinault M, Mayne-lhermite M, Brun N, Mory C, Cotte M, Doucet J, Boczkowski J, Launois P (2008) Carbon nanotubes in macrophages: imaging and chemical analysis by X-ray fluorescence microscopy. Nano Lett 8:2659–2663

Buzea C, Pacheco II, Robbie K (2007) Nanomaterials and nanoparticles: sources and toxicity. Biointerphases 2:MR17–MR71

Caurant F, Navarro M, Amiard J-C (1996) Mercury in pilot whales: possible limits to the detoxification process. Sci Total Environ 186:95–104

Chai Z, Zhu H (1994) Introduction to trace element chemistry. Atomic Energy Press, Beijing, pp 1–280

Chai Z, Sun J, Ma S (1992) Neutron activation analysis in environmental sciences, biological and geological sciences. Atomic Energy Press, Beijing, pp 1–302

Chai C, Mao X, Wang Y, Sun J, Qian Q, Hou X, Zhang P, Chen C, Feng W, Ding W (1999) Molecular activation analysis for chemical species studies. Fresenius J Anal Chem 363:477–480

Chai ZF, Zhang ZY, Feng WY, Chen CY, Xu DD, Hou XL (2004) Study of chemical speciation of trace elements by molecular activation analysis and other nuclear techniques. J Anal At Spectrom 19:26–33

Chandra S (2004) 3D subcellular SIMS imaging in cryogenically prepared single cells. Appl Surf Sci 231-232:467–469

Chandra S, Morrison GH (1992) Sample preparation of animal tissues and cell cultures for secondary ion mass spectrometry (SIMS) microscopy. Biol Cell 74:31–42

Chen T, Huang Z, Huang Y, Xie H, Liao X (2003) Cellular distribution of arsenic and other elements in hyperaccumulator Pteris nervosa and their relations to arsenic accumulation. Chin Sci Bull 48:1586–1591

Chen C, Xing G, Wang J, Zhao Y, Li B, Tang J, Jia G, Wang T, Sun J, Xing L (2005) Multihydroxylated [$Gd@C_{82}(OH)_{22}$] n nanoparticles: antineoplastic activity of high efficiency and low toxicity. Nano Lett 5:2050–2057

Chen C, Yu H, Zhao J, Li B, Qu L, Liu S, Zhang P, Chai Z (2006) The roles of serum selenium and selenoproteins on mercury toxicity in environmental and occupational exposure. Environ Health Perspect 114:297–301

Chen Z, Chen H, Meng H, Xing G, Gao X, Sun B, Shi X, Yuan H, Zhang C, Liu R, Zhao F, Zhao Y, Fang X (2008) Bio-distribution and metabolic paths of silica coated CdSeS quantum dots. Toxicol Appl Pharmacol 230:364–371

Chen C, Chai Z, Gao Y (2010) Nuclear analytical techniques for metallomics and metalloproteomics. RSC Publishing, Cambridge, pp 1–428

Chen C, Li Y-F, Qu Y, Chai Z, Zhao Y (2013) Advanced nuclear analytical and related techniques for the growing challenges in nanotoxicology. Chem Soc Rev 42:8266–8303

Cheung KC, Strange RW, Hasnain SS (2000) 3D EXAFS refinement of the Cu site of structural change at the metal centre in an oxidation–reduction process: an integrated approach combining EXAFS and crystallography. Acta Crystallogr D 56:697–704

Colvin VL (2003) The potential environmental impact of engineered nanomaterials. Nat Biotechnol 21:1166–1170

Corezzi S, Urbanelli L, Cloetens P, Emiliani C, Helfen L, Bohic S, Elisei F, Fioretto D (2009) Synchrotron-based X-ray fluorescence imaging of human cells labeled with CdSe quantum dots. Anal Biochem 388:33–39

Derfus AM, Chan WCW, Bhatia SN (2003) Probing the cytotoxicity of semiconductor quantum dots. Nano Lett 4:11–18

Devos W, Senn-Luder M, Moor C, Salter C (2000) Laser ablation inductively coupled plasma mass spectrometry (LA-ICP-MS) for spatially resolved trace analysis of early-medieval archaeological iron finds. Fresenius J Anal Chem 366:873–880

Dickson DPE, Berry FJ (1986) Mössbauer spectroscopy. Cambridge University Press, New York, pp 1–16

Dobrovolskaia MA, McNeil SE (2007) Immunological properties of engineered nanomaterials. Nat Nanotechnol 2:469–478

Drescher D, Giesen C, Traub H, Panne U, Kneipp J, Jakubowski N (2012) Quantitative imaging of gold and silver nanoparticles in single eukaryotic cells by laser ablation ICP-MS. Anal Chem 84:9684–9688

Drickamer K, Taylor M (2002) Glycan arrays for functional glycomics. Genome Biol 3:10341–10344

Durrant SF, Ward NI (1994) Laser ablation-inductively coupled plasma-mass spectrometry(LA-ICP-MS) for the multielemental analysis of biological materials: a feasibility study. Food Chem 49:317–323

Espinosa EH, Ionescu R, Bittencourt C, Felten A, Erni R, Van Tendeloo G, Pireaux JJ, Llobet E (2007) Metal-decorated multi-wall carbon nanotubes for low temperature gas sensing. Thin Solid Films 515:8322–8327

Fahrni CJ (2007) Biological applications of X-ray fluorescence microscopy: exploring the subcellular topography and speciation of transition metals. Curr Opi Chem Biol 11:121–127

Falnoga I, Tušek-Žnidarič M, Horvat M, Stegnar P (2000) Mercury, selenium, and cadmium in human autopsy samples from Idrija residents and mercury mine workers. Environ Res 84:211–218

Feigin LA, Svergun DI (1987) Structure analysis by small-angle X-ray and neutron scattering. Plenum Press, New York, pp 1–335

Feldmann Y, Jakubowski N, Thomas C, Stueweret D (1999) Part II: analytical figures of merit and first applications. Fresenius J Anal Chem 365:422–428

Fischer HC, Liu L, Pang KS, Chan WCW (2006) Pharmacokinetics of nanoscale quantum dots: in vivo distribution, sequestration, and clearance in the rat. Adv Funct Mater 16:1299–1305

Franklin NM, Rogers NJ, Apte SC, Batley GE, Gadd GE, Casey PS (2007) Comparative toxicity of nanoparticulate ZnO, bulk ZnO, and $ZnCl_2$ to a freshwater microalga (Pseudokirchneriella subcapitata): the importance of particle solubility. Environ Sci Technol 41:8484–8490

Freitas VAP, Glories Y, Bourgeois G, Vitry C (1998) Characterisation of oligomeric and polymeric procyanidins from grape seeds by liquid secondary ion mass spectrometry. Phytochemistry 49:1435–1441

Fu C, Lü T, Cao X, Liu J (2005) Pharmacokinetic research of nerve growth factor marked with ^{99}mTc in rabbit. Chin J Pract Med 4:1108–1109

Gailer J, George GN, Pickering IJ, Madden S, Prince RC, Yu EY, Denton MB, Younis HS, Aposhian HV (2000) Structural basis of the antagonism between inorganic mercury and selenium in mammals. Chem Res Toxicol 13:1135–1142

Gan H, Gao H, Zhu H, Chen J, Zhu P, Xian D (2006) X-ray fluorescence tomography. Laser Optoelectr Prog 43:56–64

Gao Y, Chen C, Chai Z (2007) Advanced nuclear analytical techniques for metalloproteomics. J Anal At Spectrom 22:856–866

Gao Y, Liu N, Chen C, Luo Y, Li Y-F, Zhang Z, Zhao Y, Zhao Y, Iida A, Chai Z (2008) Mapping technique for biodistribution of elements in a model organism, Caenorhabditis elegans, after exposure to copper nanoparticles with microbeam synchrotron radiation X-ray fluorescence. J Anal At Spectrom 23:1121–1124

Ge R, Sun H (2009) Metallomics: an integrated biometal science. Sci China Chem 52:2055–2070

Ge C, Du J, Zhao L, Wang L, Liu Y, Li D, Yang Y, Zhou R, Zhao Y, Chai Z, Chen C (2011) Binding of blood proteins to carbon nanotubes reduces cytotoxicity. Proc Natl Acad Sci U S A 108:16968–16973

Green M, Howman E (2005) Semiconductor quantum dots and free radical induced DNA nicking. Chem Comm 122:121–123

Hansel CM, La Force MJ, Fendorf S, Sutton S (2002) Spatial and temporal Association of as and Fe Species on aquatic plant roots. Environ Sci Technol 36:1988–1994

Haraguchi H (2004) Metallomics as integrated biometal science. J Anal At Spectrom 19:5–14

Hasnain SS, Strange RW (2003) Marriage of XAFS and crystallography for structure±function studies of metalloproteins. J Synchrotron Rad 10:9–15

Hazemann I, Dauvergne MT, Blakeley MP, Meilleur F, Haertlein M, Van Dorsselaer A, Mitschler A, Myles DA, Podjarny A (2005) High-resolution neutron protein crystallography with radically small crystal volumes: application of perdeuteration to human aldose reductase. Acta Crystallogr D 61:1413–1417

Heumann K, Gallus SM, Radlinger G, Vogl J (1998) Accurate determination of element species by on-line coupling of chromatographic systems with ICP-MS using isotope dilution technique. Spectrochim Acta B 53:273–278

Hoet PHM, Nemmar A, Nemery B (2004) Health impact of nanomaterials? Nat Biotechnol 22:19

http://pubs.rsc.org/en/journals/journalissues/mt#!recentarticles&all. Accessed on 23 Oct 2017

Hulle MV, Cremer KD, Vanholder R, Cornelis R (2005) In vivo distribution and fractionation of indium in rats after subcutaneous and oral administration of [^{114m}In] InAs. J Environ Monit 7:365–370

Ide-Ektessabi A, Fujisawa S, Sugimura K, Kitamura Y, Gotoh A (2002) Quantitative analysis of zinc in prostate cancer tissues using synchrotron radiation microbeams. X-Ray Spectrom 31:7–11

Iida A (1997) X-ray spectrometric applications of a synchrotron X-ray microbeam. X-Ray Spectrom 26:359–363

Iida A, Noma T (1993) Synchrotron X-ray muprobe and its application to human hair analysis. Nucl Instrum Meth Phys Res Sec B 82:129–138

Isaure M-P, Fayard B, Sarret G, Pairis S, Bourguignon J (2006) Localization and chemical forms of cadmium in plant samples by combining analytical electron microscopy and X-ray spectromicroscopy. Spectrochim Acta B 61:1242–1252

Ishii K, Matsuyama S, Watanabe Y, Kawamura Y, Yamaguchi T, Oyama R, Momose G, Ishizaki A, Yamazaki H, Kikuchi Y (2007) 3D-imaging using micro-PIXE. Nucl Instrum Meth Phys Res Sect A 571:64–68

Jackson BP, Hopkins WA, Baionno J (2003) Laser ablation-ICP-MS analysis of dissected tissue: a conservation-minded approach to assessing contaminant exposure. Environ Sci Technol 37:2511–2515

Jackson B, Harper S, Smith L, Flinn J (2006) Elemental mapping and quantitative analysis of Cu, Zn, and Fe in rat brain sections by laser ablation ICP-MS. Anal Bioanal Chem 384:951–957

James P (1997) Protein identification in the post-genome era: the rapid rise of proteomics. Quart Rev Biophy 30:279–331

Janssens K, Proost K, Falkenberg G (2004) Confocal microscopic X-ray fluorescence at the HASYLAB microfocus beamline: characteristics and possibilities. Spectrochim Acta B 59:1637–1645

Jarvis KE, Williams JG (1993) Laser ablation inductively coupled plasma mass spectrometry(LA-ICP-MS): a rapid technique for the direct, quantitative determination of major, trace and rare-earth elements in geological samples. Chem Geol 106:251–262

Jenkins R (1999) X-ray fluorescence spectrometry. Vol. 152 in chemical analysis: a series of monographs on analytical chemistry and its applications. Wiley, New York, pp 1–232

Jiang J, Sato S (1999) Detection of calcium and aluminum in pyramidal neurons in the gerbil hippocampal CA1 region following repeated brief cerebral ischemia: X-ray microanalysis. Med Electron Microsc 32:161–166

Johansson E (1989) PIXE: a novel technique for elemental analysis. Endeavour 13:48–53

Johansson E, Lindh U, Johansson H, Sundstrom C (1987) Micro-PIXE analysis of macro-and trace elements in blood cells and tumors of patients with breast cancer. Nucl Instrum Methods Phys Res Sect B 22:179–183

Kametani K, Nagata T (2006) Quantitative elemental analysis on aluminum accumulation by HVTEM-EDX in liver tissues of mice orally administered with aluminum chloride. Med Mol Morphol 39:97–105

Kang D, Amarasiriwardena D, Goodman A (2004) Application of laser ablation–inductively coupled plasma-mass spectrometry (LA–ICP–MS) to investigate trace metal spatial distributions in human tooth enamel and dentine growth layers and pulp. Anal Bioanal Chem 378:1608–1615

Kang S-G, Zhou G, Yang P, Liu Y, Sun B, Huynh T, Meng H, Zhao L, Xing G, Chen C, Zhao Y, Zhou R (2012) Molecular mechanism of pancreatic tumor metastasis inhibition by

$Gd@C_{82}(OH)_{22}$ and its implication for de novo design of nanomedicine. Proc Natl Acad Sci U S A 109:15431–15436

Kanngießer B, Malzer W, Reiche I (2003) A new 3D micro X-ray fluorescence analysis set-up – first archaeometric applications. Nucl Instrum Meth Phys Res Sect B 211:259–264

Kanngießer B, Karydas AG, Schütz R, Sokaras D, Reiche I, Rohrs S, Pichon L, Salomon J (2007a) 3D micro-PIXE at atmospheric pressure: a new tool for the investigation of art and archaeological objects. Nucl Instrum Meth Phys Res Sect B 264:383–388

Kanngießer B, Malzer W, Pagels M, Lühl L, Weseloh G (2007b) Three-dimensional micro-XRF under cryogenic conditions: a pilot experiment for spatially resolved trace analysis in biological specimens. Anal Bioanal Chem 389:1171–1176

Karanatsios J, Freiburg C, Reichert W, Barnert-Wiemer H (1988) Quantitative multi-element analysis of denitration ceramics by X-ray fluorescence spectrometry. J Anal At Spectrom 3:979–983

Karydas AG, Sokaras D, Zarkadas C, Grlj N, Pelicon P, Zitnik M, Schütz R, Malzer W, Kanngießer B (2007) 3D Micro PIXE—a new technique for depth-resolved elemental analysis. J Anal At Spectrom 22:1260–1265

Kemner KM, Kelly SD, Lai B, Maser J, O'Loughlin EJ, Sholto-Douglas D, Cai Z, Schneegurt MA, CFJr K, Nealson KH (2004) Elemental and redox analysis of single bacterial cells by X-ray microbeam analysis. Science 306:686–687

Kindness A, Sekaran CN, Feldmann J (2003) Two-dimensional mapping of copper and zinc in liver sections by laser ablation-inductively coupled plasma mass spectrometry. Clin Chem 49:1916–1923

Koeman JH, Peeters WHM, Koudstaal-Hol CHM, Tjioe PS, de Goeij JJ (1973) Mercury-selenium correlations in marine mammals. Nature 245:385–386

Koppenaal DW, Eiden GC, Barinaga CJ (2004) Collision and reaction cells in atomic mass spectrometry: development, status, and applications. J Anal At Spectrom 19:561–570

Kosta L, Byrne AR, Zelenko V (1975) Correlation between selenium and mercury in man following exposure to inorganic mercury. Nature 254:238–239

Kouichi Tsuji KN (2007) Development of confocal 3D micro-XRF spectrometer with dual Cr-Mo excitation. X-Ray Spectrom 36:145–149

Kramer U, Grime GW, Smith JAC, Hawes CR, Baker AJM (1997) Micro-PIXE as a technique for studying nickel localization in leaves of the hyperaccumulator plant Alyssum lesbiacum. Nucl Instrum Meth Phys Res Sect B 130:346–350

Kreyling WG, Semmler M, Erbe F, Mayer P, Takenaka S, Schulz H, Oberdorster G, Ziesenis A (2002) Translocation of ultrafine insoluble iridium particles from lung epithelium to extrapulmonary organs is size dependent but very low. J Toxicol Environ Health A 65:1513–1530

Laera S, Ceccone G, Rossi F, Gilliland D, Hussain R, Siligardi G, Calzolai L (2011) Measuring protein structure and stability of protein-nanoparticle systems with synchrotron radiation circular dichroism. Nano Lett 11:4480–4484

Lambertsson L, Lundberg E, Nilsson M, Frech W (2001) Applications of enriched stable isotope tracers in combination with isotope dilution GC-ICP-MS to study mercury species transformation in sea sediments during in situ ethylation and determination. J Anal At Spectrom 16:1296–1301

Landsiedel R, Kapp MD, Schulz M, Wiench K, Oesch F (2009) Genotoxicity investigations on nanomaterials: methods, preparation and characterization of test material, potential artifacts and limitations—many questions, some answers. Mut Res 681:241–258

Lazof DB, Goldsmith JG, Rufty TW, Linton RW (1994) Rapid uptake of aluminum into cells of intact soybean root tips (a microanalytical study using secondary ion mass spectrometry). Plant Physiol 106:1107–1114

Le XC, Li XF, Lai V, Ma M, Yalcin S, Feldmann J (1998) Simultaneous speciation of selenium and arsenic using elevated temperature liquid chromatography separation with inductively coupled plasma mass spectrometry detection. Spectrochim Acta B 53:899–909

Lesk AM (2002) Introduction to bioinformatics. Oxford University Press, New York, pp 1–400

Li Y-F, Chen C (2011) Fate and toxicity of metallic and metal-containing nanoparticles for biomedical applications. Small 7:2965–2980

Li Y-F, Chen C, Xing L, Liu T, Xie Y, Gao Y, Li B, Qu L, Chai Z (2004) Concentrations and XAFS speciation in situ of mercury in hair from populations in Wanshan mercury mine area, Guizhou Province. Nucl Tech 27:899–903

Li Y-F, Chen C, Li B, Sun J, Wang J, Gao Y, Zhao Y, Chai Z (2006) Elimination efficiency of different reagents for the memory effect of mercury using ICP-MS. J Anal At Spectrom 6(21):94–96

Li Y-F, Chen C, Li B, Wang Q, Wang J, Gao Y, Zhao Y, Chai Z (2007) Simultaneous speciation of selenium and mercury in human urine samples from long-term mercury-exposed populations with supplementation of selenium-enriched yeast by HPLC-ICP-MS. J Anal At Spectrom 22:925–930

Li Y-F, Wang L, Zhang L, Chen C (2010) Nuclear-based metallomics in metallic nanomaterials: nanometallomics. In: Chen C, Chai Z, Gao Y (eds) Nuclear analytical techniques for metallomics and Metalloproteomics. RSC publishing, Cambridge, pp 342–384

Li Y-F, Gao Y, Chai Z, Chen C (2014) Nanometallomics: an emerging field studying the biological effects of metal-related nanomaterials. Metallomics 6:220–232

Li Y-F, Sun H, Chen C, Chai Z (2016) Metallomics. Science Press, Beijing, pp 1–103

Liang X (1976) Nuclear magnetic resonance. Science Press, Beijing, pp 1–358

Liang X-J, Huang B, Meng H, He H, Meng J, Wang Y, Lu J, Wang PC, Zhao Y, Gao X, Chen C, Sun B, Xing G, Gottesman MM, Shen D, Jia L (2010) Metallofullerene nanoparticles overcome tumor resistance to cisplatin by reactivating endocytosis. Proc Natl Acad Sci U S A 107:7449–7454

Linse S, Cabaleiro-Lago C, Xue W-F, Lynch I, Lindman S, Thulin EN, Radford SE, Dawson KA (2007) Nucleation of protein fibrillation by nanoparticles. Proc Natl Acad Sci U S A 104:8691–8696

Liu Y, Gao Y, Zhang L, Wang T, Wang J, Jiao F, Li W, Liu Y, Li Y, Li B, Chai Z, Wu G, Chen C (2009a) Potential health impact on mice after nasal instillation of nano-sized copper particles and their translocation in mice. J Nanosci Nanotechnol 9:1–9

Liu Y, Jiao F, Qiu Y, Li W, Lao F, Zhou G, Zhao Y, Sun B, Xing G, Dong J, Chai Z, Chen C (2009b) The effect of $Gd@C_{82}(OH)_{22}$ nanoparticles on the release of Th1/Th2 cytokines and induction of TNF-α mediated cellular immunity. Biomaterials 30:3934–3945

Lundqvist M, Stigler J, Elia G, Lynch I, Cedervall T, Dawson KA (2008) Nanoparticle size and surface properties determine the protein corona with possible implications for biological impacts. Proc Natl Acad Sci U S A 105:14265–14270

Meng H, Chen Z, Xing G, Yuan H, Chen C, Zhao F, Zhang C, Zhao Y (2007) Ultrahigh reactivity provokes nanotoxicity: explanation of oral toxicity of nano-copper particles. Toxicol Lett 175:102–110

Meurer WP, Claeson DT (2002) Evolution of crystallizing interstitial liquid in an arc-related cumulate determined by LA ICP-MS mapping of a large amphibole oikocryst. J Petrol 43:607–629

Montes-Bayón M (2002) Metal speciation in biomolecules. Anal Bioanal Chem 376:287–288

Motelica-Heino M, Le Coustumer P, Thomassin JH, Gauthier A, Donard OFX (1998) Macro and microchemistry of trace metals in vitrified domestic wastes by laser ablation ICP-MS and scanning electron microprobe X-ray energy dispersive spectroscopy. Talanta 46:407–422

Mounicou S, Szpunar J, Lobinski R (2009) Metallomics: the concept and methodology. Chem Soc Rev 38:1119–1138

Muradin AK (2000) Capillary optics and their use in x-ray analysis. X-Ray Spectrom 29:343–348

Nakano K, Tsuji K (2006) Development of confocal 3D micro XRF spectrometer and its application to rice grain. Bunseki Kagaku 55:427

Nemmar A, Hoylaerts MF, Hoet PHM, Dinsdale D, Smith T, Xu H, Vermylen J, Nemery B (2002) Ultrafine particles affect experimental thrombosis in an in vivo hamster model. Am J Respir Crit Care Med 166:998–1004

Ng C-T, Li JJ, Bay B-H, Yung L-YL (2010) Current studies into the genotoxic effects of nanomaterials. J Nucl Acids 2010:Article ID 947859. 12 pages

Nicholson JK, Lindon JC (2008) Systems biology: metabonomics. Nature 455:1054–1056

Oberdörster G (2000) Pulmonary effects of inhaled ultrafine particles. Int Arch Occup Environ Health 74:1–8

Oberdorster G, Oberdorster E, Oberdorster J (2005) Nanotoxicology: an emerging discipline evolving from studies of ultrafine particles. Environ Health Perspect 113:823–839

Oberdoster G, Sharp Z, Atudorei V, Elder A, Gelein R, Kreyling W, Cox C (2004) Translocation of inhaled ultrafine particles to the brain. Inhal Toxicol 16:437–445

Ortega R, Bohic S, Tucoulou R, Somogyi A, Deves G (2004) Microchemical element imaging of yeast and human cells using synchrotron X-ray microprobe with Kirkpatrick-Baez optics. Anal Chem 76:309–314

Oughton DH, Hertel-Aas T, Pellicer E, Mendoza E, Joner EJ (2008) Neutron activation of engineered nanoparticles as a tool for tracing their environmental fate and uptake in organisms. Environ Toxicol Chem 27:1883–1887

Pan F, Tyson JF, Uden PC (2007) Simultaneous speciation of arsenic and selenium in human urine by high-performance liquid chromatography inductively coupled plasma mass spectrometry. J Anal At Spectrom 22:931–937

Paunesku T, Vogt S, Maser J, Lai B, Woloschak G (2006) X-ray fluorescence microprobe imaging in biology and medicine. J Cell Biochem 99:1489–1502

Paunesku T, Vogt S, Lai B, Maser J, Stojicevic N, Thurn KT, Osipo C, Liu H, Legnini D, Wang Z, Lee C, Woloschak GE (2007) Intracellular distribution of TiO_2-DNA oligonucleotide nanoconjugates directed to nucleolus and mitochondria indicates sequence specificity. Nano Lett 7:596–601

Provencher SW, Gloeckner J (1981) Estimation of globular protein secondary structure from circular dichroism. Biochemistry 20:33–37

Przybylowicz WJ, Mesjasz-Przybylowicz J, Pineda CA, Churms CL, Ryan CG, Prozesky VM, Frei R, Slabbert JP, Padayachee J, Reimold WU (2001) Elemental mapping using proton-induced x-rays. X-Ray Spectrom 30:156–163

Punshon T, Jackson BP, Bertsch PM, Burger J (2004) Mass loading of nickel and uranium on plant surfaces: application of laser ablation-ICP-MS. J Environ Monit 6:153–159

Qu Y, Li W, Zhou Y, Liu X, Zhang L, Wang L, Li Y-F, Iida A, Tang Z, Zhao Y, Chai Z, Chen C (2011) Full assessment of fate and physiological behavior of quantum dots utilizing Caenorhabditis elegans as a model organism. Nano Lett 11:3174–3183

Raith A, Godfrey J, Hutton RC (1996) Quantitation methods using laser ablation ICP-MS. Fresenius J Anal Chem 354:163–168

Rowe EM, Mills FE, Tantalus I (1973) A dedicated storage ring synchrotron radiation source. Particle Accelerators 4:211–227

Ryan CG (2000) Quantitative trace element imaging using PIXE and the nuclear microprobe. Int J Imag Syst Technol 11:219–230

Ryan CG, Jamieson DN, Churms CL, Pilcher JV (1995) A new method for on-line true-elemental imaging using PIXE and the proton microprobe. Nucl Instrum Methods Phys Res Sect B 104:157–165

Savchenko A, Yee A, Khachatryan A, Skarina T, Evdokimova E, Pavlova M, Semesi A, Northey J, Beasley S, Lan N (2003) Strategies for structural proteomics of prokaryotes: quantifying the advantages of studying orthologous proteins and of using both NMR and X-ray crystallography approaches. Proteins Struct Funct Genet 50:392–399

Scheuhammer AM, Wong AHK, Bond D (1998) Mercury and selenium accumulation in common loons (*Gavia immer*) and common mergansers (*Mergus merganser*) from eastern Canada. Environ Toxicol Chem 17:197–201

Scott RA, Shokes JE, Cosper NJ, Jenney FE, Adams MWW (2005) Bottlenecks and roadblocks in high-throughput XAS for structural genomics. J Synchrotron Rad 12:19–22

Seemungal D, Newton G (2001) Human Genome Project. In: Sydney B, Jeffrey HM (eds) Encycl. Gen. Academic Press, New York, pp 980–981

Service RF (2003) Nanomaterials show signs of toxicity. Science 300:243–243

Shi Y, Acharya R, Chatt A (2004) Speciation of arsenic in natural waters by HPLC-NAA. J Radioanal Nucl Chem 262:277–286

Sie SH, Thresher RE (1992) Micro-PIXE analysis of fish otoliths: methodology and evaluation of first results for stock discrimination. Int J PIXE 2:357–379
Skerfving S (1974) Methylmercury exposure, mercury levels in blood and hair, and health status in Swedes consuming contamination fish. Toxicology 2:3–23
Snigirev A, Kohn V, Snigireva I, Lengeler B (1996) A compound refractive lens for focusing high-energy X-rays. Nature 384:49–51
Szpunar J (2005) Advances in analytical methodology for bioinorganic speciation analysis: metallomics, metalloproteomics and heteroatom-tagged proteomics and metabolomics. Analyst 130:442–465
Tertian R, Claisse F (1982) Principles of quantitative XRF analysis. Heyden, London, pp 1–386
Thompson M, Walsh JN (1983) Handbook of inductively coupled plasma spectrometry. Blackie, Glasgow, pp 1–327
Tu C, Ma X, House A, Kauzlarich SM, Louie AY (2011) PET imaging and biodistribution of silicon quantum dots in mice. ACS Med Chem Lett 2:285–288
Tylko G, Mesjasz-Przybylowicz J, Przybylowicz WJ (2007) In-vacuum micro-PIXE analysis of biological specimens in frozen-hydrated state. Nucl Instrum Meth Phys Res Sect B 260:141–148
Upadhyay AK, Hooper AB, Hendrich MP (2006) NO reductase activity of the tetraheme cytochrome c554 of *Nitrosomonas europaea*. J Am Chem Soc 128:4330–4337
Van Langevelde F, Vis RD (1991) Trace element determinations using a 15-keV synchrotron x-ray microprobe. Anal Chem 63:2253–2259
Villiers C, Freitas H, Couderc R, Villiers M-B, Marche P (2010) Analysis of the toxicity of gold nano particles on the immune system: effect on dendritic cell functions. J Nanopart Res 12:55–60
Vincze L, Vekemans B, Szaloki I, Janssens K, Van Grieken R, Feng H, Jones KW, Adams F (2001) In high resolution X-ray fluorescence micro-tomography on single sediment particles, 46th SPIE annual meeting international symposium of optical science and technology, San Diego, 2001; Bonse Ulrich, U., Ed. San Diego, pp 240—245
Vincze L, Vekemans B, Brenker FE, Falkenberg G, Rickers K, Somogyi A, Kersten M, Adams F (2004) Three-dimensional trace element analysis by confocal X-ray microfluorescence imaging. Anal Chem 76:6786–6791
Walkey CD, Chan WCW (2012) Understanding and controlling the interaction of nanomaterials with proteins in a physiological environment. Chem Soc Rev 41:2780–2799
Wang Q-W, Liu W-H (1994) X-ray absorption fine structure and it's application. Science Press, Beijing, pp 1–347
Wang J, Chen C, Li B, Yu H, Zhao Y, Sun J, Li Y, Xing G, Yuan H, Tang J, Chen Z, Meng H, Gao Y, Ye C, Chai Z, Zhu C, Ma B, Fang X, Wan L (2006) Antioxidative function and biodistribution of [$Gd@C_{82}(OH)_{22}$]n nanoparticles in tumor-bearing mice. Biochem Pharmacol 71:872–881
Wang M, Feng W, Lu W, Li B, Wang B, Zhu M, Wang Y, Yuan H, Zhao Y, Chai Z (2007) Quantitative analysis of proteins via sulfur determination by HPLC coupled to isotope dilution ICP-MS with hexapole collision cell. Anal Chem 79:9128–9134
Wang B, Feng W, Wang M, Wang T, Gu Y, Zhu M, Ouyang H, Shi J, Zhang F, Zhao Y, Chai Z, Wang H, Wang J (2008a) Acute toxicological impact of nano- and submicro-scaled zinc oxide powder on healthy adult mice. J Nanopart Res 10:263–276
Wang J, Chen C, Liu Y, Jiao F, Li W, Lao F, Li Y, Li B, Ge C, Zhou G, Gao Y, Zhao Y, Chai Z (2008b) Potential neurological lesion after nasal instillation of TiO_2 nanoparticles in the anatase and rutile crystal phases. Toxicol Lett 183:72–80
Wang J, Liu Y, Jiao F, Lao F, Li W, Gu Y, Li Y, Ge C, Zhou G, Li B, Zhao Y, Chai Z, Chen C (2008c) Time-dependent translocation and potential impairment on central nervous system by intranasally instilled TiO_2 nanoparticles. Toxicology 254:82–90
Wang L, Li Y-F, Zhou L, Liu Y, Meng L, Zhang K, Wu X, Zhang L, Li B, Chen C (2010) Characterization of gold nanorods in vivo by integrated analytical techniques: their uptake, retention, and chemical forms. Anal Bioanal Chem 396:1105–1114

Wang B, Yin J-J, Zhou X, Kurash I, Chai Z, Zhao Y, Feng W (2013) Physicochemical origin for free radical generation of iron oxide nanoparticles in biomicroenvironment: catalytic activities mediated by surface chemical states. J Phys Chem C 117:383–392

Watanabe I, Kashimono T, Kawano M (1987) A study of organic bound halogens in human adipose, marine organisms and sediment by neutron activation and gas chromatographic analysis. Chemosphere 16:847–857

Watmough SA, Hutchinson TC, Evans RD (1997) Application of laser ablation inductively coupled plasma-mass spectrometry in dendrochemical analysis. Environ Sci Technol 31:114–118

Wilson RG, Stevie FA, Magee CW (1989) Secondary ion mass spectrometry: a practical handbook for depth profiling and bulk impurity analysis. Wiley, New York, pp 1–364

Wilson LJ, Cagle DW, Thrash TP, Kennel SJ, Mirzadeh S, Alford JM, Ehrhardt GJ (1999) Metallofullerene drug design. Coord Chem Rev 190-192:199–207

Wishart DS, Sykes BD, Richards FM (1992) The chemical shift index: a fast and simple method for the assignment of protein secondary structure through NMR spectroscopy. Biochemistry 31:1647–1651

Xiong Y, Ouyang L, Liu Y, Xie Q, Wang J (2006) One of the most important parts for bio-elementomics: specific correlation study of bio-elements in a given tissue. J Chin Mass Spectrom Soc 27:35–36

Xu DD, Zhong WK, Deng LL, Chai Z, Mao X (2003) Levels of extractable organohalogens in pine needles in China. Environ Sci Technol 37:1–6

Yang P, Gao F (2002) Principle of inorganic biochemistry. Science Press, Beijing, pp 1–628

Yang L, McRae R, Henary MM, Patel R, Lai B, Vogt S, Fahrni CJ (2005) Imaging of the intracellular topography of copper with a fluorescent sensor and by synchrotron x-ray fluorescence microscopy. Proc Natl Acad Sci U S A 102:11179–11184

Yang S-T, Liu Y, Wang Y-W, Cao A (2013) Biosafety and bioapplication of nanomaterials by designing protein–nanoparticle interactions. Small 9:1635–1653

Yin J-J, Lao F, Fu PP, Wamer WG, Zhao Y, Wang PC, Qiu Y, Sun B, Xing G, Dong J, Liang X-J, Chen C (2009) The scavenging of reactive oxygen species and the potential for cell protection by functionalized fullerene materials. Biomaterials 30:611–621

Yin Y, Tan Z, Hu L, Yu S, Liu J, Jiang G (2017) Isotope tracers to study the environmental fate and Bbioaccumulation of metal-containing engineered nanoparticles: techniques and applications. Chem Rev 117:4462–4487

Yoneda S, Suzuki KT (1997) Equimolar Hg-Se complex binds to selenoprotein P. Biochem Biophys Res Comm 231:7–11

Zhang P, Chen C, Zhao J, Li B, Qu L, Chai Z (2004) Correlation of mercury, selenium and other elements in the tissues of fishes from the regions at different mercury exposure level. Environ Sci 25:159–165

Zhang H, Chai ZF, Sun HB, Zhang JL (2006) A survey of extractable persistent organochlorine pollutants in Chinese commercial yogurt. J Dairy Sci 89:1413–1419

Zhang H, Chai Z, Sun H, Xu H (2007) Neutron activation analysis of organohalogens in Chinese human hair. J Radioanal Nucl Chem 272:561–564

Zhang L, Wang L, Hu Y, Liu Z, Tian Y, Wu X, Zhao Y, Tang H, Chen C, Wang Y (2013) Selective metabolic effects of gold nanorods on normal and cancer cells and their application in anticancer drug screening. Biomaterials 34:7117–7126

Zhao Y, Nalwa HS (2006) Nanotoxicology – interactions of nanomaterials with biological systems. American scientific publishers, California, pp 1–300

Zhao Y, Xing G, Chai Z (2008) Nanotoxicology: are carbon nanotubes safe? Nat Nanotechnol 3:191–192

Zhong W, Xu D, Chai Z, Mao X (2004) Neutron activation analysis of extractable organohalogens in milk from China. J Radioanal Nucl Chem 259:485–488

Zhu M-T, Feng W-Y, Wang Y, Wang B, Wang M, Ouyang H, Zhao Y-L, Chai Z-F (2009) Particokinetics and extrapulmonary translocation of intratracheally instilled ferric oxide nanoparticles in rats and the potential health risk assessment. Toxicol Sci 107:342–351

Zhu M, Li Y, Shi J, Feng W, Nie G, Zhao Y (2012) Exosomes as extrapulmonary signaling conveyors for nanoparticle-induced systemic immune activation. Small 8:404–412

Zoeger N, Streli C, Wobrauschek P, Jokubonis C, Pepponi G, Roschger P, Hofstaetter J, Berzlanovich A, Wegrzynek D, Chinea-Cano E, Markowicz A, Simon R, Falkenbergu G (2007) Determination of the elemental distribution in human joint bones by SR micro XRF. X-Ray Spectrom 37(1):3–11

Chapter 10
New Frontiers of Metallomics: Elemental and Species-Specific Analysis and Imaging of Single Cells

Javier Jiménez-Lamana, Joanna Szpunar, and Ryszard Łobinski

Abstract Single cells represent the basic building units of life, and thus their study is one the most important areas of research. However, classical analysis of biological cells eludes the investigation of cell-to-cell differences to obtain information about the intracellular distribution since it only provides information by averaging over a huge number of cells. For this reason, chemical analysis of single cells is an expanding area of research nowadays. In this context, metallomics research is going down to the single-cell level, where high-resolution high-sensitive analytical techniques are required. In this chapter, we present the latest developments and applications in the fields of single-cell inductively coupled plasma mass spectrometry (SC-ICP-MS), mass cytometry, laser ablation (LA)-ICP-MS, nanoscale secondary ion mass spectrometry (nanoSIMS), and synchrotron X-ray fluorescence microscopy (SXRF) for single-cell analysis. Moreover, the capabilities and limitations of the current analytical techniques to unravel single-cell metabolomics as well as future perspectives in this field will be discussed.

Keywords Single-cell analysis · ICP-MS · Laser ablation · NanoSIMS · X-ray fluorescence · Imaging · Intracellular distribution · Metal content

Abbreviations

ALOD	Absolute limit of detection
AP-MALDI	Atmospheric pressure matrix-assisted laser desorption ionization
CE	Capillary electrophoresis
cryo-XT	Cryo nanoscale X-ray tomography
CyTOF	Cytometry by time-of-flight

J. Jiménez-Lamana (✉) · J. Szpunar · R. Łobinski
Institute of Analytical Sciences and Physico-Chemistry for Environment and Materials (IPREM), UMR 5254, CNRS-UPPA, Pau, France
e-mail: j.jimenez-lamana@univ-pau.fr; joanna.szpunar@univ-pau.fr; Ryszard.Lobinski@univ-pau.fr

M. A. Z. Arruda (ed.), *Metallomics*, Advances in Experimental Medicine and Biology 1055, https://doi.org/10.1007/978-3-319-90143-5_10

ESI-MS	Electrospray ionization mass spectrometry
fs-LI-O-TOFMS	Femtosecond laser ionization orthogonal time-of-flight mass spectrometry
HECIS	High-efficiency cell introduction system
HPCN	High performance concentric nebulizer
ICP-AES	Inductively coupled plasma atomic emission spectrometry
ICP-MS	Inductively coupled plasma mass spectrometry
ICP-SFMS	Inductively coupled plasma sector field mass spectrometry
ICP-TOFMS	Inductively coupled plasma time-of-flight mass spectrometry
LADE	Liquid-assisted droplet ejection
LA-ICP-MS	Laser ablation inductively coupled plasma mass spectrometry
LB-HPCN	Large-bore high performance concentric nebulizer
MALDI	Matrix-assisted laser desorption ionization
μDG	Micro-droplet generator
M-DIS	Micro-droplet injection system
μFI	Micro-flow Injection
μXRF	Micro-X-ray fluorescence
NPs	Nanoparticles
PBS	Phosphate-buffered saline
PDMS	Poly(dimethylsiloxane)
PFH	Perfluorohexane
QDs	Quantum dots
RR	Ruthenium red
SC-ICP-MS	Single cell inductively coupled plasma mass spectrometry
SIMS	Secondary ion mass spectrometry
SNMS	Secondary neutral mass spectrometry
SOD	Superoxide dismutase
SP-ICP-MS	Single particle inductively coupled plasma mass spectrometry
SR-nXRF	Synchrotron radiation nano-X-ray fluorescence
SXRF	Synchrotron X-ray fluorescence
TEM	Transmission electron microscopy
TOF-SIMS	Time-of-flight secondary ion mass spectrometry

10.1 Introduction

An integrated approach of biometal sciences called metallomics is now a well-established research field aimed at understanding the biological functions of metal ions and their chemical structures in biological systems and how their usage is fine-tuned in biological species and in populations of species with genetic variations. In fact, chemistry and biology are expected to come together in investigations of biometals because neither establishing only structures nor finding only functions is sufficient (Maret 2016).

Cells are the basic building blocks of all living organisms; they provide structure for the bodies, take in nutrients, convert them into energy, and carry out specialized functions. Thus, study of these "basic units of life" can be considered one of the most important areas of research helping understand how organisms function and how cellular components work together to carry out life functions and enable organisms to meet their basic needs.

Most known biological processes depend on metals, and, as a result, the entirety of metal ions present in a cell, including, e.g., Mg, Ca, Mn, Fe, Cu, and Zn, determines its life functions; cells use the physicochemical properties of metal ions to control essential processes. Complexation with metals is a basic functionalization reaction of cell components such as proteins (proteome), metabolites (metabolome), and nucleic acids (genome). Based on their physicochemical properties, metals are recognized, classified, and localized by living cells and then distributed to organelles – such as nuclei and mitochondria – to enable them to perform their functions.

Metallomics on the level of single cells seems to be an ultimate goal in the development of the field. Although the establishment of metallomics as a stand-alone scientific field had at its beginnings the study of individual cells (salmon eggs) by Haraguchi et al. (2008), technical limitations have been hampering more advanced investigations at this level due to small size of cells of interest in biological studies; the work with tissue homogenates (or slices in elemental imaging) has been privileged. However, it has to be underlined that cell disruption during sample preparation process (homogenization) results in loss of individual cell properties and mixing cell compartments in a way that is not likely to occur in Nature. For these reasons, development of analytical strategies for characterization of metallome homeostasis at cellular level is necessary for full understanding of life processes. Moreover, studies of individual cells in their integrity undoubtedly possess a philosophical appeal of the fundamental primary approach to explore a microcosm encapsulating the essence of life.

The first step in this direction is the determination of total element contents in individual cells. It has been suggested that correlating gene mutations, physiological status, and stress response with the trace element pattern should be carried out on the single-cell level because due to the variability of cells bulk analysis may be not meaningful (Schmid et al. 2010). The technical feasibility of such determination requires a combination of high metal detection sensitivity combined with ultrafast data acquisition. Its realization, based on the concept similar to that proposed by Tanner's group (Bandura et al. 2009) in their mass cytometry scheme of inductively coupled plasma time-of-flight mass spectrometry (ICP-TOFMS) investigation of individual cells containing metal-labeled species, has been facilitated by new instrumental developments in ICP-MS. They include systems originally designed for the analysis of nanoparticles by single-particle ICP-MS as well as miniaturized on-chip sample introductions setups. Multielemental (and multiisotopic) capability of ICP-MS allows a holistic view of cell microcosms with a perspective to study isotopic distribution and fractionation variability which is expected to bring new insights into metal utilization by the Nature.

The total element analysis is complemented by the development of intracellular metallome imaging techniques such as nanoSIMS (secondary ion mass spectrometry), synchrotron X-ray fluorescence microscopy (SXRF), and, to some extent, laser ablation (LA)-ICP-MS. NanoSIMS analysis can be performed at lateral resolution down to 50 nm thus allowing imaging at subcellular level. This methodology, based on the coaxial design of the ion gun and the secondary ion extraction, allows the analysis of up to seven ions with multicollection and recently developed oxygen probe made possible the detection of most of the metals (Nuñez et al. 2018; Pett-Ridge and Weber 2012). Another powerful tool for measuring element localization is SXRF based on a spectroscopic analysis of the emission of characteristic fluorescence from a material that has been excited by bombarding with high-energy X-rays. A new (third) generation synchrotron sources, providing high brilliance at high photon energies and highly advanced X-ray focusing systems, make it possible to visualize intracellular element localization at the sub-100 nm scale. In the case of nanoSIMS, only square-element mapping is possible (depending on the implementation energy – up to a few nanometers), while in the case of SXRF, the whole depth of the cell (up to few μm) is accessible. Another, by far more robust and easier accessible option, is laser ablation ICP-MS offering, however, worse resolution of down to several μm (Gundlach-Graham and Günther 2016).

Although the ultimate goal of all-species-of-all elements on a single-cell level seems still to be distant, the first reports on single-cell metabolomics and proteomics (Rubakhin et al. 2011) open a way to the characterization of metal-containing molecules at cellular levels. The success in this field is highly dependent on the development in small-scale high-resolution molecular mass spectrometry.

The present chapter summarizes the developments in metallomics studies on the level of individual cells highlighting novel approaches and, at the same time, discussing the current limitations and presenting perspectives of future progress.

10.2 Detection and Quantitative Distribution of Metals and Metal Nanoparticles at the Single-Cell Level

In order to fully understand a biological system such as a cell, it is necessary to determine the trace metals within the system, due to their essential roles in biological or physiological functions. In contrast with stochastic average analysis by bulk measurements, the quantitative analysis of individual cells leads to a more sensitive representation of cell-to-cell variations of content of elements (Miyashita et al. 2017). In this context, single-cell elemental analysis is considered one of the major research fields in metallomics in the next decade (Haraguchi 2017a).

However, elemental analysis of individual cells must not be restricted to the investigation of intracellular metal ions. Engineered nanoparticles are already present in our daily life in a number of consumer and industrial products from where they are directly released and incorporated into the environment. As a result, analytical methods to study their uptake by biological systems and, more specifically, by cells are needed.

In recent years, detection and quantitative multielement analysis in a single biological cell has been carried out using various measurement systems. Different applications reported in literature are shown in Table 10.1.

One of the first multielemental analyses in single biological cells was reported by Haraguchi et al. (2008). In this study, two or three salmon eggs were decomposed with HNO_3 by using a microwave-assisted acid digestion method and subsequently analyzed by ICP Atomic Emission Spectrometry (ICP-AES) and ICP-MS. In the experiment, 78 elements were the target of analysis. Sixty-six elements out of these 78 were determined in salmon egg cells, whereas other 7 elements (Li, Zr, Nb, Hf, Ir, Bi) could be only detected close to their detection limit as a result of their low abundance in the sample. The other 5 elements (F, Rh, Te, Ta, Re) were below their detection limit and they could not be determined or detected in this study. However, all these elements except F were detected in a recent experiment by using High Resolution ICP-MS (Haraguchi 2017b). The results supported the concept of the extended all present theory of the elements in a single biological cell, i.e., "cell microcosm," which postulates the existence of all elements in a single biological cell (Haraguchi 2004). Nevertheless, it has to be taken into account that the size of salmon eggs is bigger than the normal biological cells of study (mm vs μm), which requires the use of higher sensitive techniques.

More recently, Umemura et al. (2017) recently reported a multielement quantitative determination (more than 30 elements) of prokaryotic and eukaryotic cells and subcellular organelles, by microflow injection (μFI)-ICP-MS. Preliminary studies performed by ICP-MS in time-resolved mode only allowed the determination of a few major elements, such as P, Mg and Fe, whereas the signal of the other elements could not be distinguished from the signal produced by the background. As a result, an alternative approach was proposed, based on traditional solution-based ICP-MS after microwave-assisted acid digestion of less than 10 mg of dry weight sample. The concentration determined by ICP-MS was normalized as number of atoms per cell or organelle.

10.2.1 Time Resolved ICP-MS Analysis

In recent years, time-resolved ICP-MS, known as Single Cell ICP-MS (SC-ICP-MS), has attracted a great deal of interest for multielemental analysis of single cells as an alternative of classical bulk analysis of large amounts of cells with a lysis, extraction, or digestion (Miyashita et al. 2014a, 2017; Mueller et al. 2014; Wang et al. 2017a). This technique is able to detect and quantify metallic elements in individual cells with high sensitivity, with the advantage of obtaining information about cell-to-cell variations instead of the average information from large cell populations obtained by conventional bulk measurements.

Table 10.1 Summary of analytical methods for quantitative distribution and imaging analysis of single cells

Type of cell	Analytical technique	Analyte	Analytical information	Ref
Salmon eggs	ICP-AES/ ICP-MS	78 elements	Elemental content	Haraguchi et al. (2008)
Bacteria (*E. coli*) Cyanobacteria (*Synechocytis* sp.) Chlorella (*Chlorella kessleri*) Organelles	ICP-MS	>30 elements	Elemental content	Umemura et al. (2017)
Mouse fibroblasts (L929) Human pancreas (SR-4-SH) Human umbilical vein endothelial (HUVEC)	Single Cell ICP-AES	Ca	Elemental content	Nomizu et al. (1994)
Bacteria (*Bacillus subtilis*)	Single Cell ICP-SFMS	U	Uptake	Li et al. (2005)
Unicellular alga (*Chlorella vulgaris*)	Single Cell ICP-MS	Mg, Mn, Cu, Cr	Elemental content	Ho and Chan (2010)
Bacteria (*Helicobacter pylori*)	Single Cell ICP-MS	Mg, Bi	Uptake	Tsang et al. (2011)
Mouse leukemic monocyte macrophage (RAW 246.7)	Single Cell ICP-MS	Quantum dots	Uptake	Zheng et al. (2013)
Human cervical cancer (HeLa) Normal human bronchial epithelial (16HBE)	Single Cell ICP-MS	Gd@ $C_{82}(OH)_{22}$ CisPt	Uptake Elemental content	Zheng et al. (2015)
Human cervical cancer (HeLa) Human lung carcinoma (A549) Normal human bronchial epithelial (16HBE)	Single Cell ICP-MS	Fe, Cu, Zn, Mn, P, S	Elemental content	Wang et al. (2015)
Ovarian cancer (CP70, A2780, CAOV3)	Single Cell ICP-MS	CisPt AuNPs Cu, Zn	Uptake Elemental content	Amable et al. (2017)
Yeast (*Saccharomyces cerevisiae*)	Single Cell ICP-MS	Mg, P, Ca, Mn, Fe, cu, Zn	Elemental content	Groombridge et al. (2013)
Yeast (*Saccharomyce cerevisiae*) Cyanobacterium (*Synechocystis* sp. *PCC 6803*), red algae (*Cyanidioschyzon merolae 10D* and *Galdieria sulphuraria*), green alga (*Chlamydomonas reinhardtii CC-125*)	Single Cell ICP-MS	C, Mg, Al, P, S, K, Ca, Cr, Mn, Fe, Zn	Elemental content	Miyashita et al. (2014b)
Yeast (*Saccharomyce cerevisiae*)	Single Cell ICP-SFMS	Na, Mg, Fe, Cu, Zn, Se	Elemental content	Shigeta et al. (2013a)
Unicellular alga (*Pseudococcomyxa simplex*)	Single Cell ICP-AES	Ca, Mg, Fe	Elemental content	Ishihara et al. (2015)

(continued)

Table 10.1 (continued)

Type of cell	Analytical technique	Analyte	Analytical information	Ref
Bovine/calf red blood	Single Cell ICP-MS	Fe	Elemental content	Verboket et al. (2014)
Human liver carcinoma (HepG2)	Single Cell ICP-MS	ZnO NPs	Uptake	Wang et al. (2017b)
Bacteria (*Pseudomonas putida* KT2440)	CyTOF	Ru, CisPt, Ag, AgNPs	Detection and quantification	Guo et al. (2017)
Mouse fibroblasts (3T3)	LA- ICP-SFMS	AuNPs AgNPs	Quantification Intracellular distribution	Drescher et al. (2012)
Mouse fibroblasts (3T3) Mouse macrophages (J774)	LA- ICP-SFMS Cryo-XT	AuNPs AgNPs SiO_2NPs	Uptake Intracellular distribution	Drescher et al. (2014)
Mouse leukemic monocyte macrophage (RAW 246.7)	LA-ICP-MS	AuNPs	Uptake	Wang et al. (2013)
Marine microalgae (*Scrippsiella trochoidea*)	LA-ICP-MS	Cu	Uptake Intracellular distribution	Van Malderen et al. (2016a)
Mouse fibroblasts (3T3)	LA-ICP-MS	AuNPs	Intracellular distribution Quantification of nanoaggregates	Büchner et al. (2014)
Paramecium	Fs-LI-O-TOFMS	Na, Mg, Al, K, Ca, Mn, Fe, Cu, Zn, Cs	Elemental composition	Gao et al. (2013)
Human regulatory macrophages (from $CD14^+$ monocytes)	LA-ICP-MS	AuNPs	Tracking single cells	Managh et al. (2013)
Alga (*Chlorella kesslerii*)	NanoSIMS	Cu	Intracellular distribution	Slaveykova et al. (2009)
Hyperaccumulator plant (*Alyssum lesbiacum*)	NanoSIMS	Ni, K, Mg, Na, Ca	Intracellular distribution	Smart et al. (2010)
Bacterium, strain GFAJ-1	NanoSIMS	As	Intracellular distribution	Wolfe-Simon et al. (2011)
Human breast adenocarcinoma	NanoSIMS	Au	Intracellular distribution	Wedlock et al. (2011)
Human liver carcinoma (HepG2)	NanoSIMS	CuNPs	Intracellular distribution	Audinot et al. (2013)
Chlamydomonas reinhardtii and *Arabidopsis thaliana*	NanoSIMS	Na, Ca, Fe, Mn, Cu	Intracellular distribution	Malherbe et al. (2016)
Marine microalgae (*Scrippsiella trochoidea*)	X-ray fluorescence	Cu, Ni, Zn, Mn	Uptake Intracellular distribution	Vergucht et al. (2015)
Fungus (*Heliscus lugdunensis*)	X-ray fluorescence	Cd	Intracellular distribution	Isaure et al. (2017)

(continued)

Table 10.1 (continued)

Type of cell	Analytical technique	Analyte	Analytical information	Ref
Microalga (*Chlamydomonas reinhardtii*)	X-ray fluorescence	Cd	Intracellular distribution	Penen et al. (2017)
Mouse fibroblasts (3T3)	X-ray fluorescence	Zn, Cu, Fe	Intracellular distribution	McRae et al. (2013)
Green microalgae (*Coccomyxa actinabiotis*)	X-ray fluorescence	Au	Intracellular distribution	Leonardo et al. (2014)
Human intestinal epithelial (HT29 – MD2)	X-ray fluorescence	Mn	Intracellular distribution	Mathieu et al. (2017)
Human fibrosarcoma (T1080)	X-ray fluorescence	Ni	Intracellular distribution	Szyrwiel et al. (2015)
Bovine ovaries	X-ray fluorescence	Se	Intracellular distribution	Ceko et al. (2015)
Pea (*Pisum sativum*) embryo	X-ray fluorescence	Fe	Intracellular distribution	Roschzttardtz et al. (2011)
Hyperaccumulator (*Arabidopsis halleri*) and non-accumulator (*Arabidopsis lyrata*) leaves	X-ray fluorescence	Cd	Intracellular distribution	Isaure et al. (2015)
Human macrophages (THP-1)	Laser-SNMS	AgNPs	Intracellular distribution	Haase et al. (2011)
Rat kidney epithelial (NRK)	TOF-SIMS	SiO_2 particles	Intracellular distribution	Hagenhoff et al. (2013)

SC-ICP-MS is based on the well-established technique of Single Particle ICP-MS (SP-ICP-MS), whose theoretical basis was outlined by Degueldre and Favarger (2003) and further developed by Laborda et al. (2014), that has shown that information about the particle size and particle number concentration can be derived from the time-resolved signals. Briefly, a cell suspension is introduced into the plasma (via conventional concentric nebulizers or other dedicated introductions systems as it will be discussed later) where droplets containing cells are vaporized and the intrinsic metals inside the cell are atomized and ionized, producing an ion plume that is detected by the ICP-MS as a short transient signal or a spike signal (also called cell event).

Each individual cell produces its own ion plume with a typical duration in the range of 100–500 μs (Miyashita et al. 2017). If the acquisition time (i.e., dwell time) of the detector is longer than the duration of the ion plume generated, signals corresponding to individual cells can be detected as high spikes, whose signal is proportional to the amount of metals in the cell. Therefore, the spike intensity distribution can be correlated with the mass distribution of the metal in the cells. On the other hand, if the dwell time is shorter than the duration of the ion plume, cells events are recorded as transient signals. In SC-ICP-MS an essential requirement is that each cell event corresponds to the signal generated by an individual cell, so that cells must be well separated in time and space in the ICP.

One of the first studies on single-cell introduction into ICP in time-resolved mode was performed by Nomizu et al. (1994). They used a previously developed airborne particle analyzer (Nomizu et al. 1992, 1993) to determine the calcium content in some single mammalian cells with diameters from 10 to 20 μm by ICP-AES. Droplets containing cells were dried in a drying chamber, and cells were directly introduced into the plasma and the calcium signal produced by each cell registered by the analyzer. For quantitative purposes, the system was calibrated with monodisperse calcium acetate aerosols generated by a vibrating orifice aerosol generator, assuming that a cell and a particle behave the same way in the plasma and thus the pulse signals derived from both produced the same pulse height if the calcium content is the same. Through this approach, they determined the calcium content in mouse fibroblasts, human pancreas, and human umbilical vein endothelial cells, with measured contents ranging from 0.057 to 0.27 pg of Ca per cell. The detection limit was reported at ~0.01 pg of Ca in a cell. However, other major elements present in cells, such as Mg, K, and Na, could not be detected due to the system that was not sensitive enough, a problem that authors thought could be overcome by using ICP-MS as detection method. In addition, the introduction efficiency of dried cells into the ICP was less than 0.1%.

The direct analysis of a single bacteria cell by SC-ICP-MS was reported for the first time by Houk's group (Li et al. 2005). *Bacillus subtilis* (*B. subtilis*), grown in a spiked uranium medium, were introduced by a microconcentric nebulizer, and the U signal incorporated intrinsically into single cells was registered by a magnetic sector instrument (ICP-SFMS) operating at the fastest integration time (4 ms). The study showed that bacteria cells behave more like large particles rather than wet droplets. U^+ spikes were observed when introducing the sample as dry and wet bacterial aerosols. On the other hand, they also showed that the U response is different for lysed and unlysed bacteria, observing a 30% increase when bacteria are lysed by sonication. The quantification of U was possible by using an inorganic U standard but applying a correction factor that has to be known.

Ho and Chan (2010) used a conventional pneumatic V-groove nebulizer to perform a single-cell analysis of a unicellular alga (*Chlorella vulgaris*, 1–6 μm) using single quadrupole ICP-MS at its lowest integration time (10 ms). This model alga was chosen as a model due to its high magnesium content. By monitoring this major element and counting the number of spikes, the number density of the cells was determined. The sample uptake rate and the algae number density were optimized to ensure that each magnesium spike corresponded to one cell. However, the transport efficiency was estimated at 0.54% due to an inefficient sample introduction system. On the other hand, they demonstrated the feasibility of a quantitative determination of Mg in the algal cells by using MgO particles for calibration. Alternatively, a semiquantitative measurement was also possible by using aqueous standards for calibration. The use of MgO particles for calibration reduces the error due to a different diffusion loss of Mg atoms coming from cells with respect to those coming from aqueous standards. Apart of the presence of a major element like Mg (10^8 atoms per cell), they were able to detect other trace elements in the algal cells, like Mn and Cu ($5x10^6$ atoms per cell). The method was also applied to a kinetic study of the sorption rate of Cr onto algal cells without separation of the cells from the original suspension.

The same group used a similar approach to obtain valuable biologically relevant info about the uptake of bismuth-based drugs in single cells (Tsang et al. 2011). The magnesium and bismuth content in *Helicobacter pylori* was monitored at the single-cell level in order to track the presence of individual cells and the metallodrug uptake. For quantitative purposes, MgO particles were used as calibration standards as in their previous work (Ho and Chan 2010). A magnesium content of about 2.9×10^7 atoms per cell was determined in wild-type bacteria, whereas the bacteria treated with the bismuth-based drug deposited nearly 1×10^6 atoms per cell. The study showed that tracking the changes of bismuth content in single bacteria cell with time provides a snapshot of bismuth uptake. Moreover, the competitive effect of ferric ions against the bismuth uptake by cells was demonstrated.

Zheng et al. (2013) applied SC-ICP-MS to determine intracellular quantum dots (QDs) in single mouse leukemic monocyte macrophage cells (RAW 246.7) after different exposure times. They used a microconcentric nebulizer to directly introduce the cells after exposure with carboxyl CdSeS of 7 nm, and the cadmium signal was monitored with a dwell time of 5 ms by using a single quadrupole ICP-MS. The method was applied to study uptake kinetics of QDs by incubating the cells for 2–12 h. A transport efficiency of 2.9% was obtained for this study. The CdS uptake was quantified by using a semiquantitative approach based on the introduction of a cadmium standard solution for calibration, assuming that ions from QDs and dissolved ions from standard solution behave alike in the plasma. A suspension of QDs of the same diameter could not be used for calibration purposes, since 7 nm is below the size detection limit of the technique. The results were validated by flow cytometry and cell digestion methods.

The same system was used to determine the uptake of $Gd@C_{82}(OH)_{22}$, a new nanomaterials antitumor agent, and cisplatin in single cells (Zheng et al. 2015). Cancer cells (HeLa) and normal cells (16HBE) were incubated with $Gd@C_{82}(OH)_{22}$ and cisplatin at different doses for 2 and 24 h and directly introduced into the ICP after treatment, monitoring the gadolinium and platinum signals. Standard solutions of Pt or Gd were used to quantify the number of atoms per cell after each exposure dose and exposure time. Metal contents ranging from 0.03 to 2.4 pg and from 0.15 to 25 pg were obtained for Pt and Gd, respectively. Results showed that $Gd@C_{82}(OH)_{22}$ has higher bioavailability.

SC-ICP-MS was used to analyze at the single-cell level the contents and distribution patterns of essential mineral elements (Fe, Cu, Zn, Mn, P, and S) in two types of cancer cells (HeLa and A549) and one type of normal cells (16HBE) (Wang et al. 2015). Analyses were performed by using a concentric PFA-ST nebulizer which resulted in low cell transport efficiency (~0.2%). In order to fulfill the criteria that each spike signal corresponds to one cell event, dwell time and cell number density were optimized. As expected, different distribution patterns of the elements among the three types of cells were observed, which led the authors to the conclusion that these discrepancies can be used to distinguish between normal and tumor cells in mixed populations. The quantitative analysis of the elements in single cells was performed by the use of standard solutions.

Amable et al. (2017) used SC-ICP-MS to investigate the uptake of cisplatin by ovarian cancer cells, by tracking the platinum content in individual cells over time, which could be used as a predictor of the effectiveness of the treatment. The platinum content increased over time, signifying increased cisplatin uptake. Moreover, the same authors used the same analytical approach to study the uptake rate of gold nanoparticles (AuNPs) at two different concentrations by three different ovarian cell lines and to determine the intrinsic copper and zinc contents in one ovarian cell line. In the later study, the use of a short dwell time (50 μs) allowed to reduce the background signal while keeping the signal-to-background ratio constant.

One of the major drawbacks of SC-ICP-MS is the high sample consumption (as a consequence of inefficient cell transport efficiency) in conventional sample introduction systems, usually based on a concentric nebulizer and a spray chamber. As we have seen in the previous studies reported, the cell transport efficiency was <5%. In order to try to overcome this problem, some groups have developed special sample introduction devices.

Groombridge et al. (2013) developed a high-efficiency cell introduction system (HECIS), which was based on a total consumption sample introduction device, to determine trace elements in single yeast cells (*Saccharomyces cerevisiae*). The sample introduction system consisted of a modified high performance concentric nebulizer (HPCN) and a 15 mL on-axis spray chamber utilizing a sheath gas glow to prevent sample deposition. Besides, a solution of NaCl at low concentration was added to the cell suspension in order to reduce cell adsorption. The modified nebulizer was able to generate aerosols below 10 μm in size by working at 10 μL min^{-1}, which leads to a lower background noise level. With the developed system, a cell transport efficiency of 75.0 ± 4.7% into the plasma was achieved for the analysis of single yeast cells. The signals of Mg, P, Ca, Mn, Fe, Cu, and Zn were obtained by single quadrupole ICP-MS at a dwell time of 10 ms. They also conducted a preliminary investigation into the potential for multielement correlation analysis within cells by using ICP-TOFMS. The time-resolved profiles of six elements (Mg, P, Mn, Fe, Cu, and Zn) were simultaneously obtained in one measurement at a dwell time of 1 ms. Results show a relatively strong correlation for the spectra between P and Zn (correlation factor, 0.69), between P and Mg (0.63), and between Mg and Zn (0.63). Despite the successful application of the HECIS, the cell transport efficiency was not close to 100%, probably due to cell loss during introduction, according to the author (Miyashita et al. 2014b). Therefore, the same group modified the HPCN in order to improve the efficiency of single-cell analysis by reducing the loss of cells, as well as to accept a wider range of size cells (from 2.0 to 6.4 μm) (Miyashita et al. 2014b). The modified system consisted of an HPCN with a large-bore center capillary tube (LB-HPCN) and a 15 mL on-axis spray chamber still with a sheath gas flow but with a longer total length and shorter inner tube length. Besides, an external ion pulse counting unit allowed improving the data acquisition frequency (0.05–1 ms). The system was tested by monitoring a wide range of biologically important elements (C, Mg, Al, P, S, K, Ca, Cr, Mn, Fe, Zn) in unicellular microbes of different sizes: yeast (*Saccharomyces cerevisiae*), cyanobacterium (*Synechocystis* sp. *PCC 6803*), red algae (*Cyanidioschyzon merolae 10D* and *Galdieria sulphura-*

ria), and green alga (*Chlamydomonas reinhardtii CC-125*). The cell introduction efficiencies determined ranged 86% to ca. 100% for microbes with cell diameter of 6.4 mm and 2.0–3.0 mm, respectively.

A sample introduction system based on micro-droplet generator (μDG) was developed by Shigeta et al. (2013a) and coupled to a sector field ICP-MS instrument, operating in fast scanning mode (shortest dwell time of 100 μs). Uniformly sized droplets down to 23 nm were achieved thanks to the use of triple pulse mode of the droplet generator. The generated droplets were injected into a low-temperature desolvation system, using He as drying gas. The system was successfully applied for the detection of trace elements (Na, Mg, Fe, Cu, Zn, Se) in single selenized yeast cells (*Saccharomyce cerevisiae*) with 100% transport efficiency (Shigeta et al. 2013b). The single yeast cells with a diameter of roughly 6 μm were embedded into droplets and introduced into the plasma, with a fixed droplet generation rate of 50 Hz, producing equidistant signals that allowed the discrimination between signals produced due to background from those produced by the analyte. Signal intensities from single cells were measured for Cu, Zn, and Se from the histograms for about 1000 cell events. In addition, they performed open vessel digestion of washed yeast cells for multielement analysis, determining absolute amounts per single cell for Na (0.91 fg), Mg (9.4 fg), Fe (5.9 fg), Cu (0.54 fg), Zn (1.2 fg), and Se (72 fg).

Ishihara et al. (2015) developed a desolvation system for micro-droplet injection system (M-DIS) in order to detect trace metals in the unicellular alga *Pseudococcomyxa simplex* by ICP-AES. In the M-DIS system, the cell solution is not nebulized but introduced at the center of the ICP as a single micro-droplet with a diameter between 30 and 70 μm. The desolvation system readily removed the water vapor by placing a cooling part to the introduction route downstream of the heating, allowing the decrease of the solvent load in the droplet and thus improving the sensitivity of the elements present in the cell. The modified M-DIS system equipped with the desolvation system was coupled to ICP-AES, and the emission spectra of Ca, Mg, and Fe contained in a single cell were recorded; however, the emission spectra corresponding to the metals Mn, Zn, and Mo could not be detected.

Verboket et al. (2014) proposed a novel droplet microfluidic sample introduction system for ICP-MS. The system is based on a liquid-assisted droplet ejection (LADE) chip made entirely of poly(dimethylsiloxane) (PDMS), which produced droplets in a stream of perfluorohexane (PFH). A membrane desolvator is used prior the injection into the plasma to reduce the droplet size. The chip was able to generate highly monodisperse droplets in the size range from 40 to 60 μm, which remain intact during the ejection and can be transported and directly injected into the ICP with >50% efficiency. The system was tested by introducing calf red blood cells suspended in phosphate-buffered saline (PBS) and monitoring the ^{56}Fe signal. However, the transport efficiency of cells was low (4.5%) as a result of losses in the transport assembly. The amount of Fe per cell was calculated by calibrating the instrument with droplets consisting of a multielement solution.

Recently, Wang et al. (2017b) fabricated a droplet chip which produced droplets with an average diameter of 25 μm, coupled online with an ICP-MS via a microflow nebulizer. The single cells encapsulated in the droplets generated by the chip remained intact during its transportation from the outlet of the droplet chip to the microflow nebulizer. The developed droplet chip was applied for the quantification of Zn and zinc oxide nanoparticles (ZnO NPs) uptake/adsorption in single HepG2 cells by SC-ICP-MS. ZnO NPs were used for the quantification of Zn in single cells. The cell transport efficiency was low (2.96%) compared with other systems, but the frequency of droplet generation was high (3–6 × 10^6 droplets per minute).

Mass cytometry, usually in combination with a time-of-flight analyzer, can offer the capability of single-cell analysis on the basis of specific metal-based cell markers. Tanner's group introduced in 2008 for the first time a specialized system for the multiparametric analysis of single cell at high throughput (Bandura et al. 2009, Tanner et al. 2008). The developed system was used for the analysis of human cells through an approach based on multiatom elemental antibody tagging. The mass cytometer was later introduced to the market as "CyTOF" by DVS Sciences Inc.

More recently, Guo et al. (2017) used CyTOF for label-free detection and quantification of Ag in single bacterial (*Pseudomonas putida*) cells of different physiological states. Cells were treated with silver ions or 10 nm silver nanoparticles (AgNPs). The amount of Ag per cell was determined via the intensity of the mass cytometric signal for Ag. The information about the amount of Ag per cell was combined with an approach using ruthenium red (RR) staining as a marker for all cells of a population and cisplatin staining for live/dead cell discrimination. They found that the treatment with AgNPs caused higher frequencies of dead cells, higher frequencies of Ag-containing cells, and higher per-cell silver quantities.

10.2.2 Other MS-Based Approaches

Apart of its capabilities to obtain information about the intracellular distribution of metal and metal nanoparticles, as it will be discussed in the next section, LA-ICP-MS is able to provide quantitative information. In this context, LA-ICP-MS allows the quantification of inorganic nanoparticles without losing spatial information of the nanoparticle distribution at the cell level, overcoming the disadvantages of ICP-MS elemental analysis after cell digestion or extraction.

One of the major problems to achieve an accurate quantification of metals and metal nanoparticles by LA-ICP-MS is a lack of suitable single-cell standards. In order to overcome this problem, different approaches have been investigated.

Drescher et al. (2012) used LA coupled to a sector field ICP-MS to study the uptake of AuNPs (with a diameter of 25 nm) and AgNPs (with a diameter of 50 nm) in single eukaryotic cells (mouse fibroblasts). The number of nanoparticle per cell after incubation was determined on the basis of a matrix-matched calibration using nitrocellulose membranes spiked with nanoparticle suspension. The limit of detection and the limit of quantification for AgNPs were 20 and 60 particles and for AuNPs were 190 and 550 particles, respectively. Results showed a strong dependency of particle uptake on concentration and incubation conditions. In the case of AgNPs, increasing nanoparticle concentration and incubation time was shown to lead to an increase of the silver intensity inside the cells, which was directly correlated with an increasing number of nanoparticles.

The same group used LA-ICP-MS for a quantitative comparison of the uptake of uncoated AgNPs and AuNPs, silica-coated AgNPs, and silica nanoparticles with a plasmonic (Au or Ag) core (BrightSilica) by single mouse fibroblasts and macrophages (Drescher et al. 2014). The use of nanoparticles with thicker shells (silica-coated AgNPs) provided the possibility to compare the uptake of relatively large nanoparticles with smaller ones (BrightSilica with thin shells) at an identical size of the plasmonic cores. Since the uptake of nanoparticles by cells is size dependent, the signal intensity obtained was about six times higher for BrightSilica(Ag) with a diameter of 57 nm than for the 126 nm silica-coated AgNPs. Results also showed a smaller uptake by fibroblast cells of BrightSilica NPs in comparison with uncoated AgNPs and AuNPs of almost the same size.

Wang et al. (2013) quantified the gold mass in single mouse leukemic monocyte macrophage cells (RAW 246.7) exposed to 30 nm AuNPs by LA-ICP-MS. For calibration purposes an inkjet printer producing dried residues of picoliter droplets was used in order to simulate matrix-matched calibration standards. Under controlled conditions, the droplets dispensed from the printer were almost identical with less than 1% mass variations. Through this calibration approach, a limit of quantification of 1.7 fg Au was achieved. A log-normal distribution, ranging from 1.7 to 72 fg Au per cell, corresponding to 9 to 370 AuNPs per cell, was obtained in single cells exposed to AuNPs for 4 h.

Van Malderen et al. (2016a) presented an interesting quantitative approach in order to overcome the limitations in precision, accuracy, and efficiency of current calibration strategies in LA-ICP-MS. For this purpose, they designed and proposed the use of high-density microarray plates for multiple matrix-matched standards. A calibration curve with good linear correlation was obtained within the range of 5–200 $\mu g\ g^{-1}$, exhibiting a detection limit of 3 fg Cu. The system was applied for the quantification at the single-cell level of Cu in the marine microalgae *Scrippsiella trochoidea* after copper exposure.

Büchner et al. (2014) studied the accumulation of AuNPs in fibroblast cells during cellular uptake, processing, and mitosis. Cells were exposed to 14 nm AuNPs with different incubation times and analyzed by LA-ICP-MS. Results showed that AuNPs form aggregates of different sizes, and the gold signal intensity can be compared for the different incubation times. As an indicator for the aggregation of nanoparticles inside the cells, the authors determined the number of spots per cell with a gold signal intensity >50,000 cps. The number of spots with gold intensity >50,000 cps showed a different trend that was attributed to a change in the intracellular particle distribution over time.

Finally, Gao et al. (2013) proposed an approach based on high irradiance femtosecond laser ionization orthogonal time-of-flight mass spectrometry (fs-LI-O-TOFMS) to determine the elemental composition of single paramecium cells, with the size of several tens of micrometers. After a simple preparation procedure, based on heating and drying the cells, a series of elements (Na, Mg, Al, P, S, Cl, K, Ca, Mn, Fe, Cu, Zn, I, Cs) were determined in paramecium at the pg per cell level. Salts of these elements were used in order to increase the element concentration. This approach allowed the measurement of nonmetallic elements like P, S, or Cl difficult to analyze with single quadrupole ICP-MS instruments. An absolute limit of detection (ALOD) at the femtogram level was achieved.

10.3 Subcellular Distribution Analysis of Metal and Metal Nanoparticles in Single Cells by Imaging Techniques

As it has already been mentioned, classical studies of biological cells consist on the digestion of a cell suspension and thus averaging over a huge number of cells. With this approach, not only the intrinsic individual variation but also the information about intracellular distribution of metals is lost. Nevertheless, each cell is subdivided into different compartments with specific reactions, and thus chemical elemental imaging is critical to elucidate the biochemical processes that take place in each of them.

The determination of the metal distribution at the subcellular level in individual cells is challenging and requires analytical methods with high sensitivity and high spatial resolution. In the last years, the advances in analytical techniques like LA-ICP-MS, SIMS, or SXRF have opened a number of possibilities for imaging at the single-cell level. Different applications reported in literature are shown in Table 10.1.

10.3.1 LA-ICP-MS

LA-ICP-MS is an established quantitative elemental analysis and mapping technique, which provides spatially resolved information on element distribution (qualitative and quantitative) in biological samples (Mueller et al. 2014). Rapid response LA-ICP-MS systems greatly benefit throughput and sensitivity, which are key parameters in 2D and 3D imaging at high lateral resolution (Van Malderen et al. 2016b). The novel capabilities of laser ablation as a sample introduction technique in ICP-MS are being increasingly applied for advanced imaging mass spectrometry.

LA-ICP-MS was used by Managh et al. (2013) for tracking single macrophage cells labeled with 50 nm AuNPs in different tissues of immunodeficient mice. Since the cell diameter (~10 μm) was much smaller than the laser spot diameter (25 μm), a single cell was completely ablated in each laser shot.

Jakubowski's group (Drescher et al. 2012, 2014) showed the capabilities of LA-ICP-MS for spatially resolved bioimaging of the intracellular distribution of nanoparticles in single eukaryotic cells in two different studies. Scan speed, ablation frequency, and laser energy were optimized in order to achieve a laser spot size of 4 μm or 8 μm so that the laser spots are widely overlapping each other and the signal of the sample is generated by the difference in the ablated area. By using this laser ablation scan mode, the lateral resolution in scan direction is much smaller than the laser spot diameter, obtaining a high-resolution image. In a first study, they used the optimized ablation system to investigate the intracellular distribution of AuNPs and AgNPs in cellular substructures of single mouse fibroblasts (Drescher et al. 2012). The subcellular mapping obtained enabled the differentiation of nanoparticles in the cytosol from those in the cell nucleus region.

In a different study, the same group combined LA-ICP-MS with cryo nanoscale X-ray tomography (cryo-XT) to investigate the intracellular distribution at the subcellular level with 3D resolution of silica nanoparticles with a plasmonic (Au or Ag) core (BrightSilica) in 3T3 fibroblast cells and J774 macrophage individual cells (Drescher et al. 2014). Taking advantage of the absorption of soft X-rays by the silica material, the study confirmed that BrightSilica nanoparticles enter cells by an endocytotic mechanism.

Büchner et al. (2014) detected Au nanoaggregates at subcellular level in single fibroblast cells by LA-ICP-MS micromapping. They achieved high sensitivity at high spatial resolution and complete cell ablation by optimizing the laser ablation parameters as previously described in the study performed by Drescher et al. (2012). By using an 8 μm diameter beam, they showed that the nanoaggregates accumulated in localized, perinuclear regions of the cells, but it was also demonstrated that reorganization occurred during mitosis.

However, these LA-ICP-MS micromapping studies were performed in fibroblast cells, which can be considered relatively large in size ($\geq$50 μm in diameter). Therefore, the application would benefit from improved spatial resolution for the investigation of the distribution of nanoparticles within smaller cells (Van Malderen et al. 2016a).

Van Malderen et al. (2016a) demonstrated the capabilities of LA-ICP-MS to study the intracellular transition metal distribution in single cells of model marine microalgae. Microimaging of *Scrippsiella trochoidea* exposed to 20 μg g^{-1} indicated an important organelle acting as preferential storage of Cu. These conclusions were corroborated by images obtained by SXRF in a cross-validation approach (Vergucht et al. 2015), showing the potential of LA-ICP-MS for subcellular level imaging.

While the use of LA-ICP-MS for single-cell imaging is just at its beginning, SXRF microscopy and secondary ion mass spectrometry have already shown that the imaging of biological trace metals at nm scale is possible (Qin et al. 2011).

10.3.2 NanoSIMS

SIMS is an analytical technique based on the use of an accelerated primary ion beam that bombards a solid surface and generates secondary ions, which are subsequently analyzed by a mass spectrometer (Boxer et al. 2009). In the case of nanoSIMS, the primary ion beam is scanned over the surface of the sample, and ejected secondary ions are analyzed by a double sector mass spectrometer (Schaumlöffel et al. 2016).

Current nanoSIMS instruments are equipped with two different primary ions sources: a thermal ionization cesium positive ion source that generates negative secondary ions (Storms et al. 1977) and a duoplasmatron oxygen negative ion source that generates positive secondary ions (Benninghoven et al. 1987). Its high spatial resolution (better than 100 nm) has opened numerous research application fields, such as the detection of chemical elements at the subcellular level in biological samples.

Slaveykova et al. (2009) demonstrated the capabilities of nanoSIMS to investigate Cu distribution in single micrometer-sized algal cells. The microalga *Chlorella kesslerii* was exposed to nanomolar and micromolar copper concentrations, representative of natural and heavily contaminated waters, and the subcellular distribution of copper in individual cells was mapped by measuring $^{63}Cu^-$ and $^{65}Cu^-$ secondary ions. A cesium source was used to generate Cs^+ primary ions that were accelerated as a tightly focused ion beam with a probe working diameter of approximately 50 nm. Results showed that the quantity of Cu in the cell wall and membrane was lower than that in the cell interior.

Smart et al. (2010) used nanoSIMS with dedicated sample preparation methodologies to study the distribution of nickel and other elements in single cells from a nickel hyperaccumulator plant (*Alyssum lesbiacum*), with spatial resolutions of better than 100 nm. A $^{58}Ni^-$ signal distribution was obtained by using the primary Cs^+ ion beam, whereas the primary O^- ion beam allowed them to obtain maps showing the distribution of the positive secondary ions $^{23}Na^+$, $^{40}Ca^+$, $^{39}K^+$, $^{24}Mg^+$, and $^{58}Ni^+$ from the same area analyzed with the Cs^+ beam. They demonstrated a significant sequestration of nickel in the epidermal cell vacuoles together with a uniform distribution throughout the cell walls. However, despite the sample preparation methodology applied in this study helped to minimize the redistribution of soluble elements, the in vivo distributions of highly diffusible ions ($^{23}Na^+$, $^{24}Mg^+$, $^{39}K^+$, and $^{40}Ca^+$) were significantly altered during chemical fixation.

Wolfe-Simon et al. (2011) used nanoSIMS to study the intracellular arsenic in a bacterium (strain GFAJ-1 of the *Halomonadaceae*) that is able to substitute arsenic for phosphorus to sustain its growth. Bacteria was grown in an arsenic-enriched medium, using radiolabeled $^{73}AsO_4^{3-}$ to obtain more specific information about the intracellular distribution of arsenic in single GFAJ-1 cells.

Wedlock et al. (2011) investigated the subcellular distribution of gold in situ in human breast cancer cells treated with an antitumor Au(I) complex. A Cs^+ primary ion beam was used for imaging the negative secondary ion $^{197}Au^-$. Different beam currents were applied to obtain primary beam diameters of ~110 nm (low-resolution images) or ~50 nm (high-resolution images). NanoSIMS maps for $^{197}Au^-$ signal in single cells after treatment showed that the subcellular distribution of gold was associated with sulfur-rich regions in the cytoplasmic, nuclear, and perinuclear regions.

The intracellular distribution of copper oxide nanoparticles (CuO NPs) in HepG2 cells was studied by Audinot et al. (2013) by nanoSIMS imaging. $^{63}Cu^-$ maps were acquired with a Cs^+ ion primary beam, and a probe diameter in the range 80–100 nm in single cells exposed to CuO NPs (25 mg mL^{-1}) for 24 h.

The detection of many trace elements at the single-cell level by nanoSIMS requires the generation of secondary positive ions with high yield by an oxygen primary ion source. However, the commonly used duoplasmatron oxygen source cannot achieve the same lateral resolution as the cesium primary source due to a larger primary ion beam size (down to 50 nm for the Cs source and 200 nm for the duoplasmatron oxygen source) (Schaumlöffel et al. 2016). In order to overcome this problem, Malherbe et al. (2016) designed a new rf plasma oxygen source that was fitted and characterized on a nanoSIMS instrument. This new oxygen ion source was applied for the localization of trace metals (Na, Ca, Fe, Mn, Cu) in individual cells of two model organisms: a unicellular alga (*C. Reinhardtii*) and a flowering plant (*Arabidopsis thaliana*). The new rf plasma oxygen primary ion source showed higher achievable lateral resolutions and higher apparent sensitivities for electropositive elements compared to the commonly used duoplasmatron source. The images produced for positive secondary ions with the rf plasma source showed similar lateral resolution to those obtained for negative secondary ions with the cesium source.

10.3.3 Synchrotron-Based X-ray Fluorescence

SXRF (also abbreviated as μXRF) is among the most powerful techniques for the investigation of metal accumulation at the cell level and can be used for imaging element distribution at high spatial resolution (<100 nm) (Roudeau et al. 2014). This technique is based in the X-ray fluorescence energy emitted by an atom that has been previously irradiated by X-ray photons produced by synchrotron radiation (Schaumlöffel et al. 2016). The irradiated atom ejects a core electron with a characteristic binding energy. In SXRF, the sample is moved by a nano-positioner stage, and spectra are recorded pixel by pixel which enables to provide images of the distribution of metals in single cells. SRXF has been widely applied in different investigation of the metal distribution at the single-cell level. Some interesting applications of SXRF will be explained below.

Vergucht et al. (2015) reported a new imaging approach based on the combination of optical tweezer technology with synchrotron radiation confocal XRF. The method allowed elemental imaging in a 2D projection mode in single marine microalgae (*Scrippsiella trochoidea*) cells in their natural, in vivo state. The elemental distributions showed that significant amounts of Mn, Fe, Cu, and Zn were detected within the Zn-exposed algae. In addition, the in vivo areal concentrations of accumulated metals in single cells could be determined in a semiquantitative way, showing larger differences (Cu > > Ni > Zn).

Elemental imaging by synchrotron µXRF was used by Isaure et al. (2017) to investigate Cd in fungal hyphae at the subcellular level. Aquatic fungus *H. lugdunensis* was exposed to 50 µM Cd and µXRF maps were obtained after 5 and 7 days. Results showed that the hyphal tip cells were depleted in Cd and that the metal was stored in older cells. S and P localization was also determined.

Penen et al. (2017) studied the Cd distribution in single microalga (*Chlamydomonas reinhardtii*) cells. Three different *C. reinhardtii* strains were exposed to 70 µM Cd and analyzed by µXRF. µXRF maps from the three strains demonstrated that Cd level was highly heterogeneous between cells.

In an interesting approach, the redistribution of transition metals in mouse fibroblast cells during the individual stages of mitosis was investigated (McRae et al. 2013). Results showed similar mechanisms for the inheritance of Zn and Cu, but not Fe. Moreover, SXRF maps provided first clues toward a mechanism that might involve compartmentalized transport and possibly a role of the Golgi apparatus.

Leonardo et al. (2014) studied the subcellular location of silver and cobalt in a newly discovered unicellular microalga (*Coccomyxa actinabiotis*) by synchrotron radiation nano-X-ray fluorescence (SR-nXRF). The microalga was exposed to different concentrations of silver and cobalt, and the subcellular distribution after exposure was investigated. They found that cobalt was homogeneously distributed outside of the chloroplast, whereas silver was localized in the cytosol at low concentration and in the whole cell excluding the nucleus at high concentration.

Mathieu et al. (2017) studied the intracellular distribution of manganese in single human intestinal epithelial cells. Cells were incubated with $MnCl_2$ or the manganese complex Mn1 mimicking superoxide dismutase (SOD), a protein involved in cell protection against oxidative stress, and manganese was mapped on cryofixed freeze-dried cells using SXRF with a resolution of 200 nm. The intracellular manganese content in a single cell was higher after incubation with the manganese complex than that in the control cell, but $MnCl_2$ was more efficiently taken up. Moreover, SXRF maps demonstrated a diffuse distribution of manganese with higher amounts where the cell is the thickest.

Szyrwiel et al. (2015) used SXRF to map the intracellular nickel in human fibrosarcoma cells. Cells were incubated with Ni^{2+} or a newly Ni^{2+}-branch peptide complex (NiA), designed to deliver Ni^{2+} to the cell nucleus. SXRF images demonstrated the nuclear location of Ni in cells treated with the NiA complex. However, in cells treated with nickel ions, effective metal uptake was not observed.

Ceko et al. (2015) determined the intracellular selenium distribution in single bovine ovaries by SXRF imaging. The approach allowed localizing Se to the granulosa cell layer of large (>10 mm) healthy follicles.

Moreover, the use of SXRF has also been reported for the investigation of metal localization in plant cells (Zhao et al. 2014).

Roschzttardtz et al. (2011) studied the intracellular location of iron in single pea (*Pisum sativum*) cells by µXRF. The results revealed that, unexpectedly, the iron concentration found in the nucleus was higher than in the expected iron-rich organelles such as plastids or vacuoles. Furthermore, they showed that within the nucleus, the iron concentration in the nucleolus is the highest encountered among the intracellular compartments. This study raised the important question of the role that nucleolar iron plays in plant cells.

Isaure et al. (2015) used μXRF to investigate the distribution of cadmium in leave cells of two plants: the hyperaccumulator *Arabidopsis halleri* and the non-accumulator *Arabidopsis lyrata*. Images were obtained after 3 weeks of cadmium exposure. In the leaves of *A. halleri*, cadmium was found inside the cell and at the rim of the cell. However, the cellular distribution in *A. lyrata* leaves could not be obtained, since the lateral resolution and the level of cadmium were too low.

10.3.4 Other MS Approaches

The intracellular distribution of nanoparticles has also been studied through other analytical approaches based on mass spectrometry techniques.

For instance, Haase et al. (2011) tested the capabilities of laser postionization secondary neutral mass spectrometry (laser-SNMS) for the study of the intracellular distribution of peptide-coated 20 nm AgNPs inside individual human macrophages cells, comparing the results with the established confocal Raman microscopy and transmission electron microscopy (TEM). Although confocal Raman microscopy is fastest and requires a much simpler sample preparation, laser-SNMS provides a much greater sensitivity and 3D resolution limits, being capable of detecting metal atoms with a spatial resolution of down to 100 nm. The images obtained demonstrated significant accumulation of silver particles under the form of aggregates inside the cells. Laser-SNMS was used in combination with TOF-MS to follow subtle biochemical alterations related to nanoparticle treatment.

Finally, TOF-SIMS 3D analysis proved to be able to localize silica micro- and nanoparticles with sizes down to 150 nm in single rat kidney epithelial cells (Hagenhoff et al. 2013). The approach followed by Hagenhoff et al. (2013) allowed the discrimination between extracellular and intracellular localization of nonluminescent, unlabeled nanoparticles, otherwise hard to localize by other analytical techniques. TOF-SIMS provided images of their prominent distribution around the cell nucleus.

10.4 Molecular Analysis and Perspectives

The development of new analytical methods has opened the door to the identification of metabolites in single cells. One of the most interesting approaches is the coupling of high-resolution separation by capillary electrophoresis (CE) with high sensitivity and selectivity detection by electrospray ionization mass spectrometry

(ESI-MS) (Týčová et al. 2017). For instance, Comi et al. (2017) developed an approach based on CE-ESI-MS to identify several amino acids as well as dopamine in single pancreatic islet cells.

On the other hand, no metalloproteomic and metallometabolomic data exist for single cells. However, in order to fully understand the cellular specificity and complexity of cells, it is necessary to measure molecular signatures with single-cell resolution (Wang and Bodovitz 2010). The future development of single-cell metallomics clearly depends on the developments in small-scale molecular mass spectrometry accompanied by custom-designed sample preparation methods. High-resolution mass spectrometry imaging, necessary for this purpose, requires a smaller region of sample for each pixel, but this sampling reduction is translated into a decrease of ionization efficiency and sensitivity (Dong et al. 2016).

SIMS is an ideal technique for mass spectrometry imaging thanks to its high surface sensitivity and submicron resolution (Passarelli and Ewing 2013). However, recent technical advances in the field of matrix-assisted laser desorption ionization (MALDI) have pushed its spatial resolution from the 20–100 μm range to the 1–10 μm range (Passarelli and Ewing 2013), reaching the point where single-cell mass spectrometry can be accomplished, although continued technology development is needed (Murray et al. 2016). For instance, Zavalin et al. (2012) developed a transmission geometry ion source which enabled to perform high-throughput MALDI MS imaging, achieving higher sensitivity. They demonstrated the capabilities of the developed method by mapping the distribution of lipids and peptides in mammalian cells with 1 μm spatial resolution opening a way to single-cell mass spectrometry imaging such as single-cell proteomic (and, in future, metalloproteomic) imaging, even though the sensitivity is still a challenge.

The analysis of large metabolites in single cells remains a challenging task. TOF-SIMS is able to provide submicrometer spatial resolution for imaging of metabolites in single cells, whereas MALDI offers a high mass resolution. However, single-cell analysis by mass spectrometry imaging is limited in mass resolution range to $m/z < 500$ in the case of TOF-SIMS and in spatial resolution (pixel size in the range of 50–200 μm) in the case of MALDI. Therefore, the analysis of metabolites and – in perspective – metallometabolites in individual cells with high spatial resolution and high mass accuracy simultaneously remains a challenge.

However, in an interesting approach, Schober et al. (2012) have already developed a high-resolution mass spectrometry single-cell imaging method. They demonstrated for the first time the combination of high spatial resolution (7 μm) with high mass accuracy (<3 ppm) and high mass resolution (R = 100,000 at m/z = 200) in imaging analysis of human cervical cancer (HeLa) cells. Analyses were performed with a high-resolution atmospheric pressure MALDI (AP-MALDI) coupled to an Orbitrap mass spectrometer. The developed method allowed them to identify larger metabolites in intact single cells thus giving a direction to future investigations of metal-binding species.

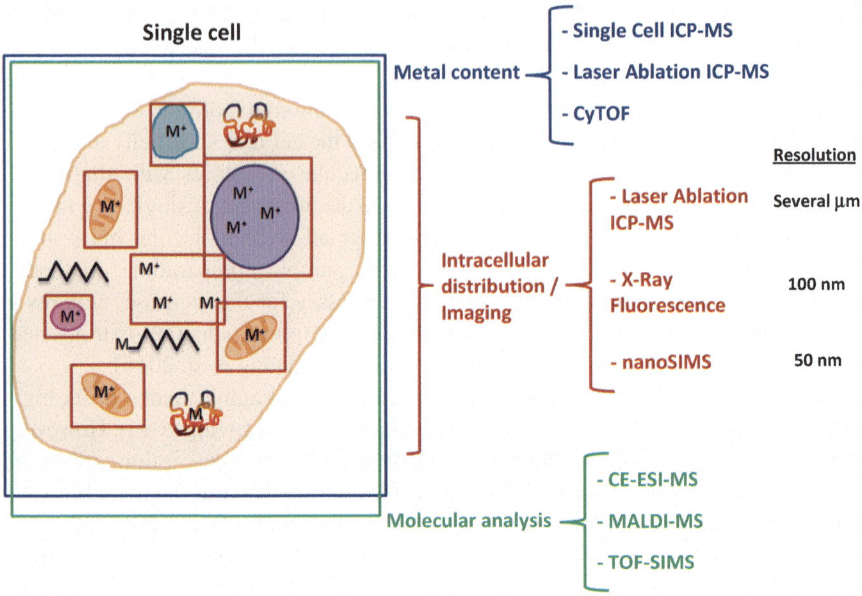

Fig. 10.1 Schematic representation of analytical information obtained by different analytical techniques at the single-cell level

Finally, and in other to sum up, the different analytical techniques discussed in this chapter, together with the relevant information obtained, are summarized in Fig. 10.1.

References

Amable L, Stephan C, Smith S, Merrifield R (2017) An introduction to single cell ICP-MS analysis. Perkin Elmer, USA, pp 3–39

Audinot JN, Georgantzopoulou A, Piret JP, Gutleb AC, Dowsett D, Migeon HN, Hoffmann L (2013) Identification and localization of nanoparticles in tissues by mass spectrometry. Surf Interface Anal 45:230–233

Bandura DR, Baranov VI, Ornatsky OI, Antonov A, Kinach R, Lou X, Pavlov S, Vorobiev S, Dick JE, Tanner SD (2009) Mass cytometry: a novel technique for real-time single cell multi-target immunoassay based on inductively coupled plasma time of flight mass spectrometry. Anal Chem 81:6813–6822

Benninghoven A, Rudenauer FG, Werner HW (1987) Secondary ion mass spectrometry: basic concepts, instrumental aspects, applications and trends. Wiley-Interscience, New York

Boxer SG, Kraft ML, Weber PK (2009) Advances in imaging secondary ion mass spectrometry for biological samples. Annu Rev Biophys 38:53–74

Büchner T, Drescher D, Traub H, Schrade P, Bachmann S, Jakubowski N, Kneipp J (2014) Relating surface-enhanced Raman scattering signals of cells to gold nanoparticle aggregation as determined by LA-ICP-MS micromapping. Anal Bioanal Chem 406:7003–7014

Ceko MJ, Hummitzsch K, Hatzirodos N, Bonner WM, Aitken JB, Russell DL, Lane M, Rodgers RJ, Harris HH (2015) X-Ray fluorescence imaging and other analyses identify selenium and GPX1 as important in female reproductive function. Metallomics 7:71–82

Comi TJ, Makurath MA, Philip MC, Rubakhin SS, Sweedler JV (2017) MALDI MS guided liquid microjunction extraction for capillary electrophoresis – electrospray ionization MS analysis of single pancreatic islet cells. Anal Chem 89:7765–7772

Degueldre C, Favarger PY (2003) Colloid analysis by single particle inductively coupled plasma-mass spectroscopy: a feasibility study. Colloids Surfaces A Physicochem Eng Asp 217:137–142

Dong Y, Li B, Aharoni A (2016) More than pictures: when MS imaging meets histology. Trends Plant Sci 21:686–698

Drescher D, Giesen C, Traub H, Panne U, Kneipp J, Jakubowski N (2012) Quantitative imaging of gold and silver nanoparticles in single eukaryotic cells by laser ablation ICP-MS. Anal Chem 84:9684–9688

Drescher D, Zeise I, Traub H, Guttmann P, Seifert S, Büchner T, Jakubowski N, Schneider G, Kneipp J (2014) In situ characterization of SiO2 nanoparticle biointeractions using BrightSilica. Adv Funct Mater 24:3765–3775

Gao Y, Lin Y, Zhang B, Zou D, He M, Dong B, Hang W, Huang B (2013) Single-cell elemental analysis via high irradiance femtosecond laser ionization time-of-flight mass spectrometry. Anal Chem 85:4268–4272

Groombridge AS, Miyashita S, Fujii S, Nagasawa K, Okahashi T, Ohata M, Umemura T, Takatsu A, Inagaki K, Chiba K (2013) High sensitive elemental analysis of single yeast cells (Saccharomyces cerevisiae) by time-resolved inductively-coupled plasma mass spectrometry using a high efficiency cell introduction system. Anal Sci 29:597–603

Gundlach-Graham A, Günther D (2016) Toward faster and higher resolution LA-ICPMS imaging: on the co-evolution of la cell design and ICPMS instrumentation Young Investigators in Analytical and Bioanalytical Science. Anal Bioanal Chem 408:2687–2695

Guo Y, Baumgart S, Stärk H-J, Harms H, Müller S (2017) Mass cytometry for detection of silver at the bacterial single cell level. Front Microbiol 8:1–9

Haase A, Arlinghaus HF, Tentschert J, Jungnickel H, Graf P, Mantion A, Draude F, Galla S, Plendl J, Goetz ME, Masic A, Meier W, Thünemann AF, Taubert A, Luch A (2011) Application of laser postionization secondary neutral mass spectrometry/time-of-flight secondary ion mass spectrometry in nanotoxicology: visualization of nanosilver in human macrophages and cellular responses. ACS Nano 5:3059–3068

Hagenhoff B, Breitenstein D, Tallarek E, Möllers R, Niehuis E, Sperber M, Goricnik B, Wegener J (2013) Detection of micro- and nano-particles in animal cells by ToF-SIMS 3D analysis. Surf Interface Anal 45:315–319

Haraguchi H (2004) Metallomics as integrated biometal science. J Anal At Spectrom 19:5–14

Haraguchi H (2017a) Metallomics: the history over the last decade and a future outlook. Metallomics 9:1001–1013

Haraguchi H (2017b) Metallomics: integrated biometal science. In: Ogra Y, Hirata T (eds) Metallomics: recent analytical techniques and applications. Springer, Berlin, pp 3–39

Haraguchi H, Ishii A, Hasegawa T, Matsuura H, Umemura T (2008) Metallomics study on all-elements analysis of salmon egg cells and fractionation analysis of metals in cell cytoplasm. Pure Appl Chem 80:2595–2608

Ho K-S, Chan W-T (2010) Time-resolved ICP-MS measurement for single-cell analysis and on-line cytometry. J Anal At Spectrom 25:1114–1122

Isaure M-P, Huguet S, Meyer C, Castillo-michel H, Testemale D, Vantelon D, Saumitou-laprade P, Verbruggen N, Sarret G (2015) Evidence of various mechanisms of Cd sequestration in the hyperaccumulator Arabidopsis halleri, the non-accumulator Arabidopsis lyrata , and their progenies by combined synchrotron-based techniques. J Exp Bot 66:3201–3214

Isaure M-P, Leyh B, Salomé M, Krauss G-J, Schaumlöffel D, Dobritzsch D (2017) The aquatic hyphomycete Heliscus lugdunensis protects its hyphae tip cells from cadmium: a micro X-ray

fluorescence and X-ray absorption near edge structure spectroscopy study. Spectrochim Acta Part B At Spectrosc 137:85–92
Ishihara Y, Aida M, Nomura A, Miyahara H, Hokura A, Okino A (2015) Development of Desolvation system for single-cell analysis using droplet injection inductively coupled plasma atomic emission spectroscopy. Anal Sci 31:781–785
Laborda F, Bolea E, Jiménez-Lamana J (2014) Single particle inductively coupled plasma mass spectrometry: a powerful tool for nanoanalysis. Anal Chem 86:2270–2278
Leonardo T, Farhi E, Boisson A, Vial J, Cloetens P, Bohic S (2014) Determination of elemental distribution in green micro-algae using synchrotron radiation nano and quantitative imaging of silver and cobalt. Metallomics 6:316–329
Li F, Armstrong DW, Houk RS (2005) Behavior of bacteria in the inductively coupled plasma: atomization and production of atomic ions for mass spectrometry. Anal Chem 77:1407–1413
Malherbe J, Penen F, Isaure MP, Frank J, Hause G, Dobritzsch D, Gontier E, Horréard F, Hillion F, Schaumlöffel D (2016) A new radio frequency plasma oxygen primary ion source on Nano secondary ion mass spectrometry for improved lateral resolution and detection of electropositive elements at single cell level. Anal Chem 88:7130–7136
Managh AJ, Edwards SL, Bushell A, Wood KJ, Geissler EK, Hutchinson JA, Hutchinson RW, Reid HJ, Sharp BL (2013) Single cell tracking of gadolinium labeled CD4+ T cells by laser ablation inductively coupled plasma mass spectrometry. Anal Chem 85:10627–10634
Maret W (2016) The metals in the biological periodic system of the elements: concepts and conjectures. Int J Mol Sci 17:1–8
Mathieu E, Lai B, Chain F, Langella P, Bachelet M, Masliah J, Seksik P, Policar C (2017) A cell-penetrant manganese superoxide dismutase (MnSOD) mimic is able to complement MnSOD and exerts an Antiinflammatory effect on cellular and animal models of inflammatory bowel diseases. Inorg Chem 56:2545–2555
McRae R, Lai B, Fahrni CJ (2013) Subcellular redistribution and mitotic inheritance of transition metals in proliferating mouse fibroblast cells. Metallomics 5:52–61
Miyashita S, Groombridge AS, Fujii S, Takatsu A, Chiba K, Inagaki K (2014a) Time-resolved ICP-MS measurement: a new method for elemental and multiparametric analysis of single cells. Anal Sci 30:219–224
Miyashita S, Groombridge AS, Fujii S, Minoda A, Takatsu A, Hioki A, Chiba K, Inagaki K (2014b) Highly efficient single-cell analysis of microbial cells by time-resolved inductively coupled plasma mass spectrometry. J Anal At Spectrom 29:1598–1606
Miyashita S, Fujii S, Shigeta K, Inagaki K (2017) Single cell analysis by using ICP-MS. In: Ogra Y, Hirata T (eds) Metallomics: recent analytical techniques and applications. Springer, Berlin, pp 107–124
Mueller L, Traub H, Jakubowski N, Drescher D, Baranov VI, Kneipp J (2014) Trends in single-cell analysis by use of ICP-MS. Anal Bioanal Chem 406:6963–6977
Murray KK, Seneviratne CA, Ghorai S (2016) High resolution laser mass spectrometry bioimaging. Methods 104:118–126
Nomizu T, Nakashima H, Hotta Y, Tanaka T, Kawaguchi H (1992) Simultaneous measurement of the elemental content and size of airborne particles by inductively coupled plasma emission spectrometry combined with the light-scattering method. Anal Sci 8:527–531
Nomizu T, Kaneco S, Tanaka T, Yamamoto T, Kawaguchi H (1993) Determination of Femto-gram amounts of zinc and lead in individual airborne particles by inductively coupled plasma mass spectrometry with direct air-sample introduction. Anal Sci 9:843–846
Nomizu T, Kaneco S, Tanaka T, Ito D, Kawaguchi H, Vallee BT (1994) Determination of calcium content in individual biological cells by inductively coupled plasma atomic emission spectrometry. Anal Chem 66:3000–3004
Nuñez J, Renslow R, Cliff JB, Anderton CR (2018) NanoSIMS for biological applications: current practices and analyses. Biointerphases 13:03B301
Passarelli MK, Ewing AG (2013) Single-cell imaging mass spectrometry. Curr Opin Chem Biol 17:854–859

Penen F, Isaure MP, Dobritzsch D, Bertalan I, Castillo-Michel H, Proux O, Gontier E, Le Coustumer P, Schaumlöffel D (2017) Pools of cadmium in Chlamydomonas reinhardtii revealed by chemical imaging and XAS spectroscopy. Metallomics 9:910–923

Pett-Ridge J, Weber PK (2012) NanoSIP: NanoSIMS applications for microbial biology. In: Navid A (ed) Microbial systems biology. Springer, Berlin, pp 375–408

Qin Z, Caruso JA, Lai B, Matusch A, Becker JS (2011) Trace metal imaging with high spatial resolution: applications in biomedicine. Metallomics 3:28–37

Roschzttardtz H, Grillet L, Isaure M, Conéjéro G, Ortega R, Curie C, Mari S (2011) Plant cell nucleolus as a hot spot for iron. J Biol Chem 286:27863–27866

Roudeau S, Carmona A, Perrin L, Ortega R (2014) Correlative organelle fluorescence microscopy and synchrotron X-ray chemical element imaging in single cells. Anal Bioanal Chem 406:6979–6991

Rubakhin SS, Romanova EV, Nemes P, Sweedler JV (2011) Profiling metabolites and peptides in single cells. Nat Methods 8:S20–S29

Schaumlöffel D, Hutchinson R, Malherbe J, Le Coustumer P, Gontier E, Isaure M-P (2016) Novel methods for bioimaging including LA-ICPMS, NanoSIMS, TEM/X-EDS, SXRF. In: Michalke B (ed) Metallomics: analytical techniques and speciation methods. Wiley-VCH, Weinheim, pp 83–116

Schmid A, Kortmann H, Dittrich PS, Blank LM (2010) Chemical and biological single cell analysis. Curr Opin Biotechnol 21:12–20

Schober Y, Guenther S, Spengler B, Römpp A (2012) Single cell matrix-assisted laser desorption/ionization mass spectrometry imaging. Anal Chem 84:6293–6297

Shigeta K, Traub H, Panne U, Okino A, Rottmann L, Jakubowski N (2013a) Application of a micro-droplet generator for an ICP-sector field mass spectrometer – optimization and analytical characterization. J Anal At Spectrom 28:646–656

Shigeta K, Koellensperger G, Rampler E, Traub H, Rottmann L, Panne U, Okino A, Jakubowski N (2013b) Sample introduction of single selenized yeast cells (Saccharomyces cerevisiae) by micro droplet generation into an ICP-sector field mass spectrometer for label-free detection of trace elements. J Anal At Spectrom 28:637–645

Slaveykova VI, Guignard C, Eybe T, Migeon HN, Hoffmann L (2009) Dynamic NanoSIMS ion imaging of unicellular freshwater algae exposed to copper. Anal Bioanal Chem 393:583–589

Smart KE, Smith JAC, Kilburn MR, Martin BGH, Hawes C, Grovenor CRM (2010) High-resolution elemental localization in vacuolate plant cells by nanoscale secondary ion mass spectrometry. Plant J 63:870–879

Storms HA, Brown KF, Stein JD (1977) Evaluation of a cesium positive ion source for secondary ion mass spectrometry. Anal Chem 49:2023–2030

Szyrwiel L, Shimura M, Shirataki J, Matsuyama S, Matsunaga A, Setner B, Szewczuk Z, Yamauchi K, Chavatte L (2015) A novel branched TAT 47–57 peptide for selective Ni 2+ introduction into the human fibrosarcoma cell nucleus. Metallomics 7:1155–1162

Tanner SD, Bandura DR, Ornatsky O, Baranov VI, Nitz M, Winnik MA (2008) Flow cytometer with mass spectrometer detection for massively multiplexed single-cell biomarker assay. Pure Appl Chem 80:2627–2641

Tsang CN, Ho KS, Sun H, Chan WT (2011) Tracking bismuth antiulcer drug uptake in single helicobacter pylori cells. J Am Chem Soc 133:7355–7357

Týčová A, Ledvina V, Klepárník K (2017) Recent advances in CE-MS coupling: instrumentation, methodology, and applications. Electrophoresis 38:115–133

Umemura T, Matsui Y, Sinnosuke S, Fukai T, Fujimori E, Kumata H, Aoki M (2017) Comprehensive element analysis of prokaryotic and eukaryotic cells as well as organelles by ICP-MS. In: Ogra Y, Hirata T (eds) Metallomics: recent analytical techniques and applications. Springer, Berlin, pp 219–237

Van Malderen SJM, Vergucht E, De Rijcke M, Janssen C, Vincze L, Vanhaecke F (2016a) Quantitative determination and subcellular imaging of cu in single cells via laser ablation-ICP-mass spectrometry using high-density microarray gelatin standards. Anal Chem 88:5783–5789

Van Malderen SJM, Managh AJ, Sharp BL, Vanhaecke F (2016b) Recent developments in the design of rapid response cells for laser ablation-inductively coupled plasma-mass spectrometry and their impact on bioimaging applications. J Anal At Spectrom 31:423–439

Verboket PE, Borovinskaya O, Meyer N, Günther D, Dittrich PS (2014) A new microfluidics-based droplet dispenser for ICPMS. Anal Chem 86:6012–6018

Vergucht E, Brans T, Beunis F, Garrevoet J, De Rijcke M, Bauters S, Deruytter D, Vandegehuchte M, Van Nieuwenhove I, Janssen C, Burghammer M, Vincze L (2015) In vivo X-ray elemental imaging of single cell model organisms manipulated by laser-based optical tweezers. Sci Rep 5:9049

Wang D, Bodovitz S (2010) Single cell analysis: the new frontier in "omics". Trends Biotechnol 28:281–290

Wang HAO, Grolimund D, Giesen C, Borca CN, Shaw-Stewart JRH, Bodenmiller B, Günther D (2013) Fast chemical imaging at high spatial resolution by laser ablation inductively coupled plasma mass spectrometry. Anal Chem 85:10107–10116

Wang H, Wang B, Wang M, Zheng L, Chen H, Chai Z, Zhao Y, Feng W (2015) Time-resolved ICP-MS analysis of mineral element contents and distribution patterns in single cells. Analyst 140:523–531

Wang H, He M, Chen B, Hu B (2017a) Advances in ICP-MS-based techniques for trace elements and their species analysis in cells. J Anal At Spectrom 32:1650–1659

Wang H, Chen B, He M, Hu B (2017b) A facile droplet-Chip-time-resolved inductively coupled plasma mass spectrometry online system for determination of zinc in single cell. Anal Chem 89:4931–4938

Wedlock LE, Kilburn MR, Cliff JB, Filgueira L, Saunders M, Berners-Price SJ (2011) Visualising gold inside tumour cells following treatment with an antitumour Au(I) complex. Metallomics 3:917–925

Wolfe-Simon F, Switzer Blum J, Kulp TR, Gordon GW, Hoeft SE, Pett-Ridge J, Stolz JF, Webb SM, Weber PK, Davies PCW, Anbar AD, Oremland RS, Hille R, Lane TW, Morel FM, Wolfe-Simon F, Davies PCW, Anbar AD, Rosen BP, Baer CD, Edwards JO, Rieger PH, Oremland RS, Stolz JF, Hollibaugh JT, Blum JS, Bindi AB, Buzzelli J, Stolz JF, Oremland RS, Takeuchi M, Makino W, Cotner J, Sterner R, Elser J, Mandelstam J, Smith PG, Pickering IJ, Holbrook S, Dickerson R, Kim SH, Oremland RS, Stolz JF, Quillaguaman J, Delgado O, Mattiasson B, Hatti-Kaul R (2011) A bacterium that can grow by using arsenic instead of phosphorus. Science 332:1163–1166

Zavalin A, Todd EM, Rawhouser PD, Yang J, Norris JL, Caprioli RM (2012) Direct imaging of single cells and tissue at sub-cellular spatial resolution using transmission geometry MALDI MS. J Mass Spectrom 47:1473–1481

Zhao F, Moore KL, Lombi E, Zhu Y (2014) Imaging element distribution and speciation in plant cells. Trends Plant Sci 19:183–192

Zheng L-N, Wang M, Wang B, Chen H-Q, Ouyang H, Zhao Y-L, Chai Z-F, Feng W-Y (2013) Determination of quantum dots in single cells by inductively coupled plasma mass spectrometry. Talanta 116:782–787

Zheng L-N, Wang M, Zhao L-C, Sun B-Y, Wang B, Chen H-Q, Zhao Y-L, Chai Z-F, Feng W-Y (2015) Quantitative analysis of Gd@C82(OH)22 and cisplatin uptake in single cells by inductively coupled plasma mass spectrometry. Anal Bioanal Chem 407:2383–2391

Index

M. A. Z. Arruda (ed.), *Metallomics*, Advances in Experimental Medicine and Biology 1055, https://doi.org/10.1007/978-3-319-90143-5

N

The manufacturer's authorised representative in the EU is Springer Nature Customer Service Centre GmbH, Europaplatz 3, 69115 Heidelberg, Germany. If you have any concerns regarding our products, please contact ProductSafety@springernature.com

Printed and bound by CPI Group (UK) Ltd, Croydon, CR0 4YY
05/10/2026
02228139-0009